SECOND EDITION

HORSE SENSE

THE GUIDE TO HORSE CARE IN AUSTRALIA AND NEW ZEALAND

SECOND EDITION

HORSE SENSE

THE GUIDE TO HORSE CARE IN AUSTRALIA AND NEW ZEALAND

Peter Huntington

Jane Myers

Elizabeth Owens

LAND
LINKS

National Library of Australia Cataloguing-in-Publication entry

Huntington, P. J.
Horse sense: the guide to horse care in Australia and New Zealand.

2nd ed.
Bibliography.
Includes index.
ISBN 0 643 06598 9.
1. Horses. 2. Horses – Australia. 3. Horses – New Zealand.
I. Myers, Jane. II. Owens, Liz. III. Title.

636.1

Available from
CSIRO Publishing
36 Gardiner Road, Clayton VIC 3168
Private Bag 10, Clayton South VIC 3169
Australia

Telephone: +61 3 9545 8400
Email: publishing.sales@csiro.au
Website: www.publish.csiro.au
Sign up to our email alerts: publish.csiro.au/earlyalert

Front cover
'Haydon Raygun', photo by Annie Minton, Horses and People Photography

Set in Adobe Minion 10.5/12; ITC Stone Sans
Cover and text design by James Kelly
Typeset by Desktop Concepts Pty Ltd
Printed by Ingram Lightning Source

Disclaimer
While the authors, publisher and others responsible for this publication have taken all appropriate
care to ensure the accuracy of its contents, no liability is accepted for any loss or damage arising
from or incurred as a result of any reliance on the information provided in this publication.

Foreword

The first edition of *Horse Sense: The Australian Guide to Horse Husbandry* was published in 1992 and went on to become a best selling reference book before going out of print.

The foreword was written by my father, the late Colin Hayes OBE and he described the book as a 'most comprehensive study on a wide range of subjects associated with equine care and well-being'.

Australia and New Zealand produce some of the best horses and horse people in the world, and it is appropriate that we have access to a homegrown horse care and management reference source, rather than relying on overseas publications written for different conditions.

Like the first edition, this book is a valuable reference guide for those with experience as well as the novice horse owner or rider. It will help develop the understanding between the horse owner and the veterinarian by explaining important equine diseases or problems, and highlighting when professional help is required.

The second edition is a fully revised and expanded book that contains a reorganisation of the existing material along with a lot of new information. There are two new contributing authors, Elizabeth Owens, who is a leading dressage rider and one of Australia's most respected equine nutritionists, and Jane Myers, an experienced equine studies lecturer and writer. Peter Huntington has provided nutrition and feeding advice to Lindsay Park Stud and Racing Stables for over 10 years, and given his expertise, it is no surprise that the feeding section is much expanded and is a comprehensive review of that important topic.

The book also contains new material on equine behaviour, horse welfare, transporting horses, buying and selling horses and educational opportunities. It also provides a host of references to other more detailed sources of information on particular topics. The drawings and photographs are excellent and illustrate points of interest particularly well.

One of the features of the first edition was its easy-to-read and practical style and this edition continues in that vein. My father's motto was 'The future belongs to those who plan for it'. With owning or caring for horses, you never know what is coming next so part of the planning process is knowledge, and this book can be recommended as a great source of knowledge in all aspects of horse care.

David Hayes

Contents

Foreword   v

Buying, selling and having a horse   1

General horse management   11

Safe handling, riding and training   39

Describing the horse   67

Feeding the horse   119

Horse health   175

Breeding the horse   237

Horse facilities   269

Transporting horses   289

Working in the horse industry   309

Bibliography and further reading   317

Appendix 1 Welfare issues covered by Codes of Practice in Victoria, Australia   320

Appendix 2 Safe riding on the road: Code of Conduct for horses on Victorian roads 323

Appendix 3 Tick clearance policy: Queensland, Australia   325

Glossary   327

Index   336

Acknowledgements

HORSE SENSE, SECOND EDITION

Fran Cleland, co-author of the first edition of *Horse Sense*

Naomi Pearson, Department of Primary Industries (Victoria)

Stuart Myers

Marguerite Bongiovanni

Julie Moran

Annie Minton, Horses and People Photography

Jane Williams, Glenormiston Campus, The University of Melbourne

John Fowler, Glenormiston Campus, The University of Melbourne

Lou Nunn, Glenormiston Campus, The University of Melbourne

Claye Cunningham, Master Farrier

Belcam Warmblood Stud, Queensland

Reg Pascoe

Steven Steele, Hayman Reese & Thretford Australasia

Jeanette Gower, Chalani Stud

Peter Turley, Woodhill Stockhorse Stud, Queensland

Sarah Hayes

Ives Cousinard, Pacific Performance Horses, Queensland

Tracy Stead, Prestonwood Training Stables, Tasmania

Dr Chris Pollit, The University of Queensland

Dr Joe D Pagan, Kentucky Equine Research, Inc., Kentucky, USA

Dr Stephen G Jackson, Kentucky Equine Research, Inc., Kentucky, USA

HORSE SENSE, FIRST EDITION
Contributing authors

Roy Butler, Department of Primary Industries, Tasmania

Royston Carr, 'Curraghmore' and Graham McCulloch, 'Grenville Stud', Tasmania

Graham Christie, Glenormiston Agricultural College

Patrick Francis, Glenormiston Agricultural College

Kerri Goff, Glenormiston Agricultural College

Chris Kimberley, Glenormiston College, The University of Melbourne

Chris Crook and Terry Clarke

Bill Dunlop, Department of Primary Industries, Queensland

Ross Mackenzie, Department of Primary Industries, Queensland

Dr Peter Ellis, Attwood, Victoria

Dr Stephen Jackson, Vice-President, Kentucky Equine Research, Inc., Kentucky, USA

Dr Joe Pagan, President, Kentucky Equine Research, Inc., Kentucky, USA

Vicki Masters, Dr Rod Graham and Ian McDuie, National All Breeds Sales, Warranwood, Victoria

Dr Michael Morris, 'Ballymore Stables' and Dr Art Meeker, 'Neptune Stud', Tasmania

Rod Patterson, as instructed by Jim Young, Farrier, Cranbourne

P Rolfe, Veterinary Officer, Division of Animal Health, Department of Agriculture, New South Wales

Australian Horse News (section on breeding and breed associations)

CONSULTANTS

Lindsay Beer, Group Education Officer, Department of Agriculture, New South Wales, Murray and Riverina Region

Ian Denney, Division of Animal Industries, Department of Agriculture, New South Wales

Greg Carroll, Manager – Legal, Department of Food and Agriculture, Victoria

Ross Dodt, District Adviser – Beef Cattle Husbandry, Queensland Department of Primary Industries

Clive Stokes, South Australia

John Glover, Principal Information Officer, Department of Primary Industries, Fisheries and Energy, Tasmania

Reg Pascoe, Oakey Veterinary Hospital, Oakey, Queensland

Ian Pickett, Morphettville Annexe, Gilles Plains College of TAFE, South Australia

Angus McKinnon, Goulburn Valley Equine Hospital, Shepparton, Victoria

Georgina Fox, Gisborne, Victoria

CHAPTER 1

BUYING, SELLING AND HAVING A HORSE

INTRODUCTION

Both the horse industry and horse ownership are flourishing in Australia and New Zealand, with an estimated contribution to gross domestic product in Australia of more than $6.2 billion each year. The horse industry has many disciplines and interest groups, but it all starts with the dream of having your own horse.

In Australia and New Zealand it is comparatively easy for that dream to come true because the cost of horses and land is cheaper than in many other countries. Keeping a horse, however, also requires knowledge and this handbook provides information from the initial stage of acquiring a horse to the level of professional horse management, written in a practical style for Australasian conditions.

COSTS OF KEEPING A HORSE

Costs will vary depending on the type of horse, the intended use, where it is kept, and the individual needs of the particular horse. You should not buy a horse unless you can afford the costs of its upkeep. If you are unsure, try leasing a horse for a while to find out if you are keen and able to sustain the financial burden or 'time share' a horse with a like-minded person or simply go for lessons or trail rides at a reputable riding school until the urge to own a horse gets too strong.

The horse

The price of a horse is determined by many factors and may not be high compared with the costs of initially setting yourself up with all of the gear and keeping the horse over the long term.

Feed

The cost of feed varies from one season to another and even within a season. It is often cheaper to purchase hay and other fodder at the time of harvest, but you must have storage facilities or the feed may spoil before it is used. A well-balanced, pre-prepared feed is often less expensive than purchasing and mixing all the ingredients required for a balanced diet.

Transportation

A float for transporting horses will cost approximately $2000–$6000 for a second-hand one and $6000–$10 000 for a new one. You must have a car or truck that can legally tow the float, and there is insurance and fuel as well.

Agistment

The cost of agistment depends on whether the property is in the city, suburbia or the country and whether you are taking 'full agistment', which would usually include the total care of the horse, or 'paddock agistment' in which you rent space in a paddock. Prices vary from $30/week for paddock only to $150/week for full agistment.

The advantage for the beginner of agisting at a centre is that there will always be other people around to ask for advice. Agistment centres usually have facilities such as round pens and arenas where you can ride your horse and many have either a resident riding instructor or a visiting instructor. It is usually easier to get a farrier to visit, because there may be several horses requiring attention, and this also applies to visits by dentists and veterinarians for routine care. Good agistment centres usually have rules and everyone is expected to contribute to the smooth running of the centre. The terms of the agistment may be set out in a written agreement that regulates the rights and liabilities of each of the parties (see the end of the chapter for useful contacts).

Cost examples

Although it is impossible to list the costs of keeping all types of horses, the following four examples are the most common uses, but prices are still only intended as a guide and the costs of feed and agistment have not been included. Many of the expenses, especially gear, can be reduced by buying second-hand items, except when safety is an issue (e.g. helmets).

Example 1. Family pony, junior pony club or trail riding

Cost of horse	$400–$2000
Equipment	
Saddle, fully mounted	$400–$1000
Saddle cloth	$30–$70
Bridle, including bit	$80–$150
Headcollar and lead rope	$30
Helmet	$60–$100
Riding boots	$70
Jodhpurs	$70
Grooming kit	$30
Veterinary costs (yearly)	
Worming	$70–$120
Visits (cuts, tetanus vaccination, teeth etc.)	$100–$1000
Farrier	
If shod	$60 per 6 weeks
If trimmed	$25 per 8 weeks
Dentist (yearly)	$60
Pony club or riding club fees	$50 plus

Example 2. Senior pony club

Cost of horse	$1000–$4000
Equipment	
Saddle, fully mounted	$600–$1200
Saddle cloth	$30–$70
Bridle, including bit	$80–$150
Headcollar and lead rope	$30
Helmet	$60–$100
Riding boots (top boots: rubber)	$80–$120
Jodhpurs, shirt, tie, jumper	$100–$150

Grooming kit	$30
Rugs x 2	$250
Veterinary costs (yearly)	
Worming	$70–$120
Visits (cuts, tetanus vaccination, teeth etc.)	$100–$1000
Farrier	$50 per 6 weeks
Dentist (yearly)	$60
Pony club fees	$50 plus
Competition fees	$20/day plus

Example 3. Show pony

Cost of horse	$2000–$10 000
Equipment	
Saddle, fully mounted	$1000–$3000
Bridle, including bit	$250
Numnah	$120
Other horse gear as per senior pony club horse	
Rider attire	
Jacket	$200–$400
Waistcoat	$90
Jodhpurs	$120
Boots	$150–$750
Shirt	$80
Tie	$15
Helmet	$60–$200
Gloves	$50
Cane	$50
Farrier, dentist and veterinary costs as per senior pony club horse	
Breed Society etc. fees	$100 plus
Competition fees	$20–30 per event

Example 4. Event or Dressage horse

Cost of horse	$1000–$20 000
Equipment	
Jumping saddle	$800–$2000
Dressage saddle	$800–$3000
Bridle, including 2 bits	$250
Breastplate	$30–$40
1 basic saddle cloth	$30
Numnah	$120
Rugs	
Canvas	$120
Under	$100
Cooling	$70
Neck	$75
Clippers	$450
or per clip	$30–$70
Grooming kit	$50
Rider attire	
Work boots	$50
Show boots	$150–$750

Jodhpurs x 2	$220
Jacket	$250
Helmets x 2	$150–$200
Gloves	$100
Shirt	$80
Tie	$15

Farrier, dentist and veterinary costs as per senior pony club horse
Competition expenses (membership of the Equestrian Federation of Australia or the New Zealand Equestrian Federation is compulsory for riders wanting to compete in 'official' events. Cost may vary, but full riding membership for an adult is between $120 and $170/annum)

BUYING, SELLING AND LEASING A HORSE

Although the information in this section is primarily for the person acquiring a first horse, if you have bought, sold or leased a horse before there are some helpful tips. The buying process can be stressful, but if you have some guidelines it can be a learning experience, even enjoyable, as you will make many new contacts and maybe new friends.

Buying a horse has usually been a case of caveat emptor (let the buyer beware), which means that if you buy a horse and it is not suitable for any reason you cannot return it to the seller. However, written contracts may now be used when buying, selling or leasing a horse.

Leasing a horse

It is possible to lease a good horse and a leased horse may assist you through a certain stage in your riding life; many good beginner's horses are only ever leased (rather than sold). Leasing is an excellent way of acquiring a horse without going through all of the buying process and usually means that you take over the care of the horse, as if it were your own, but the owner retains the right to take the horse back if they are not happy with the way it is being kept. Speaking to other people is an excellent way of finding this type of horse as they are often not advertised. The leasing of racehorses and high-level competition horses often involves a fee, but with ordinary riding horses it is commonplace that no money changes hands. However, you must ensure both parties understand the details before signing a lease agreement, which will help protect all parties involved. Standard contracts can be purchased (see the end of the chapter).

Buying a horse

The process of buying a horse should not be rushed. You need to find the right horse for your needs and if you purchase the wrong horse, not only will it erode your confidence, you also have to decide what to do with it. If the decision is made to sell, you may lack the experience to achieve a successful sale and recoup your money.

It is important to ask an experienced horseperson to assist you with the purchase of your first horse. That may be your riding instructor, who will know your capabilities, or it may be a more experienced friend; whoever you choose should know the potential pitfalls and be able to weigh up the positives and negatives about a particular horse. Ultimately, they must be able to help you decide which is the right horse for you at your current level of experience.

Timing

Many people decide that they want to learn to ride and think that they will need to acquire a horse in order to do so. In fact, it is strongly recommended that you learn to ride before buying a horse (see the end of the chapter for useful contacts).

Even after taking lessons, the horse that you buy may be different to ride than a riding school horse, which works hard and is usually quieter than a privately owned horse. If you are able to buy the horse that you have learnt to ride on (many riding schools sell horses from time to time), you may be surprised by the difference in its behaviour when it is no longer working as hard.

Your budget

You must first decide on the amount of money you are prepared to spend. Many beginners underestimate the cost of a good horse, but remember that the right horse will help you gain experience whereas the wrong horse will damage your confidence.

A good horse may seem expensive, but it takes many years of work and expense to train a horse. A horse that is less than 10 years old, good looking, well educated, very quiet and with no particular faults can be worth several thousand dollars. Other factors that add value are the level of show quality and unusual colouring or a particular breed.

What type of riding do you want to do? A beginner must first learn to ride and after you have achieved an adequate level of skill, you can then decide which discipline(s) you wish to pursue. A good horse will give you the opportunity to try out a few, such as trail riding, a novice dressage test or jumping a small course of jumps. You will be able to attend pony club or riding club rallies safely, have fun and gain experience. When you have gained more experience you may need a different type of horse or if you are lucky your first horse may still be suitable.

How far are you prepared to travel? If you are looking for a specific type of horse you may have to travel far, but finding out as much as possible about the horse before you view it can save time and expense.

If you have a limited budget you may have to compromise, but not on temperament if you are a beginner. An old horse, as well as being cheaper, is often the best horse for a beginner. Other factors such as good looks and colour are less important. Blemishes, such as scars, are only important if you are planning to show, but while you are gaining experience at local novice events they will not be relevant.

Temperament

Temperament is the most important factor; the horse must be friendly, calm and easy to handle. You must feel safe with the horse.

Performance

The better the horse's performance in any particular field, the higher the price. The potential of a young horse to perform will still influence the price.

Education

You should buy a well-educated horse, one that has good manners and well established basics. Good manners means being obedient and not pushing you around. Examples of bad manners are dragging you around, not moving over or out of the way when asked and getting nasty at feed time. Good basics mean that the horse should be easy to stop and turn, and move forward willingly but quietly. The horse should also yield (move over) to leg pressure and back up when asked.

Regardless of the education level of the horse, as the new owner you must maintain it or improve upon it because even a good beginner's horse will start to misbehave if you do not handle and ride it properly. Therefore, a novice rider should not buy a highly trained dressage horse, for example.

Age

The prime age of a horse is between 6 and 10 years, and if sound these horses will generally command the best price. Older horses in their late teens with good legs may be a better buy for the novice as they still have many years of service if well cared for and tend to be quieter because of their varied experience. An older horse will not be as supple as a young horse and this must be taken into consideration when you have the horse checked by a veterinarian.

Late-teenage horses may not pass a flexion test because they have wear and tear in the joints from both previous work and old age. This does not mean that the horse is no good, it just means that some activities may have to be restricted; however, steady work is often the best remedy for stiffness and as long as the horse is not lame, stiffness that improves after the horse has warmed up is not usually a problem.

Young horses and beginners are not a good combination. Sometimes young horses are advertised as 'very quiet' or even 'bombproof', but the seller cannot predict that the horse will always be quiet. Young horses are often very quiet at 3–4 years of age because they are still maturing, but when they are 5–6 years old the excess energy is no longer channelled into growing and their temperament may change.

Other seemingly quiet, young horses are being handled and ridden by experienced people. Less experienced riders give unclear signals until they have perfected their technique and a young horse is more likely to be confused than an older horse that has been ridden by a variety of people.

Colour
There is an old saying 'a good horse is never a bad colour'. The less experienced you are with horses, the less colour matters. Good horses are hard to find and a colour restriction will reduce the number of potential purchases. When you have learned to ride, you can select a horse based on colour if you have strong preferences.

Breed
Some breeds have a reputation for being quieter than others, but there are examples of quiet and excitable horses in all breeds. 'Hot blooded' horses such as Thoroughbreds and Arabs are thought to be more excitable than 'cold blooded' horses such as the draught breeds (e.g. Clydesdale) and although there are plenty of exceptions to this belief, an 'off the track' Thoroughbred is not a good choice for a beginner. Cross-bred horses, usually a cross between a 'hot blood' and a 'cold blood' (i.e. a 'Warmblood'), or a pony (e.g. Connemara etc.) are often very good horses, but remember that every horse is an individual. A horse is not automatically well behaved because it is a certain breed.

Horses that are registered (pedigrees) usually cost more than unregistered horses, but a pedigree is not necessarily a better horse. It is more important to know the history of the horse than to concentrate too much on the breed.

Conformation
Assessing a horse's conformation needs a trained eye and the beginner will need the help of an experienced horseperson. Certain conformation faults can be overlooked in a beginner's horse if it has a good temperament. Conformation faults in a younger horse are to be regarded with more caution because their long-term effect is unknown. Again it depends on what you are planning to do with the horse.

Action and movement
The horse should be able to move freely, without stumbling and tripping, in all of the paces. It should move straight, without hitting itself (i.e. hitting one leg with another). Some imperfections in movement can be overlooked in a beginner's horse if they are not likely to cause lameness.

The term 'forward moving' is sometimes used to describe a horse and can mean either that the horse moves freely (as opposed to sluggishly) in a calm manner or that it is 'on its toes' and ready to move faster with little or no signal from the rider. The latter type of horse is not suitable for a beginner, but may be when you have acquired the skill to control its forward movement.

Sex
Some people prefer geldings because they believe that they are more obedient, others prefer mares. It usually depends on that particular person's previous experience. There are good and bad-tempered mares and likewise geldings. A gender restriction will limit your choice.

Stallions are usually purchased for breeding or for high level competition work and even then should only be purchased by an experienced person.

Size of horse

As a general rule, when mounted, the rider's feet in the stirrups should be level with the bottom of the horse's stomach. If the rider's legs are too short or too long, the horse will not be under effective control. Many people buy a horse that is too large and strong for them. If you are buying a horse for a child, do not buy a bigger horse than the child is ready for, because the child may lose confidence.

Methods of buying

There are several methods of buying a horse, including sale registries, horse auctions, dealers, buyer's agents, private sales and by word of mouth.

Sale registries

Horse and pony sale registers keep information about buyers and sellers, and where possible will match them. A good sale register will do some of the work of finding a suitable horse, but you will still have to view it.

Horse auctions

Horse auctions range from local, livestock markets (including weekly or monthly horse auctions) to large events that are specifically designed to sell stock from studs (i.e. young, potential performance horses). The latter type of horse is unsuitable for a beginner.

Buying a horse at an auction has the advantage of enabling you to view several horses in one location, but there are many drawbacks for the inexperienced.

- It is not always possible to see the horse under saddle, or indeed ride the horse, before you buy.
- Factors such as temperament, gait, soundness etc. are sometimes difficult to judge.
- Many 'problem' horses are sold at local livestock auctions because it enables the owner to dispose of the horse without any liabilities.
- You have only a short space of time in which to make your decision to buy; once you have bought the horse, you cannot change your mind.

Horse dealers

A reputable dealer will match you with the right horse and save you time. There will be additional expenses associated with engaging professional services, so it is important to find a good dealer. Sadly, there are some people in the horse industry who will market horses to inexperienced people, knowing that the horse is unsound, unsuitable, or even downright unsafe. Good dealers are aware that if they provide good service you will use their services again in the future and you will recommend them to other people. Ask around (i.e. pony clubs, riding schools, produce stores and saddlery shops) to find a reputable dealer.

Buyer's agents

Agents act on your behalf as the buyer. Some agents inspect a horse that you have found, before you buy it, and others will find you the right horse, at the right price. This may sound expensive, however if you are busy and/or inexperienced it can be well worth the expense. Some instructors may act as the buyer's agents and similarly some dealers.

Private sale

There are lots of horses advertised and they all sound good in the advertisement, so you must find out as much as possible about the horse from the seller before you inspect it. It is always best if buyers are honest about their abilities and sellers are honest about the horse.

Word of mouth

Some owners do not want to advertise their horse and others are only concerned about getting a good home for their horse rather than a high price. Many excellent horses can only be located by word of mouth.

Questions to ask the seller

- Why are they selling the horse?
- How old is the horse?
- Is it a gelding or mare? Has the mare ever had a foal?
- Is it registered and with what society?
- How big is the horse? Has it been officially measured?
- What are the horse's accomplishments in which disciplines and when?
- Does it have any major conformation faults?
- How does the horse handle with regard to floating, shoeing, catching, lunging, riding in traffic, riding around other horses, being tied up and clipping?
- Is the horse 'forward moving' or sluggish?
- Does the horse have any behaviours such as windsucking, kicking, biting or nipping?
- What sort of temperament does the horse have? Is it ever nervous? Does it ever shy?
- Is the horse or pony currently being ridden regularly? If not, how long has it been since it was last ridden?
- What condition is the horse in? Has it been regularly wormed?
- When was it last checked by a veterinarian? Is it sound, with good clean legs?
- What level of ability and age is the present rider?
- Is a trial period possible?

If the horse seems suitable, the next stage is to view it. However, some owners may quiz you to ascertain that the horse is going to someone who is capable of caring for it and riding it, and may decide that you are not suitable for the horse at your present stage of riding.

Viewing the horse

Ask the owner if there will be someone to ride the horse for you and take an experienced advisor with you. Arrive on time and dress appropriately (boots, helmet etc.). Avoid taking young children, unless you are purchasing the horse for them.

If possible see the horse being caught. Have the horse walked and trotted on a hard level surface so you can view its natural action and gait. Together with your experienced advisor, look at the horse's teeth to check the age, do a thorough conformation check, feel the legs, look at the feet (ask the owner to pick the feet up), ask about the farrier, notice any lumps and bumps and query them.

If the horse passes your inspection, ask to see it ridden. Never get on a horse unless you have seen it being ridden first, and if you feel the horse is more than you can handle, do not ride it. The horse should be put through all its paces (i.e. walk, trot and canter) and if it is a jumper then ask for that to be demonstrated. If you still think the horse is suitable, either you or your advisor should ride the horse and where possible take it on a trail ride away from its home property in the company of one or two other horses.

If the horse meets all your requirements, you can start negotiating the sale.

The negotiation

Negotiation includes the price and method of payment, a trial period, the veterinary inspection and whether there is any guarantee with the horse. The seller may not allow a trial period because the question of who is legally responsible for the horse during this period can be difficult. If a trial is not permitted, try to negotiate at least one more opportunity to ride and handle the horse before buying it. Remember though that there will be many keen buyers for a good horse.

Veterinary inspection

A veterinary inspection (vet check) will identify health issues that you should consider before buying a horse. These include sight, internal disease, teeth, heart function, lung function, skin disease, legs and feet. The vet should carry out a flexion test on each leg, which involves the leg being held in a flexed position for approximately one minute and the horse immediately trotted out by a handler, with the vet watching for lameness (see Chapter 6).

The vet will need to know your intended purpose for the horse and it is possible that rather than proclaiming the horse as 'sound', the vet will instead give an opinion on the suitability of the horse for the intended purpose. Some vets will not do inspections because of liability claims.

Payment

You will usually be required to pay by bank cheque or cash. If paying by ordinary cheque the owner may request that the horse not be collected until the cheque has been cleared. Make sure you get a receipt for the payment.

Collecting the horse

If you have to travel a long distance, using a commercial racehorse transport company may be cheaper than collecting the horse yourself. Make sure you collect any registration papers and other documentation.

Selling a horse

Selling a horse can be just as difficult as buying, but again preparation is the key. You need to decide how you will sell and if you cannot bring yourself to sell the horse, consider leasing so that you can retain control of the animal's welfare. You can always decide to sell later.

Many of the issues associated with selling are the same as for buying, but there are some additional points. Above all, do not be dishonest about the horse because if someone is injured while riding the horse that you have falsely described, you may be liable.

- Prepare your horse for sale before you advertise. If the horse is not working, you may not get its full value because the purchaser will not be able to ascertain whether the horse will be suitable when it is 'in work' (horses can behave differently when they are unfit).
- Describe the horse correctly to attract the right type of buyer.
- Do not to rush the sale. Your horse deserves a good home.
- Plan the questions you will ask buyers before they view the horse (e.g. riding ability; are they in the market to buy, or are they just looking around and planning to buy at a later date?).
- Be prepared to tell a potential purchaser that you think they are unsuitable for the horse.
- Get legal advice about a trial period.
- Have the horse caught, cleaned and in a stable, yard or easily accessible paddock before the buyer arrives.
- Be prepared to ride the horse or arrange for someone else to do it.
- Request payment as cash or bank cheque.

WHERE TO GO FOR MORE INFORMATION

Legal contracts can be purchased at saddlery stores or can be ordered via the internet at www.horseforce.com.au. The Equine Centre website at www.equinecentre.com.au (horse health care/ guidelines and codes of practice) has information about agistment contracts for horses and the Victorian Horse Council website at www.vicnet.net.au/~vichorse/(publications) has a copy of an agistment contract. The New Zealand Ridings Clubs website (www.nzridingclubs.homestead. com/legal.html) has a copy of a lease contract.

The Australian Horse Riding Centres website (www.horseriding.org.au) lists accredited riding schools in each State. The Equestrian Federation of Australia website (www.efanational.com) has contact details of each State branch and a list of riding instructors. The New Zealand Ridings Clubs website (www.nzridingclubs.homestead.com) lists affliated riding clubs. New Zealand Horses Online (www.nzhorses.co.nz/index2.cfm) has a directory of equestrian services, which includes riding schools and instructors.

CHAPTER 2

GENERAL HORSE MANAGEMENT

UNDERSTANDING THE HORSE

A key to good horse management is understanding the physical and behavioural characteristics of the horse; that is, how and why it evolved. Many owners mistakenly believe that the horse has their best interests at heart and will not harm them, but horses have not evolved to look after humans.

EVOLUTION OF THE HORSE

The horse is a large-bodied mammal that, in the natural state, lives in a herd. It is a herbivore and the horse is also prey, so the first reaction to any threat or fright is to run. The anatomy and behaviour of the horse evolved to meet this primary 'fright and flight' instinct.

In the past, the horse's natural habitat ranged from the rolling plains in the temperate zone of the Northern Hemisphere to the drier habitat of the Middle East. Today there are no truly wild horses, although relatives of domestic horses such as zebras and wild asses still live in the wild state. Horses such as brumbies and mustangs are classed as feral because they are the descendants of domesticated horses that have either escaped or been turned loose.

The different body types of domestic horses reflect the environment of their country of origin. For example, the ancestors of heavy draught horses originated in colder, wetter parts of Europe, hence the thicker skin, coarse and more abundant hair and larger feet to cope with wetter soils, whereas lightweight Arab horses originated in hotter, drier countries requiring thinner skin, finer hair and smaller feet to cope with a stony, sandy desert-type terrain.

Horses are usually classed as grazers, which means that their staple diet is mainly grasses and herbs, but they are also browsers, eating the leaves of certain trees, berries from bushes (such as blackberries) and the flower heads of thistles.

Physical characteristics

The head and neck

The head of the horse is long, which reduces the risk of grass seeds entering the eyes while grazing, and enables the horse to see over the top of the grass (during feeding) and watch for predators. A large, long head is also needed for the large and numerous teeth required for grinding plant material in the process of digestion.

The neck is long and flexible, enabling the horse to reach the ground without having to spread its front legs apart, which would make it vulnerable to predators while it was feeding. The length of the neck also enables the horse to reach leaves in trees and also acts to balance the horse when it is in motion.

The teeth

The horse's teeth have a large surface area for grinding, are sharp and extremely strong. As the top surface of the tooth wears down from use, the teeth continue to erupt from the gums, so the roots of the teeth are very long in the younger horse and much shorter in the older horse.

The skin and hair

The skin of the horse is very sensitive. The horse has to fend off biting and bloodsucking insects, which has led to the development of a sheet of muscle just underneath the skin that stretches across the body of the horse and up the neck, enabling the horse to twitch its skin wherever an insect lands on it. It also means that the horse is very sensitive to light pressure, which can be used in the training of horses. Some horses do not respond to light pressure and are termed 'lazy', but this is a learned behaviour because the same horse will still twitch at the touch of a fly.

The horse has a long forelock (the part of the mane falling over the forehead), which helps to keep insects, debris and rain out of the eyes. Breeds that originated in wetter, colder climates usually have longer, thicker forelocks and manes.

The long, bushy tail of the horse is used for warmth in cold weather, protecting the bald area under the dock from heat loss, and as a flywhisk in summer. The tail and thighs of the mare provide protection and warmth to the foal while it suckles.

In cold weather, the horse naturally grows a thick coat for protection, which it sheds as the weather warms up, replacing it with a shorter, summer coat. Some breeds, such as Thoroughbreds and Arabs, have a particularly short, smooth summer coat and a thinner winter coat than the heavier breeds and may need extra protection in cold climates.

The body

The horse's chest (thorax) is large and deep to accommodate the large heart and lungs needed to outrun predators. The abdomen is also large to accommodate the extensive gut required for digesting plant material.

The horse does not have a shoulder joint, as in humans; instead the forelegs are attached to the body by muscles, which give strength and flexibility for locomotion and enable a cornered horse to strike out and stomp with a front leg.

Because the horse's first response to danger is flight, its body is adapted for speed and the horse has few equals for maintaining speed over long distances. A cheetah can run much faster than a horse for a short distance, but the horse can maintain a racing gallop for more than two kilometres and can maintain a fast trot/canter for much further.

A horse's metabolism can manage levels of the metabolites produced during periods of muscle exertion that would kill a human. The horse's large spleen can respond instantly to the demands of muscular effort with increased production of oxygen-carrying red cells. The muscle cells of a horse also extract oxygen from the blood very efficiently, which means energy production is very high. During locomotion, a horse is using almost 90% of its muscle mass. The horse is a natural running machine, designed to operate at the very limits of the output of the heart, lungs and muscles.

The legs and hooves

The legs of the horse raise its body so that it can move over grass rather than through it. The powerful leg muscles are located at the top of each leg with only tendons and ligaments below the knees and hocks to reduce the weight of the lower leg and thus give more speed. The upper leg bones are also angled, which gives great extension of the leg and thus a long running stride.

During locomotion, the hoof is the fastest moving part of the leg. It is small and lightweight, requiring little energy to move it. The hooves are hard, which helps the horse cope with varied surfaces. Evolution has reduced the number of toes on each leg to one and the horse has one strong, fused bone (the cannon bone) below the knee and the hock, rather than several thin bones as in the human hand or foot, resulting in stronger legs. The fetlocks, pasterns and hocks do not move sideways, only forwards and backwards, which allows the horse to move quickly over rough ground without 'knuckling over'.

The senses

The eyes of the horse are situated on the sides of the head, which gives the horse all round (360°) vision, and high up on the head so that it can see while grazing. In contrast, most predatory and tree-dwelling animals have their eyes positioned on the front of their head because they need binocular vision in order to judge distances accurately.

The rider needs to appreciate that a horse can see objects behind it and to the side that humans with their binocular vision cannot see. On the other hand, although the horse can detect objects at a distance, it is unable to focus on these as quickly as a human and if the object is not recognised, the horse may interpret it as a source of danger. The only time the horse is unable to see behind is when the head is held level and in line with the body, such as when being ridden in a collected frame. Whenever a horse is startled it attempts to put its head in the air so that it can utilise its 360° vision.

The top lip is prehensile, which means it is very mobile. It contains many nerve endings, making the lip highly sensitive, which is necessary because the length of the nose and the positioning of the eyes prevent the horse from being able to see directly under its mouth. The muzzle is covered in whiskers that also aid the horse in selecting food.

Horses rely heavily on their sense of smell and taste. Horses will usually sniff any strange feed before cautiously trying a small amount; this is because they cannot regurgitate and once ingested the food must pass through the entire digestive system before it can be eliminated. Therefore, if the horse were to eat poisonous food, it could be harmful and even fatal. Horses that have had prolonged contact with humans, and have probably been offered many different types of feed, may lose this cautious behaviour to some extent. In fact, children's ponies have even been know to accept cheeseburgers and fries (definitely not recommended!).

The ears of the horse are situated on top of the head and are funnel-shaped to facilitate the collection of sounds. They are lined with hair to protect the inner ear from debris, dust, rain and insects. Each ear can swivel 180°, giving the horse 360° of hearing without moving its head. The horse's sense of hearing is well developed, enabling it to pick up subtle signals in its environment, particularly unfamiliar sounds. Many horses become unsettled on windy days, possibly because they cannot hear or smell with the wind rushing past their ears and nostrils.

Behavioural characteristics

Eating

The horse has evolved to eat large amounts of low energy, fibrous food and will graze for up to 20 hours per day if necessary. Its natural food takes time and energy to chew and the horse has huge muscles in the jaw for this purpose. Chewing is also important for the production of saliva, which is essential for digestion.

Unlike ruminants such as cows and sheep, a horse does not digest fibrous food in its stomach. Instead, it will graze almost continuously, with fermentation of the forage occurring in the hindgut while the horse is grazing. Therefore a horse is mainly on its feet, ready to flee and is not weighed down by large quantities of undigested forage.

The horse will store body fat when food is plentiful and when grass or other plant material is not readily available (such as in winter) these reserves are utilised for energy. Therefore, when the domestic horse is not working and using energy, it will quickly become fat if grazing rich, green grass or is overfed with high-energy supplements. This can lead to serious diseases such as laminitis or colic (see Chapters 5 and 6). Horses that are confined in stables or yards and incorrectly fed may

Figure 1. Herd
instinct
(courtesy of Malissa Brown).

develop behavioural problems related to stress and boredom. Grazing is how horses naturally eat and all horses should be allowed to graze as much as possible on high volume/low energy grass. The larger the paddock, the more exercise the horse has while grazing (i.e. it uses more energy).

Socialising
Hierarchy

Horses are herd animals because there is safety in numbers. More eyes and ears means that predators are identified more quickly. In the herd, each horse has a place in the social hierarchy (sometimes called the 'pecking order').

The pecking order must be taken into account when a new horse is turned out into a paddock with resident horses. If possible, gradually introduce the horse to the group and watch it carefully to ensure that it is not injured by the other horses. The presence of a pecking order means horses must also be observed during confinement in a small yard and at feeding times to ensure that lower ranked horses are not being injured or excluded from the food.

Time budgets

Each species has a 'time budget' for its usual daily activities, which is the amount of time an animal spends on each activity. The time budget of most predators involves short periods of high activity (to catch and eat prey) and long periods of inactivity while they digest. In contrast, herbivores have to eat for most of the day, as well as watching and listening for predators, because their food is low in calories and takes a long time to chew and digest. Studies have shown that the daily time budget of feral/wild horses comprises:

- grazing: 12–20 hours
- sleeping: 2–6 hours
- loafing: 2–6 hours.

The length of time spent grazing depends on the quality of the grass available. On better quality pasture, the horse will spend less time grazing and more time sleeping and loafing. When the grass is poor, such as in a drought or during colder weather, the horse will increase the grazing time.

Adult horses usually sleep for approximately four hours each day; half of that time standing up and the other half lying down. Because of its large chest, the horse uses less energy sleeping upright than it does while lying down. Lying down rests the legs, but if the horse stretches out on its side it has to work hard to breath, which is why the horse often makes a groaning noise when lying prone.

The horse is able to sleep standing because it has an anatomical adaptation termed 'the stay apparatus' that locks the joints in place. A horse can even rest one hind leg as it nods off to sleep, but can move off quickly in an emergency. In a group of horses, one usually remains standing when the others are asleep on the ground. This horse is alert for danger (even while it is drowsing) and is a good example of how herds operate. Single horses do not have this system of shared responsibility and some nervous horses may be unwilling to lie down if they do not have another horse to watch over them.

'Loafing' describes all the other things that horses do during the day, such as mutual grooming, playing and standing in the shade swishing at flies. Mutual grooming is where two horses approach each other and use their incisor teeth to 'groom' one another. Grooming is a very important way of maintaining bonds between group members.

Playing is very important, especially in young horses, as it involves the skills required for adult life.

Standing in the shade, head to tail, swishing at flies is another common behaviour of horses, especially in hot weather. In cold, wet weather horses will stand together in a sheltered spot. If there is not enough shelter, the horses that are lower in the pecking order have to stand out in the rain.

Figure 2. Mutual grooming
(courtesy of Stuart Myers).

Communicating

Horses use 'body language' to communicate with each other, as well as sound. When working around horses an understanding of their body language is essential for safety and for appropriate training and management techniques. If the horse's signals are interpreted correctly the handler or rider may be able to forestall problems. The ears are a good indicator because they follow the movement of the eyes.

During all interactions with horses, it must be remembered that they are 'fright and flight' animals. Horse are most dangerous when they are frightened because their main concern is getting away and if you are in the way you may be injured.

When a horse is frightened it displays particularly strong body language. It usually stands still so that it can observe the danger while being less noticeable (to the potential predator). The head will be held high and the horse will be staring and have its ears pricked in the direction of the danger. The horse may then spin around and run away or if it is not able to run away it will fight. Ears laid flat back and a lifted hind leg mean the horse is planning to kick. Some horses will strike with a front leg if they are cornered and feel threatened.

A good way to learn the body language of horses is to spend time watching them interacting. Notice the number of expressions and movements a horse makes, especially with its face. Some of the gestures are very subtle, yet the other horse or horses will respond, either by moving away or with gestures in return.

Intelligence and learning

Measuring intelligence in animals is difficult because it is usually based on a human perspective. Horses can be trained to do many things because they have a very high learning ability, but they learn by association, not by reasoning (see Chapter 3 for more information on how a horse learns).

Figure 3. Playing in a dam
(courtesy of Jane Myers).

Differences between the sexes

Mares

Some mares change their behaviour when they are 'on heat' (i.e. receptive to the stallion). They tend to leave their companions, graze intermittently, and may be restless. A normally aggressive mare may become timid or submissive or a timid mare may show signs of aggression. (See Chapter 7.)

Some people consider that mares are not suitable as working horses, an opinion often held by people who have not actually owned a mare, but have heard that they are moody and unpredictable compared with geldings. Mares are capable of performing as well if not better than geldings. Indeed, in the game of polo, mares are generally preferred.

Geldings

A gelding is a castrated male horse, but its sexual behaviour can vary enormously, from non-existent to almost the same as a stallion, though it cannot impregnate a mare. The vast majority of geldings are tractable and will only show any interest in mares in the spring when hormones are on the rise due to new green grass. If a gelding seems to be showing too much sexual behaviour ask a veterinarian to check if the horse has been gelded correctly. A blood test can show whether there is still any testicular tissue present. Sometimes, one of the testicles does not descend completely into the scrotum and therefore may not have been removed at the time of gelding.

Stallions

Stallions share many of the behavioural characteristics of other classes of horses, but they also have some particular characteristics. Stallions defecate differently to mares and geldings. They back up to the dung area and defecate with the hind legs spread apart so that the dung forms a mound. Stallions are more likely to challenge authority, but this does not mean they should be treated roughly. The opposite is true. Stallions should be handled like any other class of horse, but because a stallion is less likely to tolerate mistakes, consistent, confident handling is required.

A stallion is still a herd animal and can become bored and frustrated, and consequently unmanageable, without contact with other horses. Stallions labelled as unmanageable, nasty, vicious or mean have often developed that behaviour because of poor management and handling.

Abnormal behaviour

Certain behavioural problems associated with domesticated horses have been termed 'vices', but actually are stereotypies; that is, obsessively repeated actions by the horse (e.g. rocking from one front leg to the other over and over again, which is known as 'weaving'). It is unfortunate that stereotypies have been labelled 'vices' because that suggests that the horse is deliberately behaving badly.

Sometimes the behaviour can become so entrenched that even when the environment is improved the animal still performs the behaviour, but, given time, the behaviour usually becomes less frequent and the animal only reverts to it when stressed or excited. Undomesticated horses in the wild do not exhibit these behaviours because they are busy trying to survive. Similar stereotypical behaviours are seen in animals confined in a zoo.

Stereotypes are oral (i.e. related to the mouth) or locomotory (i.e. related to movement). Oral stereotypies include windsucking/crib-biting, wood chewing and flank biting. Locomotory stereotypies include kicking stable walls, weaving, box walking and pacing up and down the fence lines.

There are many myths about these behaviours; for example, it is commonly believed that if another horse watches the affected horse, it will learn the behaviour, which is not true because horses do not learn that way. Another myth is that such horses are poor doers and lose condition, which will happen if the horse is prevented from carrying out the behaviour without its environment being improved. Furthermore if the horse is isolated because of the belief that it will teach the behaviour to the other horses, the affected horse will be further stressed and that will lead to a loss in condition.

If your horse performs a stereotypy, try to enrich the environment rather than simply preventing the behaviour from occurring. All horses benefit from:

- increased time spent grazing.
- increased fibre in the diet. If you do not have grazing available, feed as much hay as possible (grass hay rather than lucerne hay), so that the horse spends more time chewing.
- increased time spent in the company of other horses, not just on the other side of the fence. Horses will stay near the fence if they cannot be physically close and will not go and graze.

Oral stereotypies

'Crib-biting' describes the horse grabbing hold of an object, such as the top of the stable door, and gulping air. A characteristic grunt can be heard as the horse gulps. When turned out into paddocks these horses will grab fence posts and because they lean backwards while crib-biting they can loosen the posts. Some horses will arch the neck and gulp air without grabbing hold of something first; this is termed 'windsucking'.

These oral stereotypies may develop as a response to a lack of fibre in the diet, and therefore boredom through inactivity, but once established the horse will perform the action in response to any stress or excitement. A crib-biter or windsucker should be fed as much low-energy roughage (grass hay or pasture) as it wishes to eat. Turn the horse out for as much time as possible (100% is best), preferably with another horse. The companion horse should be easy going, not a bully or this will increase the level of stress/excitement for the affected horse. To protect the fences from a crib-biter, install an offset electric wire or tape (see Chapter 8) and provide a rubber-covered post somewhere in the paddock because the horse will initially continue to crib-bite. It will take time for the horse to decrease the stereotypic behaviour and it may never disappear, in just the same way that humans who bite their finger nails will revert to that behaviour when stressed.

Wood chewing is common in young horses up to approximately five years of age because they are cutting new teeth; however, it may become a habit in some horses that will then chew any wood they can get their teeth on to, including tree branches, bark, stable fittings (including doors) and wooden fencing. Check the level of fibre in the diet, because many horses will chew wood if they have insufficient roughage. There are several commercial products available for painting or spraying on railings etc. to makes the timber unpalatable and discourage this behaviour. You can also give the horses a mineralised or molasses block to chew. Hanging one in a hay net in a tree gives a bored horse something to play with and chew rather than the fencing or the tree.

Locomotory stereotypies

Horses that display locomotory stereotypies are usually responding negatively to confinement. Some horses kick at the walls of their stable while others 'box walk', which is where the horse walks around and around the stable. Remember that in the natural state, horses travel many miles each day; they have not evolved to be confined. Stallions may pace along the fence because they wish to be on the other side of it.

Horses that weave are trying to get out and once a horse has developed the habit it may weave at the gate out of the paddock. Owners sometimes interpret this as a desire to get back into the stable, but it is more likely that the horse wants to get back to where the food is. These horses usually resume weaving in the stable once they have eaten.

HORSE WELFARE

The basic needs of a horse are food and water, companionship, shelter and exercise (see Appendix 1). As the owner, you are responsible for ensuring that these basic needs are met. If you do not, the horse will be stressed, possibly resulting in ill-thrift, poor performance and difficult behaviour.

Welfare is not just ensuring that the basic physical needs of the horse are met, it is also about ensuring that the horse is not subjected to mental stress. An understanding of behaviour is essential for good horse care because you will understand what is important to the horse, which may not be the same as what you believe is important. There are many myths about horses that will only be dispelled through the education of their human handlers!

Housing

A horse can live in a paddock, stable, yard or a combination of these and will adapt well, as long as the basic principles of horse care are adhered to.

Figure 4. Over the fence
(courtesy of Jane Myers).

Ideally, a horse should be in a large, grassy paddock with other horses, a good water supply and plenty of shelter, but unfortunately this situation is not always possible. However, it is usually possible to make changes to the housing situation of a horse that will enrich its life without compromising its use.

Paddocked horses
Keeping horses in paddocks is usually less time consuming for the owner and is better for the horses. Grass is a relatively cheap form of feed and is what the horse evolved to eat. If grass is not available, roughage must be provided in the form of hay. Horses in hard work may need to be supplemented with concentrates also.

Companionship
Horses are more settled if they have a companion; they will even eat more if they can see another horse.

There are advantages and disadvantages to horses being on their own in the paddock or with other horses. Establishing and maintaining the pecking order means that horses can injure each other, although this will usually settle down after a while. Horses on either side of a fence may try to play with each another by running up and down the fence line and try to groom each other over the fence, which can be dangerous. They may also stay at the fence and not graze, rather than leave their companion.

Putting horses in together rather than having each horse in its own paddock means that paddocks can be rotated, making it easier to improve pasture (see Chapter 8 for more details). If you have limited pasture it is better to have horses in yards for part of each day and turn them out to graze together for the remainder of the day. The grazing time can be increased when good pasture is available and decreased when it is not.

Groups
It is best to group horses by age and sex whenever possible. Grouping by age can reduce the number of fights and therefore injuries. Colts should be separated from fillies after they are one year old because of the risk of pregnancy. A stallion must be safely confined either by double fencing or offset electric fencing to prevent interaction with neighbouring horses. Mares should not be paddocked next to stallions.

Stabled horses

Figure 5. In together
(courtesy of Jane Myers).

Stabled horses live in an artificial situation because they are unable to move around much or forage and graze. Ideally, the stabled horse should be turned out into a paddock or yard for part

of each day. The following guidelines are very important for all stabled horses.

- The horse must be exercised regularly. Stabled horses often develop 'filled legs' because the lymphatic (tissue drainage) system is not designed to cope with long periods of inactivity.
- The diet should comprise at least 70% fibre and ideally, a stabled horse should always have grass hay available.
- The horse should be able to see and preferably touch another horse. Modern stable design usually prevents this, so be creative and remove part of the upper half of a dividing wall so that the horses can at least groom one another.

- Try to situate the stable so that the horse can see what is happening around it. If a yard can be attached to the stable so that the horse can go outside this is even better.
- Ensure that the stable is large enough. The more time the horse has to spend in the stable the more important it is that the stable is large enough for its size.
- There should be as much fresh air available as possible. Modern stables are usually built for the convenience of humans and the horse may breath poor quality air, which can damage the respiratory system and thus affect the horse's performance.
- Keep dust to a minimum because it can irritate or infect the respiratory system. Dust and fungal spores mainly originate from bedding and hay. Some forms of bedding are better than others, but it depends on what is available to you. Poor quality hay is dustier than good quality hay and if it must be used it should be soaked just prior to feeding

Figure 6. An example of stables that are better for the welfare of the horse. The horses are able to touch each other over the walls and there is good airflow (courtesy of Annie Minton).

Yarded horses

Many of the issues associated with stabled horses are also true for yarded horses if they are confined full-time. The main difference is that the horses have access to fresh air and can usually interact with one another. Although horses are social, they should not be confined together in small areas as they may fight.

Supervision

All horses should be checked daily for health problems and if the horse is wearing a rug, that the rug is secure and not causing problems (refer to section on rugging).

Run your hands down the horse's legs, checking for swelling or heat, note your horse's respiration rate, any changes in the manure and urine, check if the horse has refused any feed, is drinking water and take a rectal temperature if the horse is not eating or appears unwell (see Chapter 6 for more details about health indicators).

Feeding

If supplementary feed is provided to groups of horses in a paddock, feed them in individual feeders placed widely apart in a large circle or triangle. Dominant horses will threaten and possible injure subordinate horses at feed time, particularly if they can force them into a fence. Alternatively, the horses can be brought in to a stable or yard each day for a feed, which ensures they will be fed according to their needs rather than their wants (e.g. the bully gets the most feed).

Foals and weanlings will usually share feed so they can be fed from trough feeders; however, check that those that need the most feed are getting it. Young horses are not able to feed in peace in the presence of older horses, which may have to be removed or tied up while the youngsters eat.

If stabled horses are receiving hard feed (in addition to their roughage) they should all be fed at the same time each day. It is preferable to split the hard feed into at least three meals, with the last meal being as late at night as possible. This rule is often ignored in many stables because the working day usually finishes at around 5 or 6 pm. The stabled horse must have enough hay to last the night, especially if the last hard feed is given early in the evening.

Exercise

Exercise is more important for confined horses than for those in a large paddock. Controlled exercise can be in the form of lunging, round-yard work, in-hand work, or driving or riding work. If the horse is ill or injured, a short walk may be allowed if the vet approves.

REGULAR HORSE CARE

The hooves, hair and teeth of the horse need regular attention to prevent health problems and to improve the appearance of the horse.

Hoof care

Basic foot care

There is an old saying, 'no foot, no horse', which still holds true because a lame horse cannot be used. Therefore, care of the horse's feet is very important.

The hoof wall grows continuously to compensate for natural wear and tear. It is also able to adapt to different surface and climatic conditions; for example, hard feet result from hard, dry ground. The shape, size and hardness of the foot, and the rate of foot growth vary between horses (see Chapter 4 for detailed anatomy and conformation of the foot).

Lameness

Any signs of lameness should be investigated immediately and any problem treated promptly. A few weeks neglect can prolong the recovery time. A minor injury may still cause pain and the horse should be rested until the lameness disappears. Your regular daily check of the horse's legs and feet will ensure any abnormalities are detected in the early stage. Never neglect cracks, wounds or punctures. Sudden, severe lameness may be a hoof abscess (see Chapter 6).

Cleaning the feet

Stabled or yarded horses need their feet cleaned daily because the continuous contact with highly acidic materials such as urine and faeces predisposes to thrush infections. A paddocked horse will usually have mud compacted in its feet, which is normal and it is usually only necessary to clean the feet before you ride.

How to pick up and pick out the feet

To lift a foreleg, stand beside the horse just in front of the shoulder, facing the hindquarters. Put your hand closest to the horse on its shoulder, then run it down the back of the foreleg to the pastern. Grasp around the back of the pastern, keeping your thumb and fingers together, lean against the horse to make it shift its weight to the other foreleg and then lift the foot.

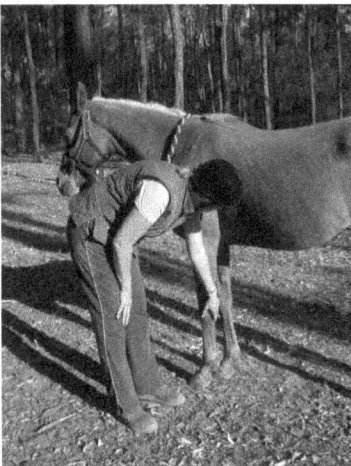

Figure 7. Picking up a foreleg
(courtesy of Stuart Myers).

Once the horse has lifted its foot, support it with your hand under the front of the hoof. Keep your body close to the horse.

To pick up a hind foot, stand level with the horse's barrel, facing the hindquarters. Place your hand closest to the horse on its hindquarters and then run it down behind the hock to the pastern. Grasp around the back of the pastern, keeping your thumb and fingers together, lean against the horse to make it shift its weight to the other hind leg and then lift the foot. If the horse goes to kick, always step back toward the head rather than toward the back of the horse. A horse that kicks when you handle its feet needs further training; if you are not confident and experienced, find someone who can help you.

To clean the foot, use a hoof pick down each side of the frog and around the sole (following the white line) from heel to toe. Clean down the centre of the frog. When you have finished picking out the foot, lower it gently to the ground.

Hoof dressings

Hoof dressings such as hoof oil should be used carefully. The oil will affect the natural ability of the foot to absorb or evaporate moisture. If the horse has good feet and the conditions in which it lives are reasonably natural it is usually best to leave the feet alone.

In very dry conditions (e.g. a prolonged drought) the hooves can become hard and brittle because there is little if any dew in the morning for the feet to soak up. Periodic soaking of your horse's feet in water and then an application of hoof oil to help retain moisture may improve the feet.

In prolonged wet conditions, applying hoof oil can control the excessive absorption of moisture and prevent the horn from becoming too soft. After application, the horse will need to stand somewhere dry until the oil dries, otherwise you will be compounding the problem because the oil will retain any moisture absorbed by the hoof. Only use hoof oil if your horse has a problem with its feet. Horses feet can usually cope, even in very wet conditions.

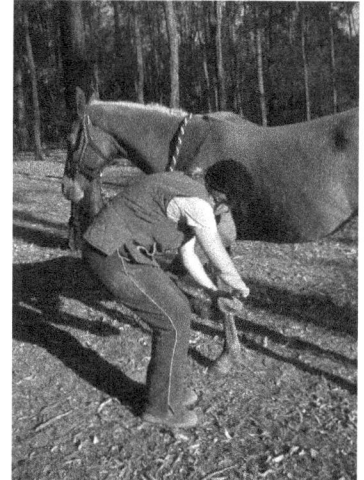

Figure 8. Picking out a foot (courtesy of Stuart Myers).

Hoof oil is applied to the wall and sole. For horses with soft soles that are prone to bruising there are sole-hardening products available. However, as with hoof oil, be aware that anything you apply to the hoof will change the balance of moisture within the hoof. See Chapter 5 for advice on how feet can be improved through correct feeding.

To shoe or not to shoe

Not all horses need to be shod, but they all need good foot care. Stabled horses that work in arenas and are turned out on grass paddocks do not usually need to be shod. On the other hand, horses that work regularly on hard, abrasive surfaces usually do need to be shod and if they are not they can quickly become lame. However, some hardy breeds of horse and pony live their lives without shoes even while working on hard surfaces. Every horse is different and you need to decide what is best for your horse and not do something just because everyone else does it.

In the unshod horse the front feet are usually the first to show signs of soreness because they usually bear 60–65% of the weight of the horse whereas the back feet only bear 40%. Horses that are only in moderate work can have just the front feet shod because the hind feet are more concave and thus cope better with rough surfaces. Many horses, and especially children's ponies, cope very well with this regimen, but if the horse starts to avoid stepping on hard or rocky surfaces then you will need to have back shoes put on as well.

An alternative to shoeing is to use hoof boots, which are slowly gaining acceptance, even with the owners of hardworking horses, and they can be very useful for a horse that is only worked periodically, but still needs foot protection when working. They are also useful as a 'spare tyre' if a horse loses a shoe. Boots must fit well or they will cause rubbing.

Trimming the hooves

The hoof wall grows downward from the coronary band and takes between 6 and 12 months to replace itself. With moderate exercise, depending on the surface, the hooves should all wear at the same rate, but hoof wear varies between horses. If a horse is not regularly exercised and lives on a soft surface the feet will grow long and need trimming every 4–10 weeks. If left too long, the hooves will splay out, crack and put strain on the tendons in the legs. Severe hoof cracks can keep a horse out of work for months so do not neglect trimming.

The aim of trimming is to balance the hoof so that:

- the structures within the hoof can work as they should
- the feet 'break over' (leave the ground) at the centre of the toe
- the feet and legs move in alignment with the direction of the body without swinging sideways and interfering

(a)

(b)

(c)

(d)

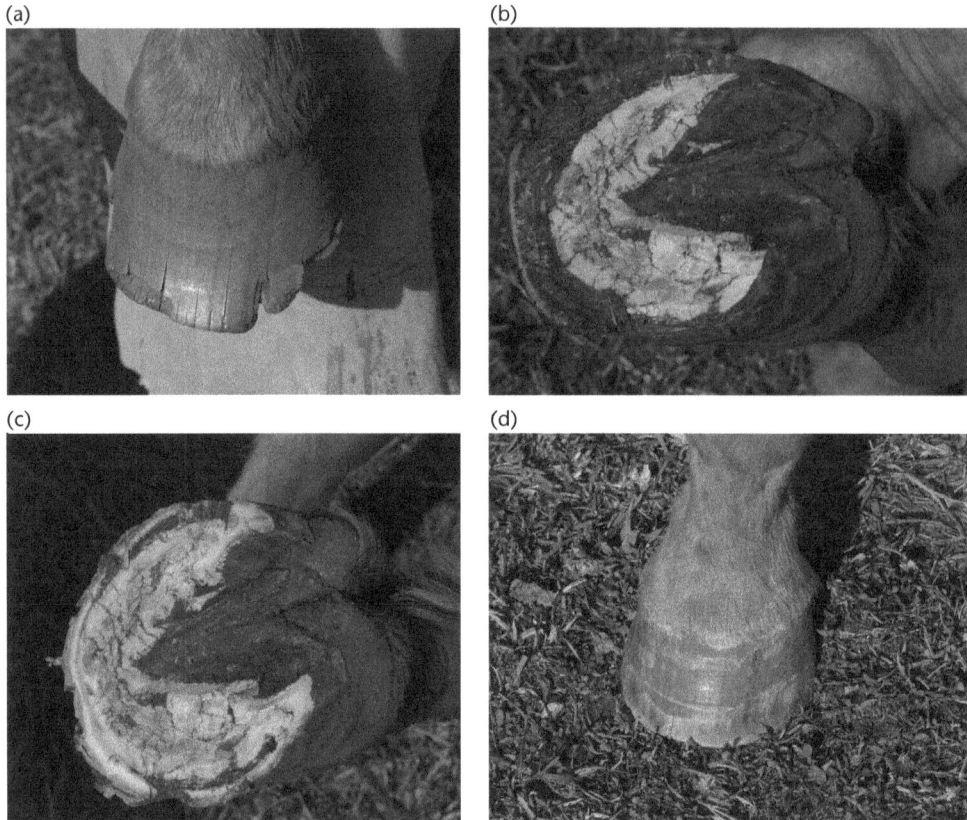

Figure 9. Trimming
a hoof
(courtesy of Annie Minton).

- the feet have enough elevation in forward extension to clear the ground properly
- the foot lands with equal force on both heels of the hoof.

Hooves that are trimmed, but not shod, have less hoof removed than is taken in preparation for a shoe. A correctly trimmed (dressed) hoof, viewed from the front while the horse is standing on a flat surface, should be level from side to side, which you can check by its relation to the vertical of the leg.

Viewed from the side, the slope of the surface of the hoof should be the same as the slope of the pastern. This is the result of a correct depth of heel and a correct length of toe. The foot should also be level from heel to toe. A correctly trimmed hind foot is described similarly except that the hind foot is naturally more pointed and oval.

It is useful to understand the principles involved in hoof trimming, but it is equally important to remember that experience and knowledge of factors affecting the horse's conformation are needed to trim a hoof correctly. Trimming the foot is the most important part of shoeing because correct trimming balances the foot.

Preparation for the farrier
Use a qualified farrier (see Chapter 10 for contacts) and be prepared.

- Catch and secure the horse.
- Clean up and dry the horse's legs and feet.
- Provide a flat, clean, well-lit, sheltered area that is easily accessible for the farrier's vehicle.
- Keep dogs and small children away while the farrier is working.

Some farriers prefer a horse to be tied up and others prefer that you hold it. If you are holding the horse, always stand on the same side as the farrier to prevent the horse swinging into the farrier.

Most farriers use machine-made shoes, but should be able to make shoes specifically for your horse, if required. The shoes should be made to fit the hoof and not the hoof fit the shoe.

Tools and equipment

The tools that the farrier uses are a rasp, a buffer (clench cutter), pull nippers, clencher, hammer, hoof nippers and hoof knife.

Figure 10. Farrier tools (courtesy of Jane Myers).

Indications of a correctly shod foot

When the feet are placed on the ground:

- the front and hind feet should be pairs, with the same size and shape and the same correctly aligned foot–pastern axis.
- clenches should be even, flat and broad. Nails should be pitched higher at the toe than at the heels and not driven into cracks.
- there should not be rasping of the wall, unless there has been flaring, except for a little below the clenches where they have been rasped smooth.
- there should be no shortening of the toes ('dumping') at the front to conceal badly fitted shoes.
- the toe clips should be low, broad and centred.
- the shoes should fit the outline of the feet and the heels should be of the correct length.

For each foot, when lifted:

- the nails should be driven home; the heads should fit their holes and protrude slightly.

Figure 11. Shoeing (courtesy of Jane Myers).

(a)

(b)

(c)

(d)

(e)

(f)

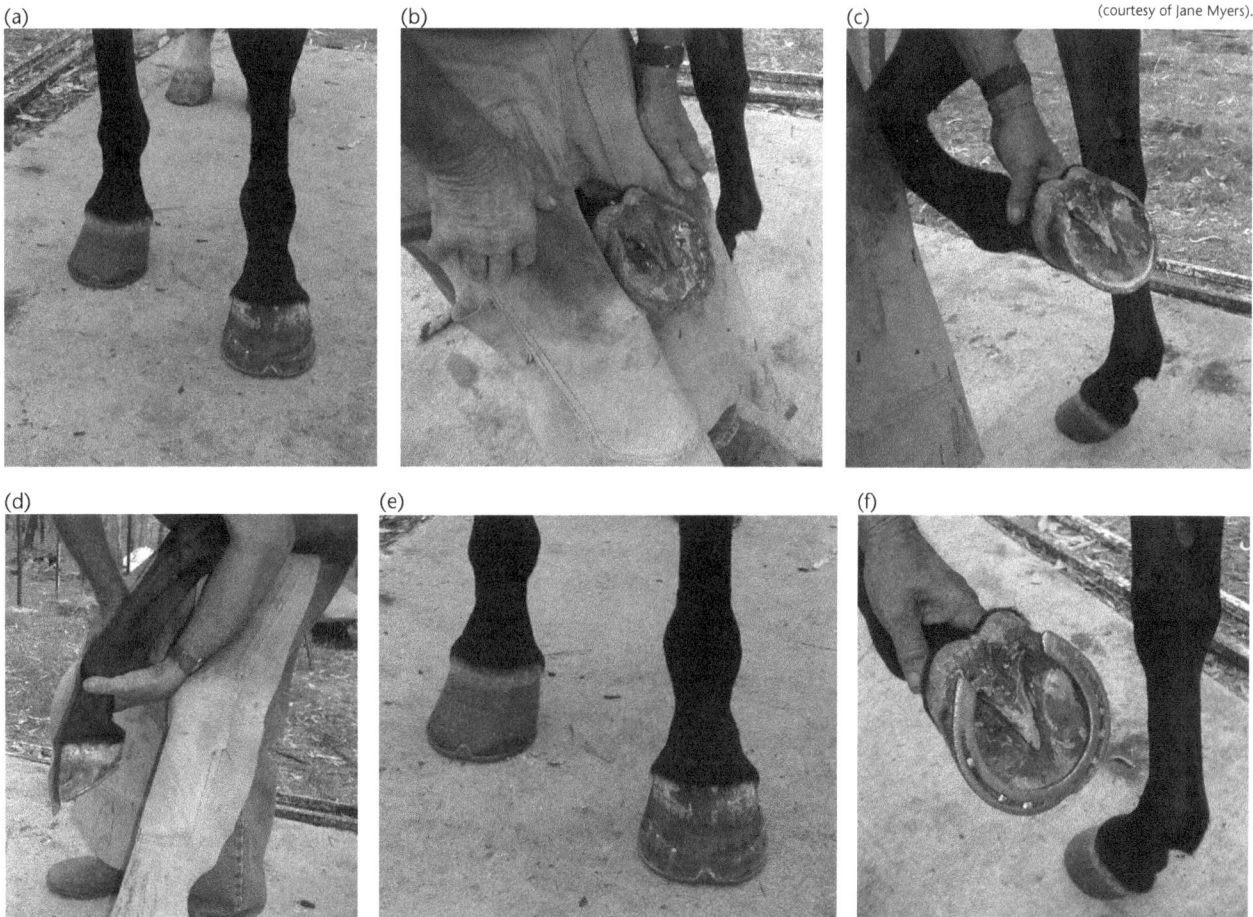

- the heels should not be opened up (opening up the heels means the bars have been cut away to make the foot wider).
- the toe clip should be centred and in line with the point of the frog.
- the frog and sole should not be excessively pared.
- the sole should be eased at the seat of the corn.
- no daylight should be visible between the foot and the shoe, which would indicate an unevenness of either the bearing surface of the foot or the foot surface of the shoe.
- the shoe should fit the foot and the heels should not interfere with the function of the frog.

When the old shoes are taken off, check how they are wearing. If the wear is even, the foot has been reduced to its correct proportions and the type of shoe is suitable. If a horse with normal action wears out the shoes more quickly than average then a wider webbed shoe rather than a thicker shoe can be fitted. Your farrier will advise.

Hoof care after shoeing

Once a horse is shod it should not go any longer than eight weeks, preferably six, without being checked by the farrier. Shoes should be reset (i.e. removed) and replaced every 4–8 weeks, depending on the rate of foot growth and shoe wear. Once shod the foot does not wear down and as it grows longer the horse's base of support changes, placing an increasing strain on the tendons of the leg. The horse may strike itself with its feet, or with loosening nails and shoes. Before replacing the shoes, the hoof should be trimmed.

Corrective shoeing for lameness

Corrective shoeing is a very skilled job. Done properly it can improve lots of conditions; done badly it can make a problem much worse. Not all farriers have the skill or experience to do corrective shoeing, so ask experienced horse people for a recommendation (see also Chapter 10).

Some of the conditions that may benefit from corrective shoeing are:

- sole bruising
- navicular disease
- pedal osteitis
- laminitis
- side bone
- spavin or hock arthritis
- ringbone
- contracted flexion tendons
- severed extensor tendons.

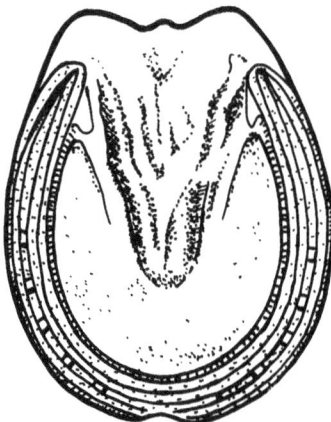

Figure 12. Correctly fitted shoe.

Flat feet and shoeing

A flat-footed horse is more prone to bruise the sole and will need shoes that lift the sole off the ground. It may be necessary to also fit pads to give extra protection, particularly if the ground where the horse lives or works is rocky.

Laminitis and shoeing

Horses that have had laminitis in the past may need to be kept shod because laminitis causes the soles of the feet to drop and the horse will need the extra clearance that a shoe gives.

Periods without shoes

All horses' feet benefit from a period without shoes, which allows the feet to grow without the damage of nails. If your horse is spelling from work, have the shoes taken off.

How to remove a shoe

There are times when you may have to take off a shoe, usually because it has twisted and become loose and if you do not the horse may be injured. It is possible but difficult to take off a shoe without specialised tools. In an emergency, you may have to use whatever tools are available in your garage however, a basic set of farrier tools will make the job a lot easier (see Figure 10).

Wear heavy duty jeans and if you have them, full-length leather chaps. A horse can cause serious damage to your legs if the nails of the shoe are displaced. Lift the horse's foot and place it between your legs with the shoe uppermost. Work the chisel edge of the buffer under each clench in turn and hammer the buffer until the clench is straightened out. When all the clenches have been straightened, take the pincers and work the jaws under each side of a branch of the shoe, near the heel. Push downwards and jerk the handles in towards the middle of the foot. When you have sprung one branch of the shoe, move the pincers to the other branch, again pushing downward and jerking inward. This direction of movement will avoid breaking the horn from the edges of the walls. When both branches have been sprung near the heels, work around each branch towards the toe until the shoe comes away. Pick up all the nails to prevent them puncturing either the horse's sole or car tyres.

Teeth care

Horses need regular dental care if they are to get the maximum benefit from their feed and thus perform well. Horses can either be born with dental problems or they can develop them during their life.

Because a horse's natural diet requires a lot of chewing and grinding, which wears down the teeth, correct dentition is essential. If the horse's ability to grind down the food is compromised for any reason, the food will not be fully digested and the horse will lose condition.

Horse aged 2–5 years should have their teeth checked prior to commencing work or at 6-monthly intervals. After the age of 5 years, all horses should have an annual dental check, more often if the horse is being fed concentrates.

Tooth problems

Sharp cheek teeth

The most common dental problem is sharp cheek teeth (the molars). The horse's upper jaw is wider (by 30%) from side to side than the lower jaw and the upper molars are also 50% wider than the bottom molars. The horse chews by grinding its bottom teeth and the food against the top teeth in a circular motion. If the wear is uneven, sharp edges and hooks will develop on the inside edge of the bottom molar and the outside edge of the top molar, cutting the cheek and tongue and making chewing very painful. The horse may reduce its intake or eat more slowly.

Shear mouth

An extreme form of sharp cheek teeth occurs when the difference in width between the upper and lower jaws is more than 50%, a condition known as 'shear mouth' because the abnormally sharp teeth resemble shearing blades. It occurs most commonly in old horses.

Figure 13. Side view of the teeth.

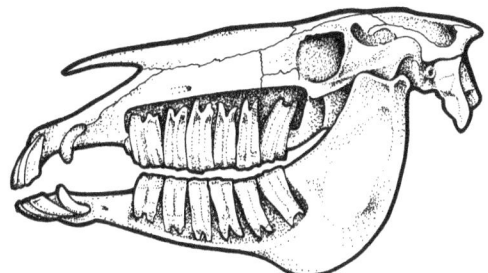

Treatment of sharp teeth consists of removing the sharp edges by rasping with a dental file, restoring the shape of the teeth as closely as possible to normal.

Wave mouth

Uneven wear can give the molar teeth a 'wave-like' profile, the biggest crest and trough occurring at the fourth molar. The teeth are worn to gum level in some cases, which causes the opposing teeth to cut into the gum and may result in infection of the jaw bone. Treatment consists of removing the sharp edges and correcting any excessively long teeth.

Step mouth

Neighbouring molars vary in length, either because the opposing tooth is missing and consequently there is lack of wear or the opposing tooth is abnormally soft. Whatever the cause, the excessively long tooth or teeth may make chewing impossible and the extra length has to be removed using a special surgical instrument.

Smooth mouth

Molar teeth have a rough grinding surface that is maintained throughout life because the denture of the tooth wears faster than the enamel. In some cases, though, the enamel and denture wear at the same rate, resulting in a smooth grinding surface.

Smooth mouth is seen most often in young horses. In old horses, it occurs when the crown of the tooth is worn down to the root of the tooth that does not have enamel. Although this condition does not cause pain, if enough teeth are affected the horse cannot chew its feed properly, resulting in poor digestion and possibly colic. The condition cannot be corrected by dental treatment and management involves feeding soft feed that requires little chewing.

Incorrect conformation of the mouth

Few horses will have perfect apposition of the front teeth (incisors), but gross defects ('parrot mouth' or 'sow mouth') are unmistakable. 'Parrot mouth' is when the upper jaw overhangs the lower jaw and 'sow mouth' is when the lower jaw is longer than the upper jaw (see Chapter 4, Figure 8). Sow mouth is less common, but if badly affected a foal may be unable to suck.

Horses with either defect will have difficulty grazing effectively and may lose weight while on pasture if not given supplementary feed. These horses will also need regular rasping of the molar teeth.

Extra teeth

Extra teeth are occasionally found, but do not need to be removed unless they interfere with chewing or the bit or because of crowding the extra teeth cause a regular tooth to grow in an abnormal direction, injuring the soft tissues.

Wolf teeth

The 'wolf teeth' are not extra teeth, but are vestiges of teeth that occurred in the ancestor of the modern horse. If present, they are usually found in the upper jaw next to the first molar and are approximately 5–10 mm in length. The wolf teeth can be removed if they are a nuisance to the horse or are loose. A loose wolf tooth may cause a horse to toss its head or be reluctant to respond to the bit. Otherwise, they can be left alone.

Figure 14. Teeth bumps visible on the jawline (courtesy of Jane Myers).

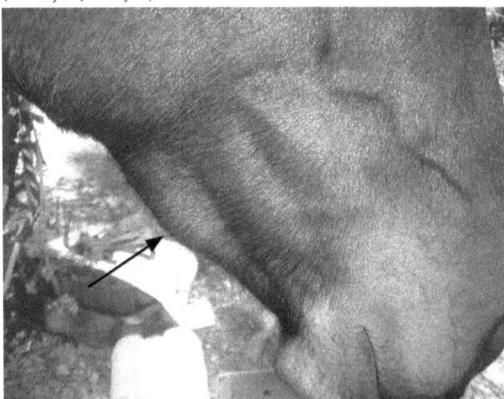

Teething

As with human babies, the eruption of teeth in young horses may cause transitory trouble. Temporary teeth may fall out prematurely, leaving a depression in the gum surrounded by an inflamed margin, which may cause difficulty with eating. Temporary teeth may not fall out, but instead form a cap over the permanent teeth. Veterinarians and horse dentists will routinely remove these because they may cause problems. Incisor caps may be present in 2–5-year-old horses.

It is not unusual for 2–3-year-old horses to have a swelling of the bottom of the jaw at the level of the second or third molar tooth. This is normal and corresponds to the eruption of the permanent molars. Be careful with the fitting of nosebands in these horses until the teeth have settled down (see Figure 14).

Broken teeth
The front teeth may be broken in an accident and the molars may be broken if the horse chews on an unexpectedly hard substance, such as a stone or piece of metal. Infection of the root and surrounding bone may follow if the fracture extends below the gum line.

Decayed teeth
Damage to the tooth and the formation of a small cavity on the chewing surface can let particles of food decompose in the cavity. Bacterial activity contributes to the destruction of the tooth, which may lead to infection of the surrounding bone. If the decayed tooth is an upper molar, the bones or the sinuses of the head can become infected and pus may appear in the nostril on the affected side. If the affected tooth is a lower molar, there may be a noticeable swelling in the lower jaw.

Crib-biting/windsucking
This behaviour will result in wear of the incisor teeth and the biting surface may be bevelled to such an extent that the horse is unable to graze because the teeth no longer meet and therefore cannot cut the grass. The horse will need supplementary feeding while at pasture (see 'Abnormal Behaviour' earlier in this chapter).

Signs of teeth problems
Weight loss
Sharp teeth, infections, cracked or broken teeth or poor mouth conformation can lead to reduced feed intake. Older horses with lost or very worn teeth may have a reduced ability to crop grass and chew feed, which can be a major cause of weight loss in an otherwise healthy horse.

Rate of chewing
Tooth problems may affect the rate of chewing. The horse may stop chewing for a few minutes, then start again.

Quidding
A horse is said to be 'quidding' when it picks up the food, forms it into a bolus, and then lets it fall from the mouth after only partly chewing it. The semi-chewed food may become packed between the teeth and cheeks instead of being dropped.

Bolting of food
A horse may bolt its food to avoid using a painful tooth, leading to indigestion and colic. Undigested food may be seen in the manure; however, this may also be seen when the horse is eating oats (in which case it is usually just the husks in the dung), or if the horse is a greedy eater and does not chew its food properly.

Unresponsiveness to the bit or head tossing
The bit may push the cheek against the sharp edge of a molar tooth, causing extreme pain. In the case of a racehorse, the animal may be unresponsive to the rein and travel wide at the turns. If a horse seems unresponsive to the bit, one of the first things to do is check its mouth. Head tossing when 'on the bit' may also be the result of teeth problems.

Excessive salivation
The horse may drip saliva, which may be tinged with blood, if sharp teeth have cut the insides of the cheeks or tongue.

Figure 15.
Hausmann's gag
(courtesy of Jane Myers).

Other signs

Other signs of teeth problems include bad breath, swelling of the face or jaw, lack of interest in hard food, yawning, loss of condition and loss of coat shine. Simply watching your horse while it is eating can tell you much about its teeth. Watch for slow eating, reluctance to drink cold water, tilting the head while chewing and moving the food around in the mouth before swallowing. Any indications of poor dental health in horses should be investigated. It is also worth remembering that some horses with chronic dental problems that are causing weight loss show no outward symptoms at all and therefore, in cases of gradual loss of condition, bad teeth should be investigated as a possible cause.

Examination of the teeth

A horse should be trained from an early age to allow inspection of its teeth in much the same way as it is taught to allow its feet to be picked up. Examining a horse's teeth is then a simple matter. However, you may still receive a painful bite or cut your hand on a sharp tooth unless care is taken.

To examine the teeth on the off side, approach the horse from the near side. Place your right hand into the horse's mouth in the space between the front and back teeth. Grasp the tongue and gently draw it out of the mouth (see Chapter 4, Figure 45). Now slide your left hand into the off side of the mouth, between the teeth and the cheek, with your knuckles toward the cheek and run your fingertips along the upper and lower teeth. As long as the tongue is held outside and against the corners of the mouth the horse will not usually close its mouth. Do not prolong the examination; 10 seconds should be long enough. Reverse the procedure to check the other side.

This examination will only indicate if the premolars are sharp. A veterinarian or horse dentist should regularly examine the horse's teeth and carry out any necessary work (see the end of the chapter for contact details of horse dentists)

'Floating' the teeth

The vet or horse dentist will usually use a Hausmann's gag, (see Figure 15) which is placed between the front teeth and stops the animal closing its mouth and biting fingers or equipment. A special tooth file (float) is used to remove the sharp edges of the teeth (see Figure 16).

Coat

Rugging

Figure 16. Floating the teeth
(courtesy of Jane Myers).

There are pros and cons associated with rugging horses. Generally, in mild climates a rug is unnecessary if adequate feed and shelter are provided. However, the factors you should take into consideration include the local climate, your budget, the work the horse does, its sensitivity to biting insects, where the horse lives (in stables, outside or both) and how much time you have available. If you decide to rug your horse, the rug must fit correctly and not rub at the shoulders or wither. The rug must not hang down on one side of the horse. You must check that all fasteners are correctly adjusted and closed to prevent the rug from slipping or flapping, which may entangle or frighten the horse and result in injury.

Under natural conditions, a horse will roll in the mud and dust and build up a protective layer of dirt and the oil secreted by the skin. The problem arises when you want to ride. The horse may be muddy/dusty/wet etc., which will affect your tack and may cause chafing if you ride the horse when it is dirty or wet. If the horse has a long winter coat and the weather is cold, working in the afternoon will cause the horse to sweat and it may not be dry before the evening gets chilly. The horse will then be cold and wet during the night if it is not rugged. If you can avoid getting the horse too sweaty you can manage without rugging.

The coat of a rugged horse will lay flatter and the horse will look smooth and shiny compared with an unrugged horse that will always look fluffy. The rugged horse will be cleaner when you want to ride. However, rugging a horse requires more work. For most owners that compete with their horses, rugging is the norm, but the horse should not be excessively rugged, which is usually for the owner's convenience and not for the horse's benefit.

Types of rugs

Heavy canvas rugs, also called New Zealand rugs, are suitable for horses kept outside during cold weather. They are heavy duty and will be nearly 100% waterproof if cared for properly. They must be well made to ensure the best fit for the horse. This type of rug is usually lined with either mixed fibres or pure wool, which is more expensive, but warmer.

Synthetic outdoor rugs are alternatives to canvas rugs, but they vary in quality according to price. The advantage of a synthetic rug is that it weighs less and is therefore easier to put on a large horse and the lighter weight is probably more comfortable for the horse as well. However, when it is sunny, they can make the horse hot, hotter than rugs made from natural fibres. Many of these rugs now have a different cut to the neckline, which brings the rug almost half way up the neck, reducing the pressure on the withers in comparison with a traditional style of rug.

Quilted (doona) rugs are used for warmth and are usually made of nylon, although those made of cotton are considered better in terms of heat regulation, but they are not as hardwearing. Doona rugs can be used either on alone for the stabled horse or under an unlined or lined canvas rug for the outdoor horse.

Woollen rugs are suitable for travelling and when attending events. If kept well-maintained these rugs look attractive and once they are worn out, they can be used as under-rugs. They are expensive, but wear well and last for years.

Sweat rugs, which look like string vests, are specifically designed to be placed underneath a cotton or woollen rug on a hot and sweaty or wet horse. When used correctly, the pockets of air trapped between the rug and the skin assist in warming up or cooling down the horse. They are not designed to be the only rug on the horse. The old-fashioned method of cooling and drying a horse by putting straw (or hay) under the rug so that the horse dries without soaking the rug is better than using a sweat rug. The straw/hay soaks up the moisture and the horse can either be left in the stable or turned out in the paddock where the straw or hay will gradually fall out as the horse dries.

Jute rugs are the stable rugs that were available before doona rugs. Even though they are considered old-fashioned, they are very useful as a spare rug and can be used on a wet horse. Racing stables often still use them as the main rug that they put on upside down while the horse dries after hosing and then turn over once the horse has dried.

Cotton rugs are useful in a variety of situations. During warm weather it is not necessary to rug a horse unless the nights are cool, but a light canvas or cotton rug will keep the horse's coat free of dust. Cotton rugs can also be used as stable rugs, travel rugs or under-rugs and will help protect against insect bites.

Neck rugs and hoods can irritate the horse if they are not fitted properly. The eye and ear holes of a hood should be big enough to enable the horse to stretch its head down without impairing its vision or rubbing the eyes or ears. The hood or neck rug must be fastened to the rug in such a way that the horse can put its head down without having to pull against the rug. Hoods and neck rugs can be made of wool, synthetic, Lycra® or canvas. Some will rub the mane out if they are worn continuously. If the horse is to wear a hood it must be checked several times a day.

Figure 17. Correctly fitted outdoor rug
(courtesy of Jane Myers).

Fitting a rug

Rugs must fit well. Outdoor rugs should be slightly roomier than indoor rugs. You should be able to run your hand from the chest, up and over the withers without feeling any tightness. The leg straps should cross through each other and the cross-over point should be roughly between the hocks and where the buttocks meet. Cross-over surcingles should be snug, but not tight. If they are too loose the horse can catch its leg in the strap when getting up and down. Never use a tight single surcingle to keep a rug in place: it does not work and the horse will be very uncomfortable.

To correctly fit your horse for a rug, measure from the centre of its chest to a point level with the buttock. Rugs are still sized in feet and inches; for example, a 14.2 h pony will usually need a 5 foot 9 inch rug. Alternatively, you can tell the saddlery store or rug maker the height of your horse and they will advise you on the size required.

Rugs and injuries

Rugs can injure your horse if they are not fitted correctly and checked regularly. They can cause:

- Pressure sores behind the withers because the rug is too large around the neck, which results in the rug sitting behind the withers. Every time the horse lowers its head it pulls the rug into its withers.
- Pressure sores on the withers. Some horses with high withers can become sore simply from the weight of the rug.
- Pressure sores at the base of the neck where it joins the chest, which can make the horse very reluctant to move. If it does move, it may move in 'fairy steps' around the paddock. Some horses get sore in this area even with a well-fitting rug, in which case a padded bib is needed under the rug.
- Sores on the inside of the thigh from the rug slipping over to one side and the leg strap pulling up into the groin.
- Sores under the belly from the cross-over surcingles and a badly fitting hood. If the fasteners for the hood do not have enough 'give' the surcingles pull tight under the belly as the horse reaches down to graze.
- Horses can get hooked on the fence via their leg-strap fasteners and injure themselves. Always clip the fasteners in toward the horse to reduce the chances of this happening.
- If any fasteners break, the rug will usually slip and the horse may panic. Panicking horses can injure themselves on fences as they attempt to get away from the flapping rug.
- Hoods and neck rugs that slip can be the cause of horrific injuries to a horse. Horses have been known to drown in dams or injure their eyes because of slipped hoods.

On a lighter note, it is not unknown to find your horse uncovered in the morning and the rug in a pile somewhere in the paddock with all the straps done up.

How many rugs?

The number of rugs needed obviously varies according to climate, but must be appropriate. In cold weather an unrugged horse raises the hairs of its coat to trap air within the coat and thus warm itself. A horse in a badly fitting, thin rug can be colder than an unrugged horse because the rug will flatten the hair and restrict the movement of the horse without providing any real warmth.

In cold weather, the correct type of rugs can help the horse maintain condition, because a cold horse uses a lot of energy just keeping warm. Cold, driving rain is the worst type of weather for horses and without adequate shelter the horse will need a rug. Regularly remove and check underneath the rugs because if they are leaking water the horse can develop rain scald and other skin problems. Heavy canvas rugs can become much heavier when wet and therefore more than one canvas rug is not recommended. It is better to put lightweight doona-type rugs under a heavy canvas rug if extra warmth is needed.

Rugs can substitute to some extent for shelter in bad weather, but shelter/shade must be provided for a horse in hot weather. Rugs should not take the place of shade. In hot weather, a rugged horse will lose condition because it will sweat to try and cool down, thus using energy and losing weight. Anything more than a thin, light coloured, cotton sheet in hot weather can lead to overheating, which is not in the horse's best interests and could become a welfare issue; even a very light level of covering can be too much in some temperatures.

Rugging a clipped horse

Horses in a moderate to heavy work programs are often clipped and will need to be rugged to substitute for the lack of a thick coat. Use a cotton doona rug under a canvas rug because even a lined canvas rug tends to rub on the thin coat and is not usually warm enough.

Rugging a stabled horse

The stabled horse is not necessarily warmer that the outdoor horse; it depends on the design of the stables and how many other horses are in the building. A stabled horse is inactive and thus can become cold. Check your horse periodically to see if it is warm enough late at night.

Checking a rugged horse

Horses that wear rugs must be checked more frequently than horses without rugs, especially if they are in a paddock. Rugs can get caught and/or slip and become dangerous to the horse. Twice daily, morning and night is recommended. Horses that wear hoods and/or neck rugs must be checked more frequently.

Rugging the itchy horse

When horses live in a group they will stand nose to tail and flick away the insects for each other. Biting insects can disturb some horses to the extent that they will panic. Lightweight, mesh rugs can assist in protecting the horse, but parts of the horse will always be exposed. Horses in Queensland may suffer from Queensland itch, which is thought to be an allergic reaction to certain biting insects. Rugging is only one aspect of the care of these horses.

Putting a rug on an inexperienced horse

Training a young horse to wearing a rug must be done very carefully. Never rug the horse for the first time and then let it loose in a paddock. Horses can gallop headlong into fences when they are frightened.

Have an assistant, who is confident and capable, to hold the horse in a small yard or stable (from which you can escape quickly if necessary e.g. slip out of through the rails). Fold the rug into a tube with the front section on the top. Then, while talking to the horse, slide the rug across the withers. Carefully unfold the front and secure it. Next, unfold the back so that the horse is covered. Carefully attach the leg straps. Leave the horse in the yard/stable to become accustomed to the rug. Always remember that when two people are handling a horse they must both be on the same side of the horse at all times (see Chapter 3 for more details on horse handling).

Most horses will be unconcerned, but some nervous horses will catch sight of the rug and jump forward, trying to run away from it, which is why you introduce the rug in a small, secure area such as the yard/stable. Give the horse some time to get used to wearing the rug (e.g. a few hours). You can then catch the horse and play around with the rug, slapping it and moving it about on the horse. Slapping the rug simulates the noise that the rug will make when the horse brushes against a tree for example. Continue to do this until the horse is not concerned about the rug. It is also a good idea to lunge the horse in a roundyard for a couple of circles in each direction to check that the horse really is unconcerned.

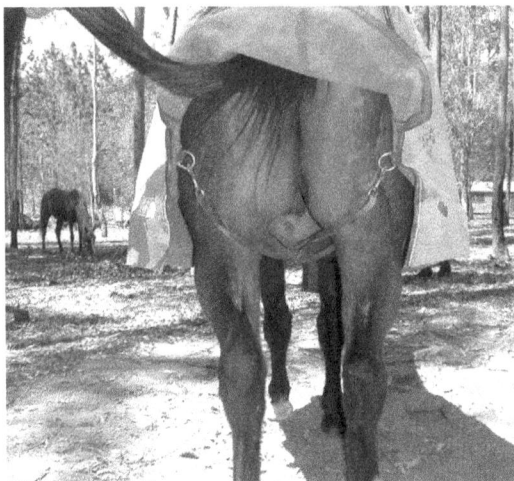

Figure 18. Leg
straps
(courtesy of Jane Myers).

Do not turn the horse out into a paddock until you are certain that
the horse is calm and accepting of the rug.

Putting a rug on an experienced horse

Throw the rug gently across the horse's back with the leg straps
fastened to themselves so that they do not swing over and hit the
horse. When removing a rug the leg straps should be fastened back
to themselves before unrugging.

There are two schools of thought about whether it should be the
chest straps or the leg straps that are done up/undone first when
putting on or removing a rug. If the chest straps are done up first
and the horse gets a fright before the leg straps are closed, the rug
can slip around the neck and between the front legs, which can
further frighten the horse. On the other hand, if the leg straps are
done up first and the horse gets a fright before the chest strap is
closed the rug will slip back and the horse may panic and kick out
because the straps have slipped into an unfamiliar place.

Whichever way you decide to fasten/unfasten the rug always do it as quickly as possible to reduce
the amount of time the rug is not completely secured.

It is a good idea to train your horse to accept unfamiliar situations with the rug to reduce the
likelihood of it panicking if something unusual happens (see Chapter 3 for instructions on training your horse).

Care of rugs

Whenever the rugs are removed they should be aired. Stable rugs need to be washed frequently
because the horses will lie down in the manure and wet bedding. Washing heavy doona-style rugs
can be difficult because they are usually too large for a non-commercial type of washing machine
and public laundrettes often will not allow horse rugs to be washed. Doona rugs can be hosed,
scrubbed and hung out to dry when the weather is fine. A cotton rug placed under a doona rug
will reduce the number of times the doona needs to be washed. You will need two cotton rugs so
that you always have one clean, dry rug.

The skin of some horses is sensitive to certain washing products, so use products suitable for
babies or a wool wash.

Heavy canvas rugs should be hosed and re-waterproofed once or twice each year after the wet
weather has ended. This is also the time to have the rugs repaired before they are put in storage.
Heavy canvas rugs should be hung rather than folded when in storage because folding can damage
the fibres along the creases.

Always check a rug for spiders and damage after it has been in storage, before putting it on the
horse. Check between the lining and the canvas if the rug is lined because this is a favourite
hiding place for spiders. Always wear protective gloves. White-tail spiders in particular are dangerous to both horses and humans.

Grooming

Grooming is the regular care of the coat, and the feet, of the horse. It is of particular importance
when an animal is being worked. The major reasons for grooming your horse are:

- to maintain and promote good health
- to ensure cleanliness and improve appearance.

The amount of grooming that a horse needs depends on its living conditions. A horse that is
constantly rugged needs to be groomed more than a horse that is not rugged.

Grooming the unrugged horse

An unrugged horse, presuming that it lives in a paddock with other horses, is able to remove dead skin and hair, massage the body and scratch itches by rolling, rubbing on objects such as trees and mutual grooming. The weather also plays a part in removing dead skin through the action of wind and rain. The unrugged horse will not need to be groomed until you ride and even then it is only the areas where the tack comes into contact with the horse that really matter.

Before you saddle up any dirt or matted hair must be removed from the saddle and bridle areas, otherwise it will cause rubbing and soreness. The horse's hooves should be picked out to remove any wedged-in stones and the feet and shoes inspected for maintenance. The rest of the horse's body can be groomed for smart looks, but does not need to be as thorough as for the horse that is rugged. In fact, the unrugged horse should not be too thoroughly groomed or bathed because it will remove the oil from the coat that 'water-proofs' the horse in wet weather.

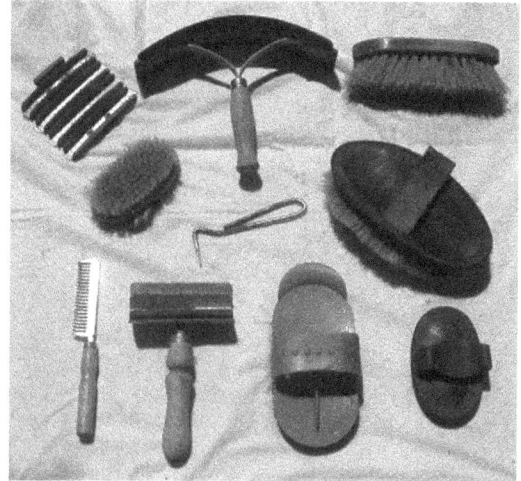

Figure 19. Grooming tools. Clockwise from top left: curry comb, sweat scraper, dandy brush, body brush, rubber curry comb, plastic curry comb, another metal curry comb, mane comb, small face brush and hoof pick (in the centre) (courtesy of Jane Myers).

After exercising in cold weather, the horse should be cooled down and dried with a towel before being turned out into the paddock. After exercise on a warm day, the sweat can be removed with a sponge or a hose (no detergent) and the horse returned to the paddock to roll in the dust.

Grooming the rugged horse

Irrespective of whether it lives mainly in a stable or in a paddock the rugged horse must be groomed frequently. Rugs cause a build-up of the dead skin that is shed daily and which in the unrugged horse is sloughed off by natural means. Also the rugged horse needs the massage that is part of daily life in the more natural-living horse. It is a good idea to allow the rugged horse some time each day without the rug so that it can roll and get dirty (e.g. when the horse is sweaty after work). The horse can then be cleaned and rugged again. A good tradition in racing stables is to let the horse roll in a sand roll after exercise, then hose it down and put on the rugs.

How to groom

A basic grooming kit usually consists of a dandy brush, body brush, plastic and rubber curry comb, sweat scraper, hoof pick and a mane comb. A dandy brush has firm bristles for removing dried mud. For very thick mud or for removing dead hair, you can use a rubber or plastic curry comb. Softer brushes, such as the body brush, are for removing dust. A mane comb, even though it is usually supplied with a kit, is not very useful and a human hairbrush works better for the mane and tail.

Figure 20. Pulling the mane (courtesy of Annie Minton).

A metal curry comb is a useful addition to a basic kit. Swipe it across the softer brushes when grooming and it will clean them. Periodically bang the metal curry comb on a wall to clean it. A cloth is another useful addition as it can be used to give the horse a final polish after grooming. The feet should be picked out as part of the grooming routine.

Hose or groom?

Grooming takes longer than hosing so many horses are now hosed after work instead of being groomed, which is a pity because the horses do not get the massaging effects of grooming. A sweat scraper is used to scrape the water off the horse.

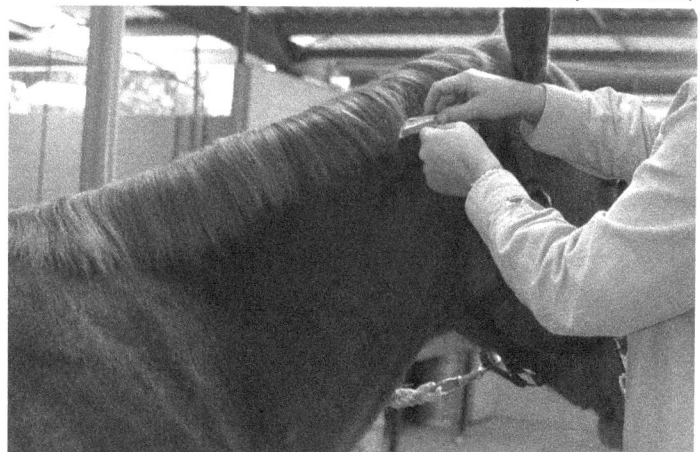

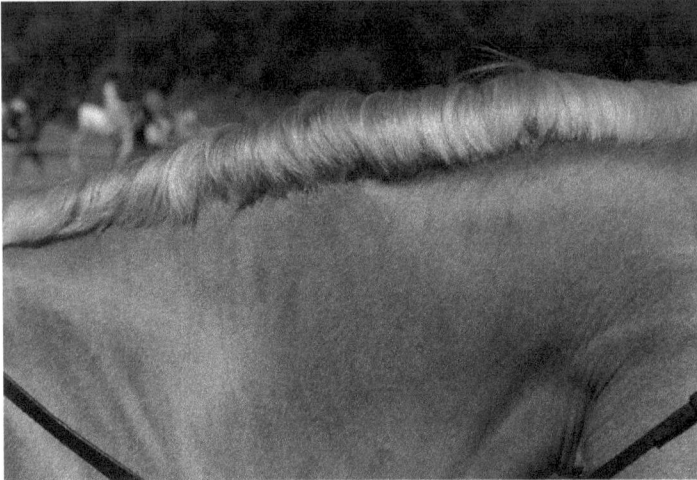

Figure 21. A pulled mane (courtesy of Annie Minton).

Preparing for competition

Events such as shows, horse trails, dressage and show jumping often require horses to be fully groomed, plaited and trimmed.

Pulling the mane

To improve presentation and ease of plaiting, long or thick manes are kept pulled. A pulled mane looks neat and does not interfere with the reins while you are riding. If your horse has a very long thick mane, pull it gradually rather than all at once because it may hurt the horse to do the entire mane. Some horses do not mind having their mane pulled; others are very sensitive and will not tolerate it and for those horses you can achieve the 'pulled' look with special trimming scissors and cutting combs available from a saddlery store.

To pull the mane, first brush and then comb it out. Starting at the poll, comb a narrow hank of mane almost all its length. Grasp the longest hairs below the comb with one hand, then holding them firmly, backcomb the section of mane, wrapping the longest hairs around your forefinger or the comb, pulling them out with one brisk pull. Comb out and repeat up and down the mane until it is all approximately 10 cm long.

Plaiting the mane

Divide the mane into sections, plait each section and put a rubber band around the end of each plait to secure it, double the plait up and secure it with the rubber band, then double the plait up again or roll it into a secure ball. Take a second rubber band and fasten it around the entire plait as many times as possible. Do not leave the plaits in any longer than necessary as the skin in that area can get sore from the tension on the mane (anyone who has worn their hair plaited will understand this).

Braiding the tail

Tails can be pulled, trimmed or braided. Pulling is neat; the hairs on either side of the dock are pulled out until the dock looks narrow and neat. Often people do the same with scissors, which is kinder for the horse and will look attractive if done correctly. Some professional show people now do this rather than pull the tail. If you can do a neat job a braided tail looks very nice. A tail braid is not unlike a French plait. Hair is taken from the side of the tail into a central braid. Remember, practise makes perfect. The circulation in the tail is poor so do not leave the braid in for any longer than a few hours or the tail may be damaged and in any case it is uncomfortable for the horse after a while.

Figure 22. Plaiting the mane (courtesy of Jane Myers).

(a)

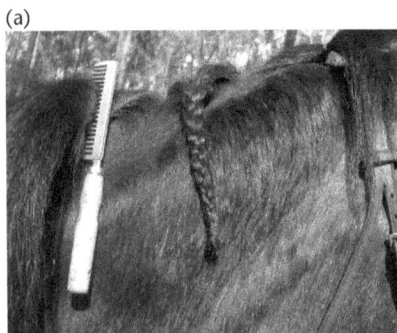

(b)

Trimming excess hair

Trimming the horse will make it look neat, but you must be careful because some of the hair you may regard as scruffy is essential to the horse. You should never trim the inside of a horse's ears (even though many people do); you can neaten the look of the ear by holding it closed and only trimming the hair that sticks out beyond the edges of the ear.

The whiskers around the eyes and muzzle should never be removed (again, many people do), because they are an essential sense organ for the horse. The hair at the back of the heels can be trimmed with scissors or clippers, as can the hair along the lower jaw line.

Clipping

Clipping is the shaving away of the horse's coat with a clipping machine (clippers). Horses are usually clipped in winter when the extra hair growth occurs and some horses are clipped in summer to help them keep cool. Many endurance riders clip their horses all year round. The clipped horse will need extra rugs for warmth.

There are various types of clip and the style that you choose depends on the work the horse is doing and the facilities you have available. A clipped horse may need to be re-clipped every few weeks; you may be able to clip at the beginning of winter and then halfway through the season depending on how fast the hair grows and how important it is for the horse to look smart.

There are various reasons for clipping.

Figure 23. Braided tail
(courtesy of Annie Minton).

- Prevent excessive sweating, especially in winter when the coat is long and thick.
- Prevent the horse getting cold because a heavy coat hold sweat and dries slowly in cold weather.
- Keeping the horse clean.
- Maintaining condition on a horse that would otherwise sweat it off.
- Enable the horse to dry more quickly after working.
- Enable the horse to work harder/faster and longer without getting too hot.
- Make the horse look smart.
- Save time on grooming.

Figure 24. Clippers
(courtesy of Jane Myers).

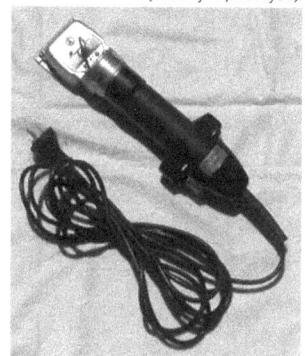

You can buy a set of clippers (approximately $400–$500 or buy second-hand) and learn to clip yourself or you can get a professional groomer to do it for you. A good idea is to have someone else clip your horse while you watch and learn, or maybe even pay for a lesson.

Types of clip

There are standard clips or you can customised to your requirements.

A full clip removes all the long hair including from the legs and head. Some sensitive horses may get a sore back from the pressure of the saddle in which case the hunter clip is better.

A hunter clip is a full clip leaving the saddle area and the legs with hair. Leaving the legs with hair protects them and keeps them warmer.

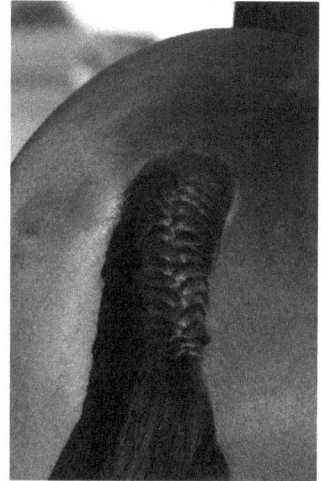

Figure 25. Hunter's clip

Figure 26. Blanket clip

Figure 27. Medium trace clip

The blanket clip leaves the shape of a blanket on the back of the horse because that area does not sweat as much as the underside of the horse. With this clip the horse does not get as cold as quickly as a full clipped horse if it has to stand still during a period of work.

The trace clip is a blanket clip with half of the neck left unclipped. The head hair can be removed and either clipped from the cheeks back or left on. Another variation of the trace clip is a chaser clip where the clipping ends at the stifle.

You can clip down the front of the neck and chest only, or over the top of the front leg, which will assist the horse with heat loss while it is working. It is useful style of clip in the winter for ponies that usually only work on weekends.

PREPARING THE HALTER AND SALES HORSE

The preparation of halter and sale horses really starts with the feed bucket (see Chapter 5), but a quality coat and athletic appearance are crucial for a professional presentation. The exercise program used to get horses fit may differ between farms and indeed even between horses.

Exercise

The most useful tool in a sales prep or training operation is a covered, round yard (minimum 16 m diameter). The ground surface must absorb concussion effectively and should be at least 15 cm deep to minimise the occurrence of splints and other exercise-related blemishes. An open-topped, round yard is an acceptable alternative in some parts of Australia and New Zealand, if the drainage is excellent.

Free lunging or loose-line lunging can be used for the exercise program, but it is critical that the session is controlled. If a lunge line is used it should remain loose and pressure should not be put on the horse's head. When the head is pulled to the centre of the circle, undue pressure is put on the inside leg and there is a much greater chance of 'popping' splints. In any case, young horses should always wear splint boots while being lunged. It is best to start horses on the lunging program at the walk, but allow some trotting if the horse wants to. The initial session should be 5 minutes in both directions. It is crucial that horses spend the same amount of time moving in both directions every time that they work. Over a period of one week you can work weanlings up to 7 minutes in both directions and yearlings up to 10 minutes in both directions. Always start with 1 minute of walking in each direction and then move to a long trot. A square, two-beat trot is safer and easier on the legs than a canter. Older horses may be worked longer as their fitness levels increase and body condition dictates. Some cresty horses benefit from being worked in a neck sweat to lose fat whereas a ewe-necked horse can be lunged in a roller to build up the muscles of the neck. (See Chapter 3 for instructions on lunging.)

Other exercise includes hand-walking, ponying, swimming or a mechanical horse walker or treadmill. The traditional method of choice for getting Thoroughbred yearlings fit is hand-walking, but the main fitness with this method occurs in the handler rather than the horse! However, there are some horses that for one reason or another cannot have a more rigorous exercise program. You need to teach a horse to stride out while walking for the best possible presentation in the sale pen. Horses should be walked 20–60 minutes at a brisk walk and, where possible, up and down hills. One positive aspect of hand-walking is that the horses are really taught to lead!

Lead ponies are very effective for fitting sales and halter horses if there is an appropriate place to 'pony' the horses. Pony the horse at the walk and the trot, and if the pasture or paddock is big enough one yearling can be exercised in a straight line, which avoids the lateral stress on the legs that may occur when some other methods are used. Swimming is not feasible for most preparation programs, but is effective for sales or show horses that may have borderline soundness.

Mechanical horse walkers (treadmills) are becoming popular because they are labour-saving devices that allow the yearling to be worked for longer than if they were hand-walked. One disadvantage is the tendency for horses that have been exercised on a walker to drag when being led. Treadmills are great tools if used judiciously. Horses can be exercised very effectively on the treadmill at the walk and trot, and the newer, high-speed treadmills are reasonably easy on a horse's legs. Gradually build up the session length, starting with just a few minutes. One observation concerning the use of the treadmill is the tendency for horses to roll their shoulders rather than breaking cleanly over and bending their knees.

As with any exercise method, you must watch for changes in the feet and legs, which may indicate an impending problem with soundness or a blemish. Common problems that necessitate reducing the work load or ceasing it completely include splints, windgalls, thoroughpins, swelling in any joint, but particularly the fetlocks, knees, hocks and stifles, active physitis, foot soreness, tendonitis or any sign of lameness. It is essential that the horses are exercised every day if they eating a lot of supplementary feed. Tying up is not just a problem affecting performance horses; a significant number of halter horses tie up because of their large starch intake.

Another precaution for horses being prepared for halter competition and sales is heat stress because most of these horses will be prepared in the summer months. Horses that are not sweating when worked hard should cease exercise immediately, especially in areas where anhydrosis is common (see also Chapter 5 and 6).

During and after an exercise session is an ideal time to work on conformational deficits. Horses with thick, cresty necks should be exercised in a neck sweat and then tethered in the stable to cool down. The shape of a horse's neck, and therefore its balance, can be improved significantly by using a sweat. Likewise, there are horses that may benefit from a throatlatch sweat or even a full-shoulder sweat. Horses that are particularly coarse in the throat would benefit from wearing a throat collar continuously (make sure the mane is protected from the collar). Another useful tool is a set of side-reins and a roller. Horses with a thin, weedy neck, a ewe-neck or appear to have an inverted neck (i.e. thin on top and a belly to the neck below) should be exercised in side-reins, which will make the horse arch its neck and can significantly change the appearance of the shoulder and neck. When the side-reins are first used they should be adjusted loosely and only after the horse has worn them for a few sessions should they be tightened and the horse be made to really work.

Grooming

An important and mostly neglected part of preparing sales and halter horses is grooming. Horses must be groomed vigorously on a daily basis to achieve a glossy, rich coat, although coat conditioning feeds are also part of the preparation. Immediately following exercise is a good time for an initial grooming. If you don't sweat while grooming the horse then you are probably not doing a good job! The best tool for the job is a small (the size of your hand) flexible, rubber curry comb. The horse should be thoroughly and vigorously curried all over and then a medium-soft brush should be used followed by a rub with a rag. Give the horse a daily bath with plain water and use a mild soap only once per week. Manes should be washed and unruly ones braided or banded to make them lay smoothly. Tails should not be brushed unless plenty of conditioner is used first and the tail is completely dry. Once the horse is fit, the only time the tail is brushed thoroughly is on the morning of the show. Yearlings and weanlings that are turned out in groups should have their tails painted with an unpalatable product to prevent them chewing on the tails (e.g. Cribbox). There is nothing that detracts more from the balance and symmetry of a yearling than a chewed-off tail.

Sales and show weanlings should be rugged as soon as night temperatures drop below 10°C (see earlier for more details about rugging horses). The rug and hood make the hair lay flatter, as well as keeping the coat short. One disadvantage of rugging the weanling is that it is often turned out after the sale without a rug and without a thick coat to withstand cold weather. It seems that hair growth and shedding for the weanling are inherited characteristics, as is hair quality. Artificial lights may be used in some preparation programs. Day length should remain at a constant

15 hours if lights are used. Eventually, however, the effect of the artificial lighting declines and the natural cycle of hair growth resumes.

Regular foot care is also essential for both types of horses. By convention, Thoroughbred sales horses are shod in front and left barefoot behind, with the exception of weanlings, which are sold barefoot, and two-year-old in-training horses, which have all four feet shod. Even though supplements may help some horses with bad feet (see Chapter 5), nothing can take the place of regular trimming well before the show or sale.

Each trainer has an individual preparation program. The important thing is to design your program and stick to it, bearing in mind that modifications may be necessary to suit the needs of each horse. Attention to details is the key. There is not a magical feed ingredient that can turn a sow's ear into a silk purse! Good breeding, good feed and hard work will beat poor genetics, poor preparation and illegal drug use (steroids) every time. Remember, you are preparing a future athlete, not fattening a lamb for market.

WHERE TO GO FOR MORE INFORMATION

The Victorian Department of Primary Industries website (www.dpi.vic.gov.au) has information, Codes of Practice and guidelines on horse welfare. The New Zealand Ministry of Agriculture and Foresty has a website (www.maf.govt.nz/mafnet) containing information on animal welfare, and Codes of Practice and Codes of Recommendations and Minimum Standards.

The Equine Dental Association of Australia has a website: www.horsedentist.com.au

The Australian Farriers and Blacksmiths Association has a website, www.afba.org.au/index.php, from where you can contact the relevant State secretary to find out about farriers in your area.

NZ Horses Online (www.nzhorses.co.nz/index2.cfm) has a directory of equestrian services, including farriers and horse dentists.

For some articles on foot care and shoeing, go to www.farmilo.com.au/horse.htm, and for teeth, go to www.horsecalendar.com/index.html

The Equine Behaviour Forum is a not-for-profit organisation devoted to the study of equine behaviour and the information it provides is for all levels of expertise: www.gla.ac.uk/External/EBF

CHAPTER 3

SAFE HANDLING, RIDING AND TRAINING

HANDLING

Horse-related activities can be dangerous, but the level of risk can be minimised by following some safety guidelines.

Remember that even the best-trained horse is still a large, strong animal, especially when it is stressed or frightened. In some respects the most dangerous horses are those that are regarded as quiet because it is easy to become complacent around them.

Correct (i.e. safe) handling involves learning a general set of rules that are applicable for all horses, keeping in mind that each horse is an individual.

General safety rules of handling

- Safety rules apply to everyone. There is not one set of rules for beginners and another for experienced people. The only difference is that an experienced horse person has developed skills and judgement that a beginner does not yet have.
- In all cases, the safety of humans comes first and then the horses.
- Wear your helmet while handling a horse, especially a young horse.
- Always make sure there is another person present, even as an observer, particularly when the horse is experiencing a new or unusual procedure/activity e.g. drenching with a worm paste, or clipping. If anything goes wrong, the second person can at least seek help.
- Handle horses firmly. A horse's skin is sensitive, so stroke or scratch (just in front of the withers is a good place) and avoid patting. Patting the head is unpleasant for the horse because its head is mainly hollow (because of the large sinuses).
- Never tease a horse, especially with food. When horses are eating leave them alone.
- Discipline must be instant. If you miss the opportunity you will have to wait for another. The horse has no understanding of delayed discipline (see later section on 'Training' for how horses learn).
- Never kneel or sit on the ground by the horse; squat so that you can get away quickly if necessary.
- Never go under the belly of a horse (children like to do this).
- Never chase a horse thinking that you (and the horse) are playing a game. You do not want to encourage the horse's natural flight instinct. Never slap the horse's rump when you release it into the paddock.
- Whenever two people are handling a horse at the same time they must always both be on the same side of the horse. Otherwise, as the horse swings away from one person it swings into the other. The person holding the horse is the one who must keep changing sides to assist the other person who is attending to the horse.

Figure 1. Tidy
equipment
(courtesy of Annie Minton).

- Do not give horses titbits by hand unless it is part of reward training. If you feed a horse for no apparent reason it can become annoyed when the food runs out, which leads to pushing and shoving with the nose and then to nipping.
- Avoid being in a small, tight space with a horse (e.g. stable, yard or crush). It can crush you against a wall if it panics.
- Incorrect feeding can become a safety issue. Horses that are overfed and under worked can be dangerous because they are bored and full of energy. Unless your horse is working hard (e.g. endurance, eventing, hunting) then you should feed a diet that is very high in fibre, such as older grass pasture and/or hay (see Chapter 5 for more advice on correct feeding).

Safe facilities

Many accidents are caused by tools and equipment left lying around. This includes shovels, rakes, brooms and buckets etc., as well as riding gear. Every item should be stored as soon as it is no longer needed. Hoses should be hung up after use.

Doors and gates should either swing both ways, in the case of paddock gates, or should open outwards in the case of yard gates and stable doors (sliding doors are best), to prevent the horse getting trapped in the opening. Doors, and especially gates, should be closed behind you as you go through.

Headcollars and other riding gear should be stored when not in use, including lead ropes left tied to a post. A person or horse can be tripped up by a hanging headcollar.

Have first aid kits for both humans and horses, as well as clearly displayed emergency telephone numbers (doctor, vet, fire etc.). Consider doing a first aid course.

Electric sockets should be placed out of reach of inquisitive horses. Never walk a horse over an electrical extension lead, which in any case should never be on the floor.

Safe handling

A horse should be either tied securely or held by a capable person when being handled. A horse should wear a headcollar and lead rope as the minimum for safe handling. Never work around an unsecured horse(s).

Figure 2. Tidy
tackroom
(courtesy of Annie Minton).

Do not clean a stable or a yard while the horse is in it. Remove and secure the horse elsewhere before you do your cleaning. Be careful whenever picking up manure around horses because you are not concentrating on the horses and you have to move around them.

When moving around a horse, the safest place is either very close to the horse or out of range of a bite or kick, which is approximately 3–4 metres away (remember that a horse can run backwards and then kick). The greatest impact from a hind kick is at its extremity. If you walk at arm's length behind a horse and it kicks out, when the leg contacts your body it will be at full power. If you are in close contact with the horse and it decides to kick, the kick will be more of a shove. Also, if you are touching the horse you will feel it tense its muscles or shift weight, which may be a sign that it is about to kick. If you suspect that a horse is about to kick, move towards the shoulder. Remember, a horse can strike out directly in front with a foreleg or bite. Standing near the shoulders is generally the safest place.

By staying close to and in physical contact with the horse and speaking to it, the horse will know where you are and will be less likely to be startled by a sudden movement. Keep talking to the horse so that it knows where you are; remember, it cannot see you when you are right behind it unless its head is held high. Never walk behind a horse that has its head down. Never stand directly behind a horse (even when washing the tail) or directly in front. These are dangerous places because they are where the horse can see you the least.

In other words, stay out of kicking range whenever possible. When accustomed to the body language of horses you will be able to read the warning signs that a horse will invariably give before biting or kicking, such as laying the ears back or picking up a leg.

Regular handling and exercise
Handle your horse on a regular basis. Unless it lives in a stable or yard a horse does not have to be ridden or exercised every day, but it does needs regular handling. If riding your horse infrequently is your only option, you need to learn the basic skills of getting your horse to focus on you and to reinforce that you are the leader each time that you do ride. If you are not comfortable handling your horse while you are on the ground you certainly should not be riding it.

Sensitive areas of the horse
The most sensitive areas of the horse are the nose, ears, feet, legs and flanks. Predators would grab the nose to pull down a horse in the wild, so horses are naturally wary of anything touching this area. The nose is supplied with many nerves, making it even more sensitive. The horse cannot see directly above its head because of the position of its eyes, so it uses its ears to sense if there are objects above the head. Hence, a horse will sometimes panic if its ears are grabbed or if they brush on a low surface (e.g. the roof of a float). Injury to the feet and legs would compromise the horse's survival in the wild, so it is naturally wary about allowing you to handle them. The flanks are a ticklish area for most horses, but especially for mares because it is an erogenous zone (i.e. it is gently sniffed by the stallion to test the mare's receptiveness to mating).

Children and horses
Small children should be kept under control at all times around horses because they can get excited at seeing horses, but do not know the dangers. Horses do not always recognise children as small humans and can get frightened if they are not used to them. Children tend to be noisier than adults. Children must be taught the correct way to approach horses and how to move safely around them.

Children like to feed horses (as they do all animals), but it is not a good idea to let them. If you must allow a child to feed a horse, put the feed in a bucket and supervise while the child is holding the bucket. Never feed someone else's horse without permission.

Never leave young children in prams around horses or unattended in a horse facility. The child is placed in great danger if a horse panics at the sight of them or the pram.

Do not allow children to lead, handle or ride large horses. Never place a baby on a horse, or carry it in front of you while you ride.

Dogs and horses
Dogs that are not used to horses can get very excited or frightened by them and begin barking and/or attacking. Horses that are not used to dogs can become frightened because a dog is a natural predator of the horse. Some horses are very defensive with dogs and will bite or strike with a foreleg or kick with a hind leg.

Figure 3. Approaching the horse (courtesy of Stuart Myers).

Horses and dogs should be allowed to slowly become accustomed to one another. If you own a horse that is frightened of dogs you will need to train it to overcome that fear before you can go riding down a road (see later section on 'Training'). Dogs will often run to the front of a property and may even come out on to the road. You do not want to be caught between a barking dog and traffic on a horse that is panicking.

In the paddock

Never take food (titbits) into a paddock with a group of horses because they will crowd around you and you may be injured as they try to get the food. Never feed titbits over the fence to a group of horses. They will crowd around and may fight each other for the food. You are in danger, even on the other side of the fence, and the horses may injure each other by fighting.

Do not 'play' with horses that are loose in a paddock. You may be knocked over or kicked and you certainly cannot outrun them.

Catching and leading
Approaching a horse

Always approach a horse at an angle, never directly from the front or the rear where the horse cannot see you. Speak to the horse as you approach so that it is not startled. Look for the movement of an ear towards you, which will show that the horse is aware of your presence. Walk confidently towards the horse; do not creep up on it. Your first contact with the horse should be on the shoulder or neck, rubbing rather than patting. Place your arms around the horse's neck and slip the lead rope over the neck to secure the horse. Then put the headcollar or halter on, moving it gently over the horse's nose and ears.

Facing up

The procedure for approaching a horse is the same whether it is in a stable, yard or paddock. However, if the horse is in a stable or yard and it is facing away from you, it must be made to 'face up'. Never enter an enclosed area with a horse that is facing away from you. The horse can swing around and knock you over or kick you.

From a safe distance (out of kicking range and behind the door or gate) command the horse to turn around. The command can either be a noise, such as clicking the tongue, or a spoken command to 'face up' (if the horse has been trained to respond). If a verbal cue does not work flick the horse lightly on the rump with a lunge whip or stock whip until it turns around. Verbally praise the horse and then walk up and secure it with the headcollar or halter.

Figure 4. Leading the horse (courtesy of Stuart Myers).

Leading a horse

A horse is usually led from the left (near) side, using the right hand to hold the lead just behind the clip (not around the clip).

Walk level with the shoulder of the horse, allowing your hand to move with the action of the horse's head. If you hold on tightly and restrict the horse's natural movement of its head while it is walking it will start to resist. For turning, push the horse around the turn rather than pull the horse around you (otherwise the horse can walk over you).

If the lead rope is long, fold the excess across the palm of your (left) hand; never wrap the lead rope around any part of your body. A knot in the end of the lead rope will prevent it slipping through your fingers. Never let the rope drag behind you.

A horse is larger and stronger than a human. If it will not move forward, do not try to pull it; stand at the horse's shoulder and use hand pressure to urge the horse forward. Try also flicking the extra lead rope behind you toward the horse's flank or use a dressage whip.

Before going through gateways and doorways stop the horse so that you can go through first. Always open doorways and gateways fully so that the horse does not bang its hips as it goes through.

Even though most horses are used to being led from the left side and most people prefer to use their right hand for holding the rope, it is a good idea to teach a horse (and yourself) to lead from the other side in case of an emergency.

Problems when leading

If a horse becomes excited or frightened while you are leading it, stay calm and try to keep control of the horse. Try to keep the head of the horse turned towards you and keep a firm hold of the lead rope. Putting your elbow into the horse's neck will help to keep it bent towards you and as long as the neck is bent in your direction you have the advantage. If the horse straightens its neck or bends it away from you, you have lost your leverage and the horse has the advantage.

There are occasions when you may have to let go; for example, if you are in danger of being dragged. Hanging on to the horse in such a situation will probably result in you being injured. Avoid potentially dangerous situations (e.g. leading a horse on a road) if you are not experienced enough to handle the horse if it starts to get excited or frightened.

Tying up a horse

Some general safety rules of tying up horses are:

- Always tie up the horse in a safe place that has a non-slip surface, does not have obstacles and is not near other horses.
- Tie the horse to something strong and secure. A solid post is usually the best. Do not tie to the rail of a fence because the horse can easily pull off the rail. Never tie a horse to a wire fence because if the horse pulls back it may pull the whole fence with it, resulting in terrible injuries.
- Tie the horse at the level of its eyes and use a quick-release knot so that the horse can be released quickly in an emergency. Allow approximately 50 cm to 1 m of rope between the horse's head and the tie. If the rope is left too long the horse may get a leg over it.
- If the horse has a tendency to pull back, tie the lead rope to a piece of strong twine and then secure the twine to the post or ring. The twine should break or can be easily cut in an emergency.
- Always use twine when tying up to a hitching rail, unless it is well made and strong (many are not), because if the horse pulls back and dislodges the rail, you and/or the horse will be hit with the rail and very likely injured.
- Never tie up with the bridle reins. Use a headcollar and rope, and always untie the horse before removing the headcollar.
- Never tie up with a rope halter. If the horse pulls back the narrow rope used for the halter can damage the horse behind the poll.
- Never tie a horse to a float that is not attached to a car or truck. Never tie up to a single float even if it is attached to a vehicle because a horse can pull it over.
- Never leave a horse unattended while it is tied, especially at a show where children may be running around.
- Keep your fingers out of any loops while tying a horse up. You may lose a finger(s) if the horse unexpectedly pulls back.
- Have a sharp knife handy for quickly releasing the horse.

See the later section on training difficult horses for how to deal with a horse that pulls back.

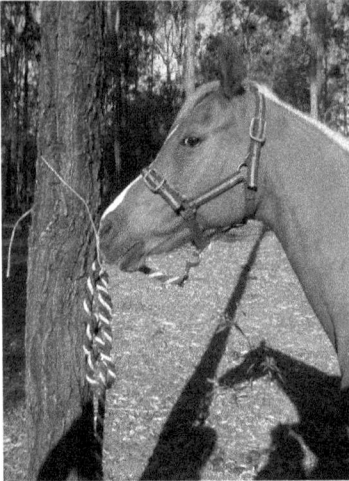

Figure 5. A horse
tied up safely.
(courtesy of Stuart Myers).

Releasing a horse

To safely release a horse into a paddock, open the gate and lead the horse through. If the gate does not swing outwards you will need to take the horse further into the paddock before releasing it (described later).

After going through into the paddock, close the gate, but do not fasten it (you may need to get out of the paddock quickly). Turn the horse so that it faces the gate before you release it and once you have let the horse go, leave the paddock immediately. Some horses kick out in play as they start to gallop off and you could be badly injured if you release the horse while it is facing into the paddock.

If there are other horses in the paddock, it is possible to get squashed in the corner of the paddock where the gate is if they come up behind the horse that you are releasing. If this situation is likely to occur take someone with you who can hold a lunge whip and keep the other horses away. That person must stand away from the horse that you are releasing.

Another method of releasing, especially if there is more than one horse to be released, is to lead the horses out into the paddock, closing and fastening the gate first. After going about 20 m into the paddock turn the horses to face the gate. Each handler should undo the headcollar, but keep the rope around the neck of each horse. Appoint a leader who can give the command to release the horses at the same time when everyone is ready. Once the horses are released the handlers should walk directly out of the paddock.

To safely release a horse into a stable or enclosed yard, always turn the horse to face the entrance so that you can release the horse without having to walk out past its hindquarters.

RIDING

Teaching you how to ride is outside the scope of this book, but there are certain safety issues related to riding horses that are not always emphasised in books on that topic. Therefore, some safety guidelines are presented in this section.

General safety rules of riding

- Safe riding means being in control. Never ride a horse that you cannot control.
- Ride in an enclosed area until you feel confident.
- Do not ride alone if you are inexperienced.
- Do not chew gum while riding (it can be inhaled and you may choke) and avoid carrying anything in your pockets that may be uncomfortable and distract you (such as a wallet) or could injure you if you fall (such as a pen).
- If you carry a mobile phone (when going out for a trail ride), make sure your horse is accustomed to the noise of it ringing. (See the later section on training for details of how to habituate your horse to noises, dogs etc.) Fastening your mobile phone to the outside of your upper arm is a relatively safe way of carrying it. If you fasten it to your saddle and you become separated from your horse, you will have lost the phone (and the horse won't use it!).
- Never smoke while riding or handling horses. Smoke only in designated areas. Horses and fires do not mix!
- Never consume liquids while riding. You must be in control of the horse at all times. If you need a drink of water, dismount first. Never drink alcohol while riding or handling horses.
- Do not allow a horse to graze while you are mounted. You do not have control of the horse and if it suddenly lifts its head and takes off, you can be moving very quickly before you have time to gather the reins.
- Do not ride where there are horses on the loose.

- Do not ride in an area that opens onto a busy road.
- Do not ride near steel posts (star pickets). You may be impaled if you fall on one or your horse can be badly injured if it runs into one.
- When returning from a ride, walk the horse for the final kilometre or two before home. Never let the horse run towards home because it will soon become a habit and you will not be able to stop the horse.

Riding a new horse

While you and your horse are getting used to each other, proceed slowly. Initially, it is safest to ride in an enclosed area so that you can be sure that you can control the horse. Remember that the horse has to get used to new surroundings as well as a new owner, whereas you only have to get used to the horse. Even if you do not have regular lessons, some instruction is recommended to prevent problems occurring during this 'getting to know you' stage. When you do go on your first trail ride, do not go alone. Take a riding partner with you or have someone follow you on foot.

Safe mounting

- Check all your gear, especially if someone else has tacked up the horse for you (see 5-minute check-up at the end of this section).
- The girth should be checked just before and just after you mount. It should then be checked again after riding for 10 minutes or so.
- Mount a horse out in the open (not in a building or under any fixture).
- Use a mounting block; this is better for you, your horse and your saddle.
- Make sure your toe does not dig into the belly of the horse as you mount.
- Sit down in the saddle gently.
- Train the horse to stand still before, during and after mounting (for approx. 1 minute).

Safe dismounting

- Make the horse stand still.
- Take both feet out of the stirrups to dismount.
- As soon as you dismount, run the stirrups up and loosen the girth.
- Bring the reins over the horse's head.

Riding alone

Do not ride alone if you are inexperienced. Sometimes it is difficult to find a riding partner, but if you are not experienced it is dangerous to ride on your own. An experienced rider gives clear signals to the horse about who is in control, but an inexperienced rider gives unclear and confusing signals that can upset the horse. Many horses that are nervous on their own, may be more confident with another horse around. Riding with a partner, preferably someone with more experience, means that there is a horse that can lead the other (e.g. past a scary object). Also, if there is an accident you can get help. If you must ride alone, take a phone and a first-aid kit, and always tell someone where you are going and what time you expect to be back.

Figure 6. Riding with a partner (courtesy of Margaret Blanchard).

Riding in groups

When riding in a group, certain rules must be followed so that everyone is safe.

- Never start riding until all riders in the group are mounted and ready.

- Designate a leader (who should be the most experienced rider), even if it is a group of friends. The lead rider should always inform/ask the others before changing pace. No one should pass the leader. The chosen leader should be on a quiet horse.
- The last rider should also be experienced and mounted on a quiet horse.
- If anyone has to dismount, everyone should halt and wait for them.
- If anyone is getting left behind, stop or slow down and wait for them to catch up.
- Be careful in wide, open spaces, such as a paddock. Horses can get out of control more quickly than on a narrow track because they will try to overtake each other.
- Ride abreast only if the horses are very accustomed to each other. Otherwise, stay a full horse length behind the horse in front to avoid being kicked. Each rider should be able to see the heels of the horse in front.(See Appendix 2 for specific instructions for riding in a group on the road.)
- Do not start to canter up a hill while others are still coming down the previous hill. Remember that if you canter up every hill you come to, then soon you will have no choice because you will have taught the horse that a hill means going into a canter.
- In a large group it is safer to keep to a walk because horses get excited when moving fast in a group.
- Never chase a runaway horse. If someone loses control, follow at a steady pace.
- Ask permission before passing another horse and rider. Pass at a steady pace. Never overtake while cantering as this may make the horses gallop.
- If you suspect/know that your horse kicks, ride at the back.
- Take a first aid kit and a mobile phone. Take a headcollar and lead rope (the horse can wear it underneath the bridle).
- Cross a road only when and where everyone can cross at the same time.
- Always ride at the pace of the least experienced/most nervous rider. Look after each other and the ride will be fun for everyone.

Riding on roads

- Only ride on the road if there is no alternative, and even then, only if you have control of your horse and it is safe.
- Expose your horse to as many sights and sounds of traffic as possible before riding on roads.
- When riding on the road, remember that you are subject to the same rules as cars. Keep to the correct side of the road, the left side, and give hand signals when turning.
- If your horse is young, or inexperienced with riding in traffic, find someone on a steady, quiet horse (a 'schoolmaster') to go with you to give your horse confidence.
- Avoid riding after dark. No matter how much care you take in making yourself visible, many drivers do not show respect for horses and riders and there are too many accidents. If you must ride in darkness, travel slowly. Remember that your horse will not like the fast-moving, oncoming headlights. Use lights, wear a reflective vest, fluorescent boots and put a reflective stripe on your helmet.
- It is a good idea to wear bright/reflective clothing even in daylight while riding on the roads.
- Always be courteous to drivers who slow down and/or move over, give them a wave or a nod.

See Appendix 2 for the Victorian Horse Council's Code of Conduct for Horses on Victorian Roads.

Improving your riding skills

Lessons and clinics

Beginners should learn how to ride correctly at a riding school or pony club or with a private instructors. In fact, everyone can benefit from lessons. You may say 'but I only want to trail ride' or 'I am only a weekend rider'. Trail riding is much more hazardous than riding in an arena and riding infrequently means it is even more important that you can control horse. A good instructor will help you gain strength, a balanced seat, and confidence (see the end of the chapter for contact details).

You can also improve both your riding skills and general control of the horse when you are not riding (i.e. ground-work techniques such as lunging) by attending clinics with your horse. Clinics are held on a on a regular basis around the country. You will be taught how to get your

Figure 7. Having a lesson
(courtesy of Stuart Myers).

horse to move, where you want and when you want, so that you are safer while riding. Even if you do not have a horse, you can go and just observe ('fence sit') and you will still learn a lot about handling and riding.

Learning to ride a horse is a long-term prospect; it takes time to be able to ride well and it is an ongoing process. Any experienced rider will tell you that the more you learn, the more you realise there is to learn! Learning to ride is fun, whatever stage you are at in your riding life.

Clothing

Wear comfortable clothes and protective gear (helmet, boots, back protector).

Helmets

Helmets have saved the life of many people involved in riding accidents; it is not worth the risk of riding without one. In Australia it is law that people under 18 wear a helmet where ever they ride (at home or out on the roads). Helmets should comply with the latest standards (currently AS/NZS 3838, EN 1384 or ASTM F1163), which will be marked on the helmet. The helmet should be less than five years old from the date of manufacture. If a helmet has any damage at all or has been knocked against a hard surface (i.e. dropped or worn in a fall) it should be destroyed and replaced with a new helmet. Your helmet should fit your head snugly and not slip. The safety harness should always be fastened. Helmets should always be worn when riding and should also be worn when handling horses (especially young horses).

Boots

Riding boots must have a heel and a smooth sole, and preferably not be a lace-up style. The heel is to minimise the risk of your foot slipping forward through the stirrup in the event of a fall. Any type of boot with ridging on the sole is dangerous because the boots can become caught in the stirrups. It is extremely dangerous to ride or handle horses in any other type of shoe or boot (e.g. running shoes, walking boots, work boots, pumps or sandals) or with bare feet.

Steel toe-cap boots are only safe if they are approved for the weight of a horse. If a horse stands on your foot and the metal cap bends into your toes you will be in a lot more trouble than if you had not worn the boots. Strong leather boots are best.

Back protection

There are many different types of lightweight, comfortable back protectors. Back protectors are compulsory for some events, but they are also a good idea in many other riding situations; for example, if you are trail riding where it is rocky.

Riding gear

Buy good quality gear

It is a big temptation to buy cheap gear when you are outfitting yourself and your horse for the first time. To the inexperienced, all the gear looks the same; it is hard to understand why a saddle can cost as little as $200 or a much as $3000 plus. There is a huge range in quality and although you do not need to buy the most expensive gear, buying the cheapest is false economy.

First of all, work out your budget. If you cannot afford good-quality new gear, consider buying second-hand, which is better than buying new, cheap, poor-quality gear. For example, if you have allocated $500 for a saddle, a new one will not be very good quality, but you can purchase a good-quality second-hand, leather saddle that will last for generations. A cheap saddle will not last long and will not have a good resale value when you decide to sell it because you have realised that you need a better quality saddle!

Cheap gear tends to have the following characteristics:

- The leather is not as strong; it stretches or snaps easily. If your reins snap while you are riding, it is similar to the brakes failing in a car.
- Fewer stitches are used and the cotton is poor quality, which means the stitches rot more quickly and the gear will have to be restitched.
- The metal parts are usually made of a composite metal rather than steel, which means they can break under pressure.
- Webbing is often made from cotton rather than nylon, which means it will rot and break more readily; check where the girth points are attached to the saddle.
- Parts are often not fastened on correctly, the stirrup bars for example.
- The horse and rider may get sore because the gear does not fit well.

Cheap gear is cheap because it has cost less to make, by using cheap (unskilled) labour and materials, and taking less time to assemble it.

Synthetic gear

Australia was the first country to manufacture synthetic gear (Wintec, made by Bates Australia; see the end of the chapter for contact details) (see also Figure 17). It has improved over the years and is a very good compromise if you cannot afford the more expensive leather versions. In fact, some top-level equestrians are using it. The price is lower because the material used is less expensive than leather, but it is strong, hardwearing and the gear has a good resale value. The quality is far superior to leather gear in the same price range. You can buy saddles, girths, driving harness and bridles. Other advantages are that rain or seawater will not affect it and it does not need as much care as leather.

Correctly fitted gear

A horse will not work well if the gear does not fit properly. Buying gear that will fit correctly can be difficult until you know what to look for. If you are buying your tack from a saddlery store, the staff should be able to advise and your riding instructor should also be able to help.

Figure 8. Basic parts of the bridle.

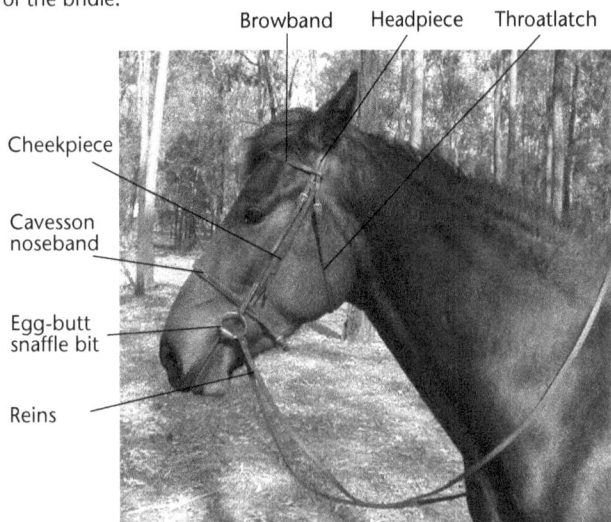

Bridles, saddles and other gear
The bridle
The bridle is a standard piece of equipment allowing the rider to communicate with and control the horse. Most bridles have the same basic parts: bit, reins, headpiece, cheek pieces and throatlatch, and browband. The bit is the major tool of communication and control, the reins allows the rider to manipulate the bit, the headpiece and cheek pieces hold the bit in place, the throatlatch stops the bridle from slipping forwards and the browband stops the bridle from slipping backwards.

A bridle can also have a noseband, which does not have a specific function unless it is fitted so that the horse cannot open its mouth or cross its jaws (see later).

There are various types of bridles, ranging from the simple English-type bridle through to the full double bridle that is used in higher levels of dressage. The traditional Australian bridle is called a Barcoo bridle and is constructed differently to an English bridle.

The bit

The bit is very important because it is your means of communicating with the horse while you are riding. Avoid using cheap metal-plated bits because the plating will flake off over time and as they wear they develop sharp edges. Good-quality bits come in a variety of materials, but the most common is stainless steel. Bits are usually sold in three sizes: pony, cob and full. A correctly fitted bit should not pinch the sides of the horse's mouth (too short) nor slide from side to side when using the reins (too long). The cheek pieces of the bridle should hold the bit so that it sits snugly in the corners of the mouth. Opinion varies as to how high the bit should sit. It should not be slack, but should not be so tight that it makes the horse 'grin'. Some horses will get their tongue over the bit if it sits too low, which becomes a bad habit that is hard to stop.

There are various styles of bit, some of which are severe. Always aim to use the least harsh bit (as long as you still have control) and seek advice before using any of the more severe types.

The reins

There are many different types of reins on the market and you will develop your own preference. Some people like thick, soft reins, some like rubber-covered reins and some like thin, leather reins. It all depends on what type of riding you are doing and the size of your hands.

Nosebands

The simple cavesson noseband goes around the horse's face, approximately halfway between the nostril and the eye, and does not have a specific function. If it is fitted correctly you should be able to place two fingers between the noseband and the horse. Some riders fit the noseband very tight so that the horse cannot open its mouth and evade the bit. It is better to train the horse to accept the bit than to rely on a very tight noseband.

The Hanoverian noseband is a cavesson with an additional strap that goes around the horse's mouth, in front of the bit, to prevent the horse opening its mouth or crossing its jaws. The cavesson strap of the noseband has to be snug or it will be pulled downward by the other strap. The Hanoverian noseband is designed to be a transitional training tool, not a permanent part of the bridle.

There are other forms of noseband that are no longer popular, such as the drop noseband, which has largely been superseded by the Hanoverian style because it is easier to get a correct fit. Most nosebands other than the simple cavesson are designed to keep the mouth closed. Think about why the horse is opening its mouth before you force it shut. It may have sharp teeth, or be trying to evade a heavy-handed rider. Be especially careful with a young horse (up to the age of 5 years), which may be teething the permanent molars and have large lumps under the jaw line (tooth bumps) until the teeth are finally through.

Figure 9. Hanoverian noseband and loose ring snaffle bit. (courtesy of Jane Myers).

Putting the bridle on

Putting on a bridle requires practise on a horse that is experienced and quiet before you attempt to bridle a young, inexperienced horse. It is very easy to make a horse difficult to bridle by being clumsy (which everyone is until they have perfected their technique).

1. Check that the bridle has been put together correctly.
2. Carry your bridle on your shoulder when approaching the horse.
3. Untie the lead rope of the horse and put it over its neck.
4. Stand at the side of the horse and undo the noseband of the headcollar (Figure 11a).

Ring snaffle

Egg-butt snaffle (a) without cheeks (b) with cheeks

Fulmer snaffle

Racing snaffle

Double-jointed 'Dr Bristol' snaffle

Double-jointed snaffle with alternative joints (a) and (b)

Half moon snaffle

Rubber or vulcanite straight bar snaffle

Straight bar metal snaffle

Gag snaffle

Bridoon

Egg-butt bridoon

Double-jointed bridoon

Bridoon with cheeks

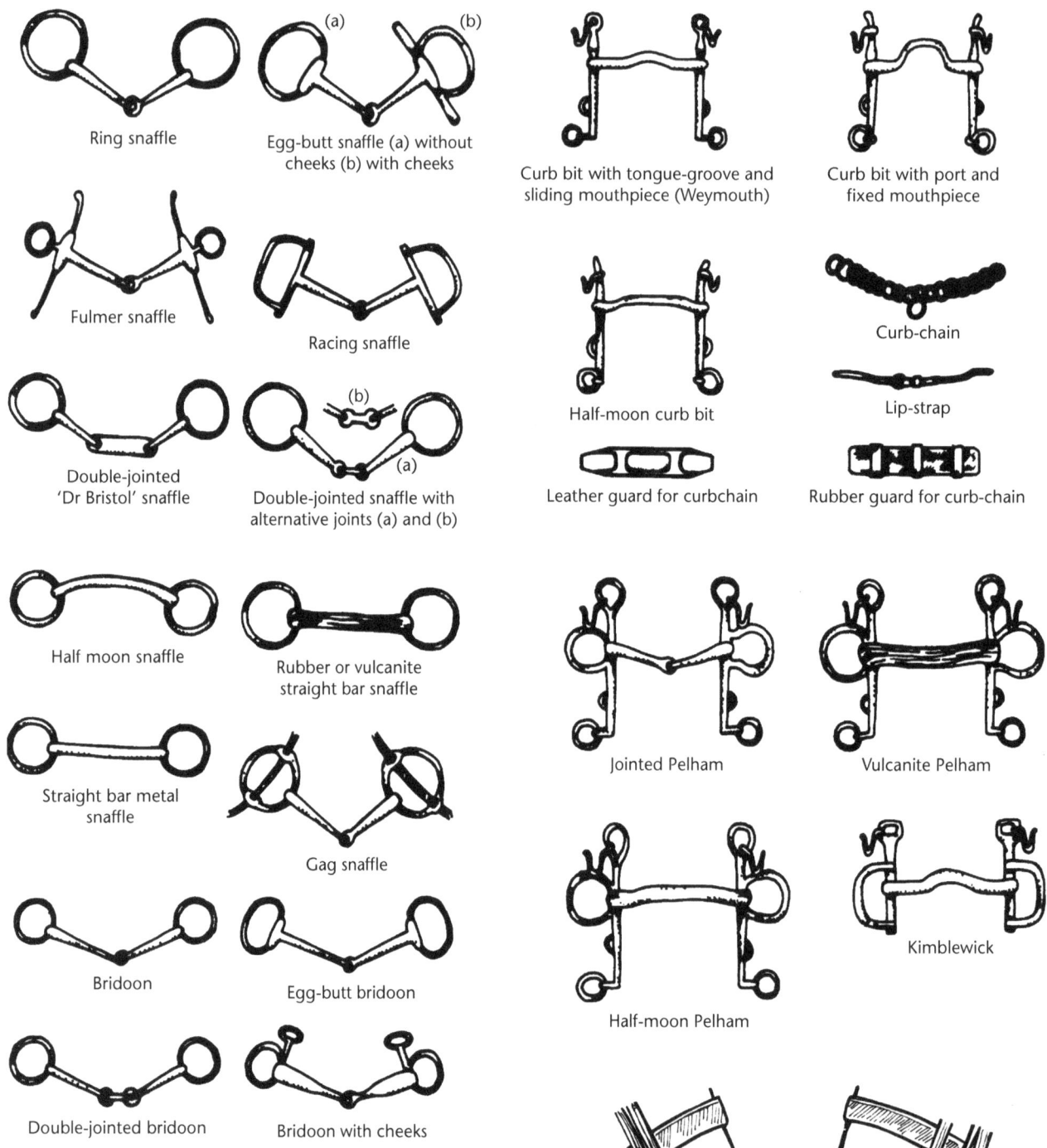

Figure 10. Types of bit.

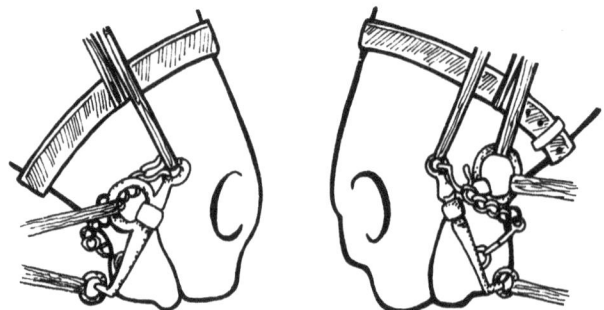

Curb bit with tongue-groove and sliding mouthpiece (Weymouth)

Curb bit with port and fixed mouthpiece

Half-moon curb bit

Curb-chain

Lip-strap

Leather guard for curbchain

Rubber guard for curb-chain

Jointed Pelham

Vulcanite Pelham

Half-moon Pelham

Kimblewick

Detail of the Pelham and double bridle bits

5. Take the bridle off your shoulder and put the reins over the neck.
6. Put your right arm under the horse's head, put the bridle around its face and grasp the cheek pieces with your right hand (Figure 11b).
7. Slip your left thumb into the side of the mouth and press on the bars.
8. Put the bit in the mouth, being careful not to bang the teeth.
9. Lift the headpiece up over the head with the left hand, putting one ear and then the other under it (Figure 11c).
10. Tidy the mane and forelock.
11. Make sure the bit, browband and noseband, if there is one, are in line.
12. The bit should be just tight enough to cause a slight dent in each side of the mouth (ideally only two or three creases in the corners of the mouth; it should never pinch) (Figure 12).
13. Fasten the throatlatch; it should be free enough to allow three fingers width against the throat, but not so loose that the bridle can easily slip forward over the horse's ears (Figure 11d).
14. If the bridle has a cavesson noseband, fasten it so that two fingers can fit between the noseband and the jaw. If the noseband is a drop or Hanoverian style it should be fitted so that only one finger can be inserted between the horse's jaw and the noseband.

Once the horse is bridled, do not tie it up by the reins. A horse can pull back and break its jaw. Make the reins safe by taking them over the head, then over the neck and twist them through themselves, looping the end through the throatlatch (Figure 11e). You can then put the headcollar back over the bridle and re-tie the horse (Figure 11f).

Removing the bridle

Undo the throatlatch (and noseband if there is one), take the reins up to the headpiece and then lift the bridle over the horse's ears (Figure 13). Allow the horse to drop the bit, being careful not to bang the teeth.

Martingales and breastplates
The running martingale

The running martingale (rings) can be a useful piece of gear if used properly. It prevents both the horse and the rider bringing the head up. The fitting of the martingale is very important. It must not be fitted so short that it causes the reins to deviate from making a straight line from the horse's mouth to the rider's hands. The martingale should only operate when the horse lifts its head too high. Rubber stoppers must be used on the reins or the rings will get caught in the billets/buckles of the reins (where they join the bit). When this happens the horse is likely to panic at the sudden downward pull on the mouth.

When unsaddling with a martingale or breastplate, unthread the loop that the girth goes through (between the front legs) before you try to remove the saddle.

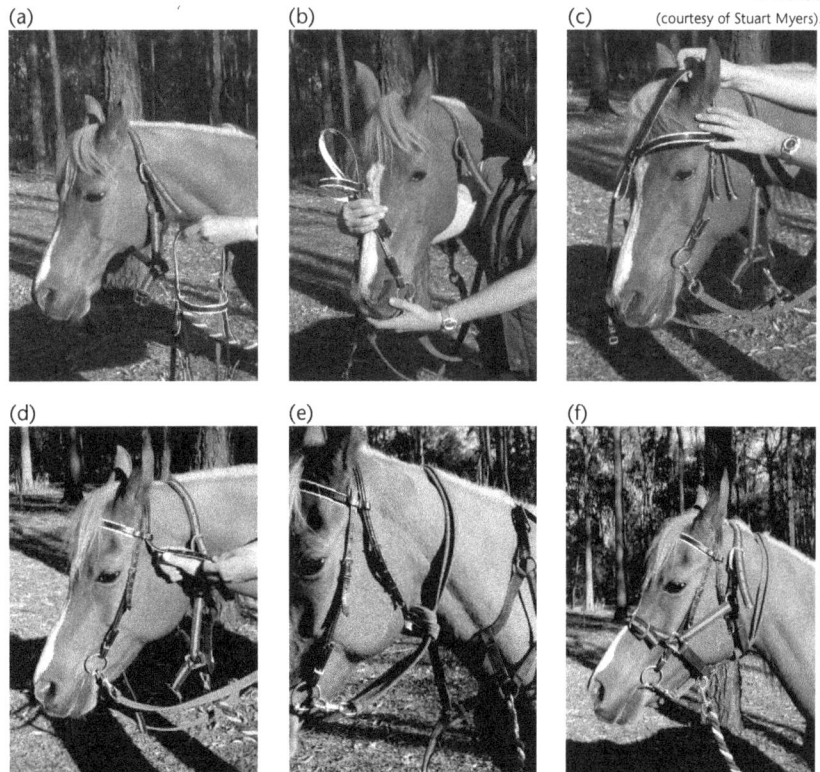

Figure 11. Bridling a horse (courtesy of Stuart Myers).

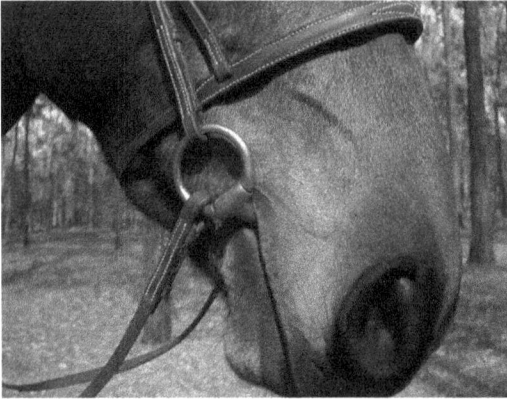

Figure 12. Correctly fitted bit (courtesy of Jane Myers).

Stockman's breastplate

A breastplate that is very popular with stockmen and trail riders is the 'stockman's breastplate' that prevents the saddle slipping back while going up steep hills. It consists of a leather neck strap that is attached to D rings at the front of the saddle, and another strap that runs between the front legs from the bottom of the neck strap to the girth. Martingale rings can also be attached to this breastplate. The breastplate should lay flat without strain when the horse is standing in a normal position. Martingale rings can be attached to a stockman's breastplate.

Breast-girth/breastplate

The 'breast-girth' or breastplate is usually made out of leather or webbing and can have elastic inserts. Sometimes the whole breastplate strap is elastic. It fits across the chest of the horse and attaches to the girth strap under the saddle flaps. It is prevented from falling downward by a strap that runs over the front of the withers.

A breastplate will prevent the saddle from slipping backwards and is frequently used on racehorses, hunters, show jumpers and eventers. The breastplate must be correctly fitted so that movement of the neck or shoulders is not restricted. It must not be so tight that it puts pressure on the windpipe. An elastic construction is good because it is much less restrictive while remaining very effective. The breastplate can be covered in sheepskin for extra comfort. This type of breastplate can also become a running martingale, using an attachment.

Figure 13. Removing the bridle (courtesy of Stuart Myers).

The saddle

There are many different types of saddle, each suited to different horse activities. Either a stock saddle or dressage saddle is a good type to begin with. Both types hold you in a good position in the saddle and are comfortable. Often an all-purpose saddle is recommended, but because it is a compromise between two very different types of saddle (dressage and jumping) it tends to be not much good for either.

All-purpose saddles are satisfactory for when you first start jumping, but when you advance you will need a more forward-cut saddle (i.e. a jumping saddle). An all-purpose saddle is a good for pony club because children tend to get involved in all the different activities. You have the choice of buying an all-purpose saddle or buying more than one saddle. Try out different styles of saddles before you buy. Some saddleries have models that you can try out at home.

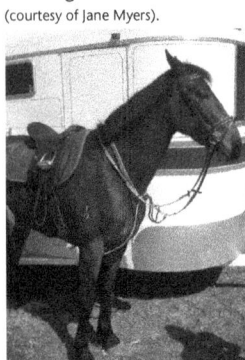

Figure 14. A bridle with a Hanoverian noseband and a martingale (courtesy of Jane Myers).

Stirrup irons

Stirrup irons should be made from stainless steel (cheap stirrup irons have been known to either snap or squash a foot in an accident) and there are various styles of safety irons available.

A correctly fitted iron should be only 1 cm wider than your foot on both sides. Any larger and your foot can slip through more easily, any smaller and your foot can be trapped. If your foot becomes trapped in a stirrup iron when you fall off, you will be dragged. Safety irons reduce this risk. There are also various types of attachments that can be fixed to an iron to prevent the foot slipping forward.

Stirrups should be 'run up' whenever they are not in use; otherwise they can swing and hit the horse, catch on projections or a horse can catch its back foot in one if kicking forward at a fly.

Girths

The girth strap can be made from different materials, but the modern types are made of synthetic material, have elastic inserts and do not go hard after being caked in sweat. These types of girth are called 'anti-gall girths' because the horse is much less likely to gall (develop pressure sores in the girth area) with their use.

Ideally, the girth should fit the horse when it is half way up the girth points on either side; that will allow for expansion or reduction over time. When you first do up the girth do not make it too tight. Gradually tighten it as you are preparing to ride. It should not be any tighter than is required to hold the saddle in place as you mount. In fact, if you tend to pull the saddle over when you mount then you need to improve your technique and/or mount from a block, which will be easier for you, the saddle and the horse. You will need to check the girth, and pull it up a bit more, once you are mounted and then again after riding for a short time (10 minutes).

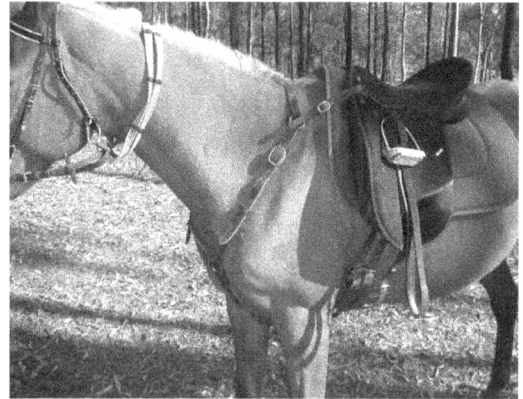

Figure 15. Stockman's breastplate (courtesy of Jane Myers).

Saddle blanket/cloth

The saddle should be used with a saddle blanket or cloth underneath. A good saddle blanket is made from natural fibre such as wool or cotton. Avoid synthetic fibres because the horse generates a lot of heat under the saddle. Make sure that the saddle blanket lies smoothly without creases. Pull it up into the gullet of the saddle before you do up the girth to prevent the blanket pressing on the withers and to allow air flow when the rider is in the saddle. Keep your saddle blanket clean. Cotton should be washed regularly; wool can be brushed once dry and washed occasionally.

Fitting a saddle

Some saddleries now have trained saddle-fitters who will come and fit a saddle to your horse. If you are buying a saddle from another source do not buy before you have tried it on your horse. You may have to pay a deposit or the full price on the proviso that you take it straight back if the saddle does not fit.

Stand your horse square on a level surface. Have someone hold the horse while you fit the saddle. Put it on the horse without a saddle blanket or cloth underneath to check the fit.

- The saddle should sit behind the withers and there should be room for you to put two or three fingers sideways between the withers and the saddle.
- The panels should mirror the shape of the horse's back, putting even pressure along the back. To check this, run your hand under the panels from front to back. If the panels are the wrong shape for the horse there will be more pressure in the middle, which will

Figure 16. Basic parts of a saddle (courtesy of Jane Myers).

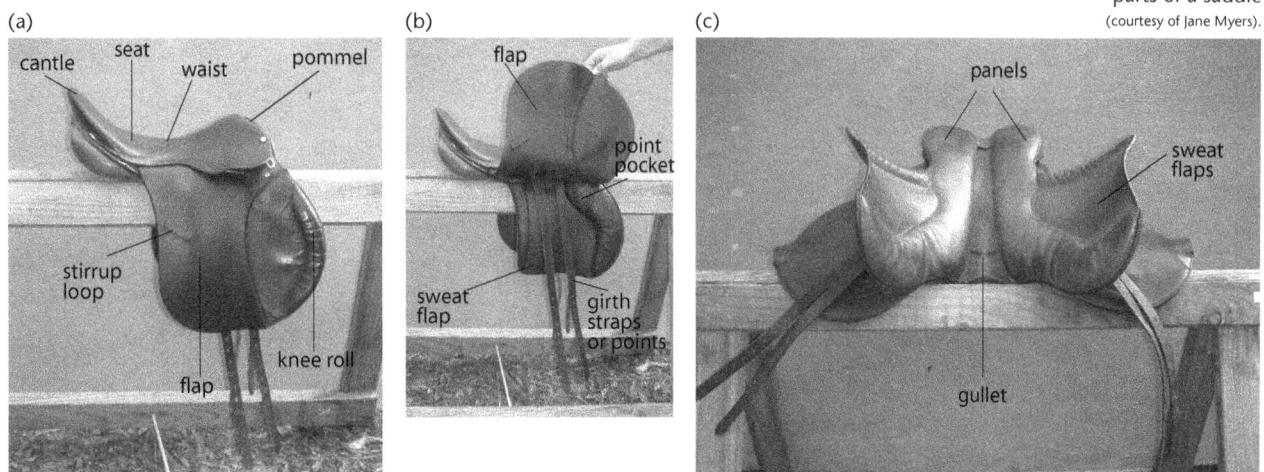

(a) cantle, seat, waist, pommel, stirrup loop, knee roll, flap

(b) flap, point pocket, sweat flap, girth straps or points

(c) panels, sweat flaps, gullet

(a)

(b)

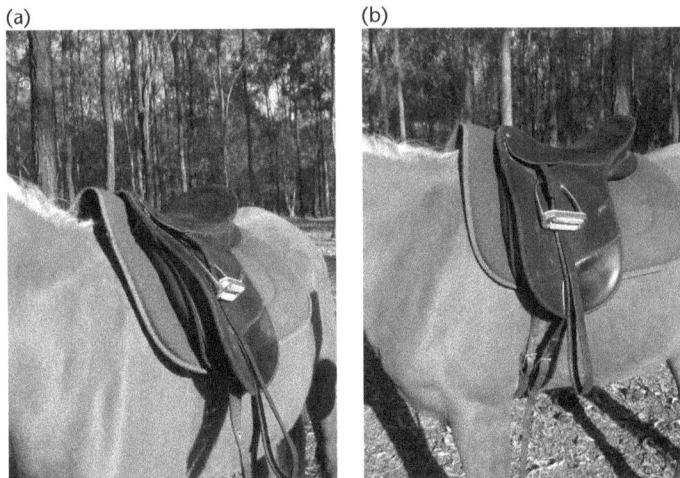

Figure 17. A Wintec dressage saddle with the stirrups 'run up'. Note how the saddle blanket has been pulled up into the gullet of the saddle before the girth is fastened (courtesy of Stuart Myers).

cause the saddle to rock, or less pressure in the middle, which will cause the saddle to bridge. Both will cause many problems later after you have ridden in the saddle for a while.

- The gullet should be the same width as the horse. To check this, run your hand between the saddle and the horse from the withers down the shoulder. There should be no increase or decrease in pressure as you go down.
- The cantle should be slightly higher than the pommel to put you in the correct riding position, both for your sake and the horse's.

If all these points are satisfactory, girth the horse up and sit on the saddle (remember to put your bridle on first). Go through all of the above checks while you are mounted (you may need an assistant to help you). If the saddle passes all the checks again, try riding in the saddle, but you should now put a saddlecloth underneath. The saddle should feel comfortable, it should not tip you forwards or backwards, and the horse should move happily without resistance.

Older saddles have stirrup bars underneath the skirt that can be either up or down. It is now regarded as safer for the bars to be in the down position at all times or you run the risk of the stirrup leather not coming off in the event of a fall (which can mean being dragged if your foot is caught in the stirrup).

Other gear

Surcingle

A surcingle is a webbing or leather strap that passes over the saddle and does up around the horse's girth. It should be used whenever there is only one girth strap and buckle (i.e. a stock saddle) and when entering into athletic activities such as the cross-country phase of horse trials. It is a safety precaution in case the girth breaks. It can be also used to hold stirrups and gear in place when lunging.

Leg boots

Boots are used to protect horses' legs from damage from jumps, obstacles on the ground, or other hooves. They can protect against damage from interference (see Chapter 4). There are many different types of boots and they are usually named after the problem they protect against (e.g. brushing boots).

Figure 18. Different types of boots.

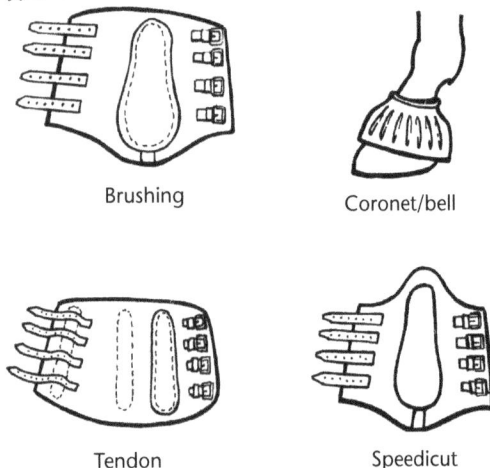

Brushing

Coronet/bell

Tendon

Speedicut

Halters and headcollars

Halters and headcollars are for catching, holding, leading, and tying up a horse. They come in many different styles and materials and, ideally, each horse should have its own correctly sized and fitted headgear. A properly fitted headcollar fits snugly, with the noseband five centimetres below the bony part of the horse's cheek. Headcollars with adjustable nosebands are better than those without because horses vary in the circumference of the nose.

Never leave a headcollar or halter on an unattended horse. They can get caught on fences, tree branches and other obstacles. Also, horses scratch their heads with their back feet and can get caught up in a headcollar or halter.

Five minute safety check

Check all of your gear for:

- rusty/bent buckle tongues
- torn buckle holes that are extending into each other, particularly in the stirrup leathers and the girth points
- worn stitching: check this by holding the two ends and trying to pull them apart; if you can, they need restitching
- leather parts that are dry, cracked, torn or stretched.

Check your bridle for:

- correct fit: bridles can change shape over time
- damaged billet hooks.

Check your bit for:

- sharp edges
- worn rings
- cracked bars.

Check your stirrup irons for:

- rusting, cracking or bending of the irons.

Check your girth for:

- frayed webbing
- perishing elastic
- ingrained dirt and sweat.

Check your saddle for:

- spreading or breaks of the head; protruding metal
- cracks and breaks in the waist of the saddle
- bends, breaks and loose or broken rivets in the stirrup bars
- lumps or bumps in the lining that could rub and irritate the horse's back.

Any other equipment/harness:

- check for damage as you would on a bridle or saddle.

Maintaining your gear

Store your gear in a clean, dry area. Hang the saddle over a wooden or metal bar with the stirrups run up so that it keeps its shape and the saddletree is protected. Hang the bridle on a peg by the headpiece so that it is ready to use next time. If your gear is leather, clean it frequently. This is a good time to inspect for defects, which should be mended immediately. Synthetic gear can be washed from time to time with warm soapy water. Keep all of your gear in good order; if you look after it, it will look after you.

TRAINING

In order to train a horse well, you must try not to label the horse with human terms such as lazy, stubborn, nasty etc. It is better to accept that the horse is misbehaving because it has not had the correct training, in which case the horse can be trained or retrained, rather than thinking 'This horse will never get any better because it is stubborn, lazy etc.'

All horses require training to some extent throughout their life. Training is an ongoing process, so when you take on the responsibility of horse ownership you also take on the responsibility of being that horse's trainer. You must ensure your horse has good manners. Ask your self 'If anything happened to me would anyone want my horse?' If the answer is 'No', then you need to train your bad-mannered horse to be a well-mannered horse. The main ingredients for creating a well-mannered horse are patience, persistence and knowledge.

How horses learn

Horses learn by association and not by reason, unlike humans who can and do reason. Even though humans have such a large brain, they often find it difficult to accept that other animals think and act differently to them. Learning to treat a horse as a horse is the first and most important step in becoming a horse trainer. Understanding the mind of the horse and accepting that it is different from a human's will lead to a better relationship between horse and handler because there will not be unreasonable expectations of what the horse can do and achieve.

A horse is not a problem solver because herbivores do not require problem-solving skills in order to eat grass: grass does not run away! Whereas a carnivore requires considerable skills in order to react to the varied actions and responses of its prey.

Most of a horse's brain activity is taken up with just keeping its legs underneath it and mobilizing them in sequence. This is an unconscious act; a foal is able to get up within two hours of birth (an essential prerequisite to the survival of a prey animal). The horse is renowned, however, for its learning ability. A task, once learned, will remain with a horse for life. As a trainer of horses, this is worth remembering because a horse will forget neither the positive nor the negative training experience, and the negative experience is likely to bring an unwanted response.

There are several types of learning that a horse (and the informed trainer) will employ. These include conditioned response learning, trial and error learning, and habituation.

Conditioned response learning

Conditioned response learning uses the horse's natural responses and connects them to a signal. Putting a certain bit in a stallion's mouth whenever he is brought out to serve a mare will shortly bring the conditioned response that the stallion becomes ready to serve a mare whenever that bit is used. This technique is also used in training. Saying 'trot' to a horse on the lunging reins, and initially reinforcing it by the movement of the whip, will shortly bring the response that the horse will trot on command, even without the movement of the whip.

Trial and error learning

Trial and error learning is commonly and successfully employed by good trainers. The horse will attempt various responses to a stimulus such as pressure from a hand on its hip. For example, the trainer wants the horse to move away from the pressure, but the horse does not know at first what it is supposed to do, so it will try various responses (trial and error); it usually moves towards the pressure before trying moving away. The trainer immediately removes the pressure when the horse makes the correct response. The release of pressure must be instantaneous and consistent if the horse is to learn the lesson well. A problem for inexperienced riders is that they produce less responsive horses because their excessive or inconsistent commands are not moderated when the horse produces the desired response.

Habituation

The process whereby a horse is slowly exposed to something that it is afraid of (by nature horses are afraid of most things initially) is called habituation. The exposure to the feared object or situation is increased gradually over time until the horse tolerates sights and sounds and sensations that previously would have frightened it. Police horses are habituated to accept many things that naturally frighten them, such as heavy city traffic and people in crowds. This training method is invaluable because it means a horse can be habituated to accept almost anything.

Breaking in and training horses

There is not one tried and true system for breaking in or training a horse. However, most of the successful methods involve pressure and the release of pressure. An experienced horse trainer can interpret the body language of the horse ('feel') and knows when to act ('timing'). As you improve your technique you also will also develop 'feel' and 'timing'.

Breaking in and training horses are ancient practices and many of the 'new' methods are actually old methods that have been 'repackaged'. In the past, horse breaking was shrouded in mystery; hence the mystical term 'horse whisperer' that has been around for more than a century. The horse breakers believed that if they shared their secrets they would lose business. In the present day, many horse trainers have realised it is better to share their knowledge with as many people as possible than keep it to themselves and have set themselves up as professional horse educators, or to be more precise 'horse owner' educators, touring the country, or even the world, doing clinics.

Going to clinics

If you have never trained a horse before (or you would like to learn more) go to a clinic, preferably with your horse or you can 'fence sit' if you prefer. The investment of time and money is well worth it. Some clinics are specifically for starting horses under saddle and are usually for one week, during which time you will start training your own horse in a safe environment and will learn a lot in the process.

Breaking or starting?

Some methods of breaking in horses were, and still are, rough and cruel. For that reason, many trainers now distance themselves from the term 'breaking' and choose to call the process 'starting' or something similar. However, there are many well-respected trainers who use very gentle methods and still call it 'breaking' because this is a generic term that everyone understands and does not necessarily mean that their methods are rough. If you are looking for someone to break/start your horse, it is the technique that matters, not its name.

When is the horse ready?

Most horses are considered ready for 'breaking in' at 18 months, although in non-racing breeds breaking in is often delayed until the horse is more mature. Groundwork, as opposed to ridden work, can be still be done with the young horse because if done correctly it will make it easier when the horse is first ridden. The actual process of breaking in or starting usually takes between 12 and 15 weeks, depending on how much the horse has been handled beforehand. A horse that has been constantly handled since it was a foal could be trained in the basics in 6–8 weeks. However, remember that there is no real end-point to the process. The initial stage of breaking in flows into the next stage and so on.

Figure 19. Yielding to the pressure on the hip (courtesy of Stuart Myers).

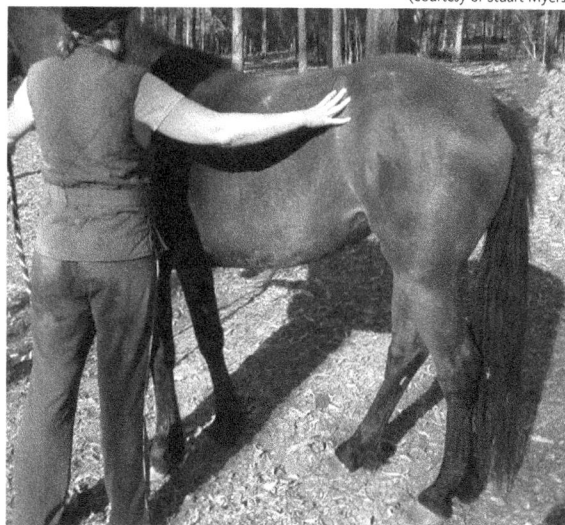

Training methods
Yielding to pressure

All horses should yield their body when asked. In the young horse, this can start as groundwork and later on will be part of the training to be ridden. All young horses should be taught to yield and if the older horse does not, then it should be taught also. Initially the horse should learn to move forwards, backwards and move the shoulders and the hindquarters sideways in both directions. To teach a horse to yield, pressure is applied until the horse makes the correct response and then the pressure is immediately released (trial and error learning).

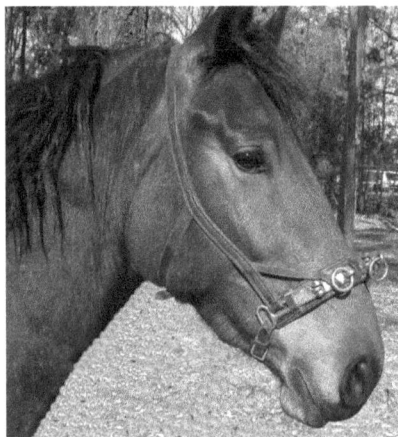

Figure 20. A lunging cavesson. The bridle can be fitted over this if using side reins (courtesy of Jane Myers).

For example, if the horse steps forwards while you are applying pressure on its chest to make it go backwards, you must keep the pressure on until the horse steps back. The horse will eventually step back because it is trying out various responses to make you release the pressure. As soon as the horse yields in the right direction the pressure must be removed.

Horses like to be comfortable, so usually you do not need to apply much pressure to get the horse to react. Examples of pressure are the fingertips, taps from a whip, flicks from a rope; even the voice can be a form of pressure.

As you become familiar with teaching the horse to yield, you can use your imagination to create many different ways to make the horse yield to pressure. A horse that yields rather than fights is much safer both for itself and its human handlers. For example, if a horse has been taught to accept ropes around its legs and will yield the leg when asked, it should not panic if it ever gets caught up in a fence or if a rug inadvertently slips. A horse that has been taught to yield its head will lead anywhere you ask, including onto a float. It will also tie up without pulling back because instead of fighting the pressure from the headcollar it will come forward (yield).

Different horses require different types and levels of pressure. In addition, the more educated the horse becomes the less pressure should be needed to get a response. You will develop your own skills and technique and if you attend clinics you will probably use the form of pressure favoured by that particular trainer.

Lunging

Lunging is a method of training based on the use of repetition to create a habit of obedience. Lunging also strengthens a horse's muscles and improves its impulsion, tracking (gaits), bending (flexibility), and balance and rhythm. Lunging is a very useful skill that most horse people employ at some stage. Lunging is also useful for:

- training young horses and retraining the spoilt horse; it is an effective aid to the teaching of obedience to aids (voice, rein, whip) and it develops correct muscles and improves balance before the horse is ridden.
- exercising a horse that cannot be ridden e.g. if the rider is ill.
- further training of the advanced horse, particularly in progressions and advanced dressage work in hand.
- training the rider, particularly with exercises to improve seat and balance, but should only be carried out by experienced instructors.
- exercising and reinforcing obedience in the fresh, lively horse. Do not allow this kind of horse to jump around on the lunge because it will reinforce the behaviour.
- allowing the handler to watch the horse and assess its movement.
- working a new and unfamiliar horse before being ridden.

Lunging equipment

All gear and equipment must be safe, well maintained and correctly fitted. The horse will need to be fitted with:

- a lunging cavesson that fits snugly over the nose. Many are not well made, and sheepskin or padding is often required under the jaw to prevent rubbing. The cheek straps of the cavesson must be kept clear of the eyes by fastening the throat latch snugly to prevent slipping and rubbing.
- a simple snaffle bridle – if you are going to be using side reins – without a noseband over the top of the lunging cavesson. The bridle reins must be taken off or fastened up.
- side reins, which are optional, and preferably should be about 1.5 metres long, adjustable and contain elastic inserts. The side reins must never pull the horse's head into position. They should allow the horse to be comfortable at the halt and should be

shortened gradually. If you have never used side reins, start by attaching only one (the outside rein). Horses can panic and flip over at the restriction caused by side reins if they are not used correctly.

- a strong roller or saddle, to which side reins can be attached.
- a lunging rein, which is attached to the middle ring on the cavesson with a swivel hook. A good lunging rein is made from heavy cotton or canvas and is approximately 10 metres long.
- protective leg boots, especially on young horses, which are prone to hitting themselves when working in circle until they learn to balance.

Figure 21. Lunging
(courtesy of Stuart Myers).

The handler must also be equipped correctly with gloves, a helmet and strong boots with a grip. It is best not to wear riding boots for lunging because they do not have grip, which you may need if the horse pulls away and you have to stand your ground. The handler should also carry a whip. A good whip is lightweight, balanced and has a long thong. The whip is simply a training aid, used mainly to encourage forward movement. It should never be used to punish the horse. When you are not using the whip, do not place it on the ground, but hold it under your arm with the end of the whip behind your body.

Where to lunge

An enclosed area, such as a round yard or small paddock, is essential, especially until the horse has learned to lunge. The surface of the lunging area should be even, flat and not slippery. Be aware that moving in deep sand, such as in a sand rolling yard, is very hard work for the horse. The diameter of the lunging area should not be less than 20 metres. There should not be any obstacles or loose horses in the area.

For how long?

The period of work depends on the individual horse's constitution, conformation and soundness, and mental capacity to learn. However, lunging, with its constant circling, can be hard on a young horse's joints if it is done for too long, so sessions should be kept to just a few minutes for this class of horse. A horse should never be lunged for more than 45 minutes and even then this should be gradually built up from an initial lunging period of 10 minutes as the horse's muscles develop. If it is necessary to lunge the horse for this length of time in order to make it quiet enough to ride then you need to re-evaluate the management of the horse. Consider whether the horse is confined for too long and/or being fed a diet with an excess of energy. Lunging the horse for long periods actually increases the level of fitness of the horse, which in turn may make it even livelier!

Voice commands

The use of the voice is a very important training aid. You can use your voice when lunging to make the horse speed up or slow down. Use the same words for each command at every lesson and only use a few words to prevent confusion. Don't 'chat' to the horse. Words must be distinct with varying tones; for example, sharper words to go faster and slower words to steady the pace.

Starting off

If you have never lunged a horse before you will need to learn with an experienced horse before you try to teach a young horse. Learning to handle a lunging rein and whip, and control the horse, all at the same time, is quite complicated.

To begin, take up the lunging rein in the same hand as the direction of travel and the whip in the opposite hand. The lunge rein should always be taut and untwisted, maintaining an even, firm contact between the horse's nose and the trainer. The rein should be folded across the palm of the hand and the loops should not be long enough to trip you up. The whip is used to control the forward movement of the horse.

Step away from the horse and ask it to walk on, using the whip initially towards its shoulder to push it out into the circle. Once the horse is circling, point the whip towards the hocks to keep the forward movement. There should now be a triangle, with the horse as one side, the rein and whip as the other two sides and you as the apex. Stand still and swivel on a heel or walk a small circle if the horse needs encouragement to start.

The whip should be used in a circular motion in the direction of the hocks. Do not punish the horse with the whip. If the horse will not go forward the whip can be used to 'flick' the horse. Flick towards the back leg.

If you step too far to the front of the horse the horse will turn towards you. If this happens step towards the back of the horse again and push the horse forward with the whip. Practise changing the gait (transitions) using your voice commands. Changes of gait are a very valuable part of lunging; in fact, there is no benefit in the horse trotting around and around without frequent changes of gait.

The horse must learn to be calm and to go forward without rushing. Eventually the horse should be relaxed and stretch down and learn to balance itself. Rhythm, regularity of pace and suppleness will develop, together with the muscles in the neck, back and hindquarters.

To finish, halt the horse on the circle and make it stand. Then approach the horse, gathering the rein in as you go.

Training difficult horses

If you are having problems with your horse, you may need professional help from either an experienced instructor or a horse trainer. Many problems should be sorted out as soon as possible before they become a habit. Most problems arise from a lack of communication between the horse and the handler. The handler is the one with the bigger brain so it is the handler who must make sure that the communication is clear. If you are confused, imagine how confused the horse is!

Uncooperative horses

Some horses that are difficult to handle (those that are usually labelled 'stubborn' or 'aggressive' etc.) often appear to be taking charge. The horse is not making a conscious decision, but is performing a learned behaviour; that is, the horse has learned that certain behaviours (e.g. dragging the handler back to the gate) tend to produce certain responses (i.e. the handler gives in and goes home). The horse needs to be retrained so that it learns that a particular cue, such as a pull on the lead rope, means that it must follow and not pull away.

Nervous horses

Nervous horses are often unwilling to cooperate because they have either been punished in the past for making the wrong response, making the horse fearful, or they have not yet had enough training to be confident around people. These horses are sometimes labelled as 'stupid' because they tend to panic and behave in what is an irrational (to humans) manner. The nervous horse needs careful, consistent handling. It is usually possible to improve this behaviour with time, explaining what is wanted by constant repetition, and offering rewards when a task is finally achieved. Nervous horses require a lot of patience (as do all horses).

Common handling problems

Always go through the following checklist if you are having problems with your horse.

- Does the gear fit properly? If gear is too tight and pinching or rubbing the horse then it will start to object sooner or later.
- Does your technique need improving, both as a handler and/or as a rider? Most people have room for improvement.

- Has the horse had its teeth checked recently?
- Is the horse being fed correctly? If the horse is under- or overfed, then problems will result.
- Is the horse confined too much? A horse that is inactive for too long will be keen to move and may play up simply because it has too much energy.
- Is the horse just too much for you? The horse may be at a stage beyond what you are capable of handling at the moment.

Hard to bridle

If a horse is difficult to bridle, check that it is not getting its ears pinched when the bridle is put on. Often a horse will start to resist if this has happened a few times. Let the bridle out by a couple of holes so that it easily goes over the ears and then re-adjust. Sometimes it only that the browband is too tight (short) and all that is needed is a longer browband.

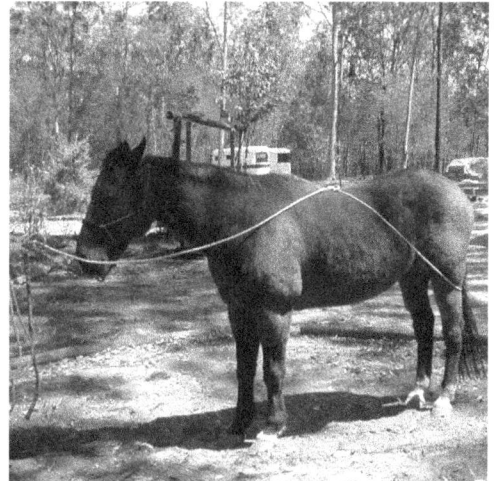

Figure 22.
Breeching rope
(courtesy of Jane Myers).

Some horses will not open the mouth for the bit, usually because someone has previously banged the teeth when bridling. This becomes a vicious cycle as the horse clenches its mouth and makes it more likely that the teeth will be banged again. If the horse will not open its mouth from thumb pressure on the bars, wrap a piece of bread around the bit (squash it on firmly with your hand) to both protect the teeth from being banged and encourage the horse to take the bit.

The horse that lifts its head up high to avoid the bridle should be trained to lower the head when pressure is placed on the headcollar. Initially this should be done without trying to put the bridle on. Put steady downward pressure on the headcollar until the horse lowers the head slightly, and then release the pressure. Keep training until the horse will lower its head right down to the ground. You can even teach the horse to lower the head from the pressure of your hand on the top of the neck. Once this is well established the bridle can be reintroduced and the horse will now be much easier to bridle.

Pulling back

A horse that pulls back is a horse that panics when tied up. The horse will pull back violently until something breaks, which is a dangerous situation for all concerned. Horses do this because they have not been taught to yield the head and therefore to tie up. Horses that pull back need expert handling because even those in which the behaviour is not too ingrained can get a lot worse if the situation is handled badly. Every time a horse pulls back and breaks free it is reinforcing the behaviour, which means it becomes harder to correct.

The solution lies in teaching the horse to yield to pressure on the headcollar. The horse has to be taught to come forward from pressure and initially this can be done with a 'breeching rope' and without actually tying the horse up.

Pressure should be put on the headcollar and the breeching rope to get the horse to come forward. As soon as the horse yields you must praise it and give a reward. Do this repeatedly until the horse comes forward every time you ask. Make sure you keep rewarding the horse when it comes forward. Do this in different situations, including where the horse will be tied up.

The next stage will require the help of an experienced person if you are not experienced. A neck collar is used to tie the horse up to a very solid post (a rubber tyre can also be used around the post, so that there is a little give in the rope).

Thread the rope of the neck collar through the headcollar, which will keeps the head forward. Have the breeching rope on the horse and go through the same procedure of getting the horse to come forward. For safety you should be on the other side of a very strong fence and the breeching rope can be threaded through the rails. Have a knife ready to cut the rope if necessary. If this stage

(a)

(b)

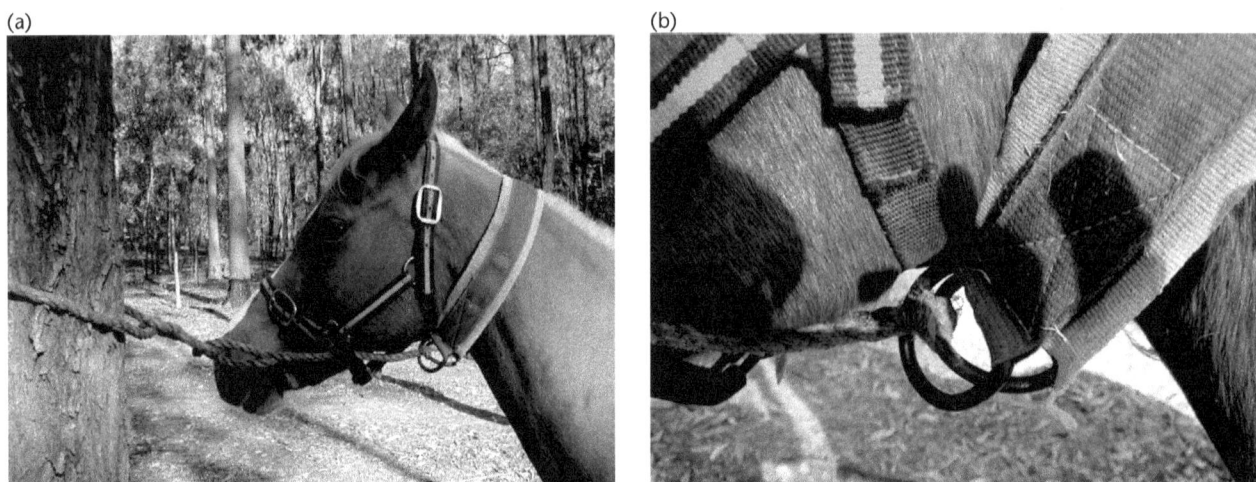

Figure 23. Neck collar
(courtesy of Stuart Myers).

is not done properly the horse will become worse. Retraining this behaviour will take time, patience and skill. If you do not have them, then send the horse to a professional.

Nipping when saddled

Most horses develop this habit because the saddle has been girthed up too tight and too fast. The horse then begins nipping whenever the saddle is put on. Tie the horse up short enough so that it cannot reach you with its teeth. Saddle the horse, putting the saddle well forward and slide it back so the hair underneath lies flat. Use an elastic girth, which is more comfortable. Do not pull the girth up straight away, ease it up until it is comfortably snug, then go up hole by hole until it is tight enough to mount.

This process can take 10 minutes or more. You can be doing other things such as getting ready to ride in the meantime. Use a mounting block so that you do not have to have the girth extra tight while you mount. Once you are in the saddle, check the girth and tighten if necessary. Check the girth again after riding for 10 minutes or so.

Nipping and biting when handled

A horse bite can do a lot of damage. Be careful when handling young horses, which are inclined to nip in play. A sharp slap and a loud 'No!' will generally be sufficient to teach a young horse that it must not bite.

Feeding titbits by hand often starts this behaviour. If the horse noses around your pockets and you then give it a titbit the horse has learned that in order to get you to feed it, it must nose around your pockets! This can then turn to nipping when the tit bits run out and the horse gets frustrated, and is the reason why many children's ponies nip, because children often freely feed ponies. Feeding as a reward for correct behaviour as a training aid does not lead to nipping, as long as the horse is not allowed to demand food and fed on demand.

Be aware that horses groom each other with their teeth. Some horses will try to groom you, especially if you scratch them in an itchy area. The horse is simply reciprocating and is not being nasty. If your horse does this just push the head away as you scratch and the horse will get the message.

Always wear loose clothing around horses that nip or bite to reduce the chance of your body getting bitten.

Never tease a horse that bites. Some horse bite because they are stabled for long periods and are frustrated. With some horses, biting is actually a game and if you try to hit the horse it will simply swing its head out of the way and come back at you. Colts play this game with each other, so it is not surprising that some horses do this.

With an older horse that bites, try to understand why it is doing it. Go through the checklist of the possible reasons for bad behaviour. Unfortunately, old habits do die hard, so even when the cause is removed an animal will still revert to habits that have worked in the past. With an older horse, it will no doubt have been smacked for biting in the past and as that has obviously not worked try to reward the horse for good behaviour (ears forward, not biting) rather than punishing it for bad behaviour.

As a temporary measure if you must work with a biter, attach a Hanoverian noseband and do it up reasonably tightly to prevent the horse opening its mouth wide enough to bite hard.

Kicking

Kicking is a very dangerous habit that includes kicking forward towards the belly (cow kicking), kicking backwards with one leg, or with both legs (double barrelling). The horse may be kicking because it is frightened and feels cornered or it may be confident and have learned that kicking is how to get people to move away.

If you have to work with a kicker be very careful. Never go behind the horse and keep it away from situations where it is likely to harm someone or another horse.

Kicking forwards is sometimes an attempt to get at a fly on the belly and if you are in the way you will get kicked. However, some horses do it in response to pressure around the belly, so they will do it when being girthed up and this must not be allowed to become a habit. Reprimand the horse and push it back for a step or two.

Kicking backwards is a more dangerous habit as the leg has more power in this position. Some people would deal with this habit by smacking the horse with a whip, but you must be very careful. If the horse is nervous and is kicking because of fear it will be even more nervous if you hit it with a whip. Even with a confident horse, if you use a whip and the horse kicks out again you will have to use the whip again and again until the horse stops. This can turn into a real fight. A horse that kicks is not a very well-educated horse and therefore needs more training. If you are not a confident handler you will need professional help with this problem.

Difficult to clip

Many horses do not like the feel of the clippers moving over them or the noise that they make, but if they are introduced to clippers carefully most will come to accept them. Horses that have been clipped before and react badly when you try to clip them may have been nipped or nicked by the clippers in the past. In either case the horse must be habituated to the clippers.

There are several ways to do this, but one method is to first get the horse to accept the clippers when they are not turned on. Run the clippers over the horse's body in the same direction as the hair so that they do not grab the hair and further worry the horse. When the horse has accepted the silent clippers, they can be turned on at a distance and very gradually moved towards the horse. This process may take several days with some horses (horses that have had a bad experience will take a lot longer than a horse that has had no previous experience).

Try to use lighter, quieter clippers for the head. The head of the horse is largely hollow and the clippers vibrate, making the horse feel very uncomfortable. It may be necessary to use some form of restraint, but spend time trying to get the horse to relax before you overly restrain it.

Bucking and rearing

Some horses will buck if they feel fresh, so if a horse is known to be fresh it should do groundwork for sufficient time to reduce its energy levels. Groundwork also reminds the horse about yielding rather than resisting (i.e. bucking and rearing). A horse that bucks for no apparent reason, however, cannot be dealt with by the casual rider and should be given to a professional trainer for correction. The same applies to a horse that rears. Some horses just lift the front end a few inches off the ground, which is bad enough, but others flip themselves over backwards, which is extremely dangerous and even many professionals will advise that the horse is too dangerous to work with.

Hard to catch

If a horse is hard to catch it can be infuriating, but this problem can be solved. Horses that are hard to catch are often good to ride because if there is a choice of horses to ride the best horse will always be picked. Consequently this horse is used more than the others and it learns that whenever it sees a human coming with a headcollar there will be work.

Run through the checklist of possible reasons why the horse is misbehaving and make sure you have not given the horse a reason to not want to be caught.

Put the horse in a small yard or stable and teach the horse to 'face up' (see earlier in this chapter). Use a reward, such as a carrot, if you want the horse to learn quickly. Make sure you do this thoroughly until the horse will face up on a verbal command, then move to a larger area (a round yard is good) to see if the horse will still face up on a verbal command. If it does not, move back to the smaller area and keep training the horse. Reward the horse with a carrot when it does face up and move back to the larger area. Continue like this until the horse will face up in a variety of situations, including the paddock.

Initially, you may have to separate your horse until it has learned to be caught. Other horses can get in the way and it is not safe if they discover that you have carrots. Never feed the other horses in the paddock; as well as being dangerous, in no time at all you will have all of them following you around (apart from your own hard-to-catch horse!). As soon as the horse has learned to be caught it can be put back in with the others.

Some horses are only hard to catch when the lush, green grass comes through in the paddock. Because the horse is getting good feed, it plays hard to get. The horse may have to be put in a less improved paddock initially, but you will need to teach the horse to be caught in any situation.

It is tempting to leave a headstall on a hard-to-catch horse, but this is dangerous because many horses have been injured as a result of catching the headstall on a branch or post. If you must leave a headstall on the horse, sew an elastic insert into the headpiece so that it will come off or break in an emergency.

Is the horse right for you?

Finally, use your own judgement and be realistic about your abilities. Do not handle or ride a horse that frightens you. If you and the horse are not well matched you will not be happy until you feel in control. Sometimes lessons will sort out the problem, but sometimes the horse is not for you at your stage of riding and you should sell it (see Chapter 1) and get a more appropriate horse before you lose your confidence.

Figure 24. Nose twitch (courtesy of Jane Myers).

Restraint methods

Sometimes it is necessary to restrain a horse to perform a job that the horse may resent and resist; for example, being injected with antibiotics or having a wound dressed after an injury. Sedation is an option that can only be carried out by a veterinarian and even then, if the horse can be restrained some other way it is preferable.

Remember that whenever two people are handling a horse at the same time they must both be on the same side of the horse at the same time.

If a yard or stable is available it can be used to restrain the horse. Back the horse into a corner and hold the front of the horse firmly. The horse is then wedged and as long as it is not frightened the corner will stop it from moving around. However, if you are working with a frightened horse it is safer to be in an open area so that the horse can move if the restraint gets too much for it. Sometimes just allowing the horse to move in a circle is enough for the horse to relax. Every situation is different so you will need to assess each new case.

Picking up a foot

Picking up one of the front feet is a method of restraining a horse. The person holding the foot up should stand beside the leg in case the horse suddenly pulls its foot forward.

Twitching

A nose twitch is a common method of restraint, as is tinker's grip on the neck.

A nose twitch is a loop of rope that is threaded through the end of a piece of poly pipe. The top lip is put into the loop, which is twisted until it is tight around the lip.

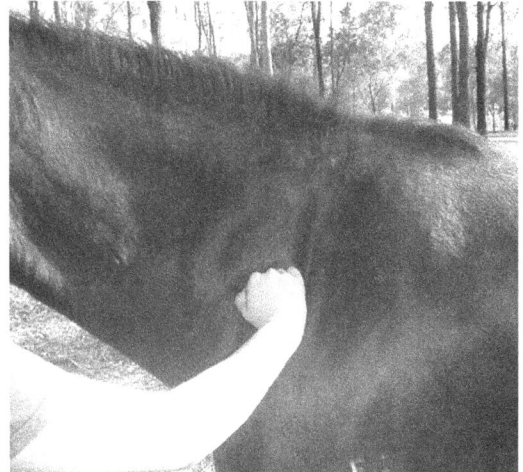

Figure 25. Tinker's grip
(courtesy of Stuart Myers).

Opinion is divided about whether this is a cruel practice. The top lip has many nerve endings and is very sensitive. It is believed that when the twitch is applied the body releases natural painkillers that result in the horse becoming quiet and still. Certainly most horses appear to go into a trance-like state when a twitch is applied. These horses can be injected etc. and then the twitch is released; the whole process is quick and relatively free of stress.

If a horse is already in a fearful state when the twitch is applied it is not as effective. It can even have an opposite effect because the fear hormones (such as adrenalin) that are already circulating in the body will override the pain-killing hormones. Therefore, it is better to twitch the horse before it gets upset about the injection or whatever procedure it must undergo.

Once the twitch is on, it can be tightened if the horse is still moving around. Work quickly and remove the twitch as soon as possible. As soon as it is removed rub the top lip to get the circulation going.

A tinker's grip is a helpful restraint for a fractious horse. Grasp a loose fold of skin from the neck where it joins the shoulder. Squeeze and twist the fold of skin to restrain the horse for a short time.

Horses that object to being twitched and cannot be otherwise restrained will need the vet to give them a tranquilliser.

WHERE TO GO FOR MORE INFORMATION

The Australian Horse Riding Centres website (www.horseriding.org.au) lists accredited riding schools in each State. The Equestrian Federation of Australia website (www.efanational.com) has contact details of each State branch and a list of riding instructors. The New Zealand Ridings Clubs website (www.nzridingclubs.homestead.com/index.html) lists affilated riding schools and has information on New Zealand riding and road safety tests.

The Australian Horse Council website (www.horsecouncil.org.au) has links to many other useful sites.

Information about the Wintec range of riding gear is available at www.wintec.net.au

CHAPTER 4

DESCRIBING THE HORSE

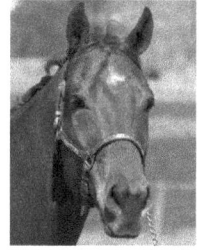

It is important for the horse owner to be familiar with the anatomy and points of the horse, correct conformation and gait, and correct identification of the horse. You can then converse with veterinary surgeons, trainers, farriers and other owners about the particular problems of your horse. You need to know what is normal so you can recognise abnormalities.

ANATOMY AND POINTS OF THE HORSE

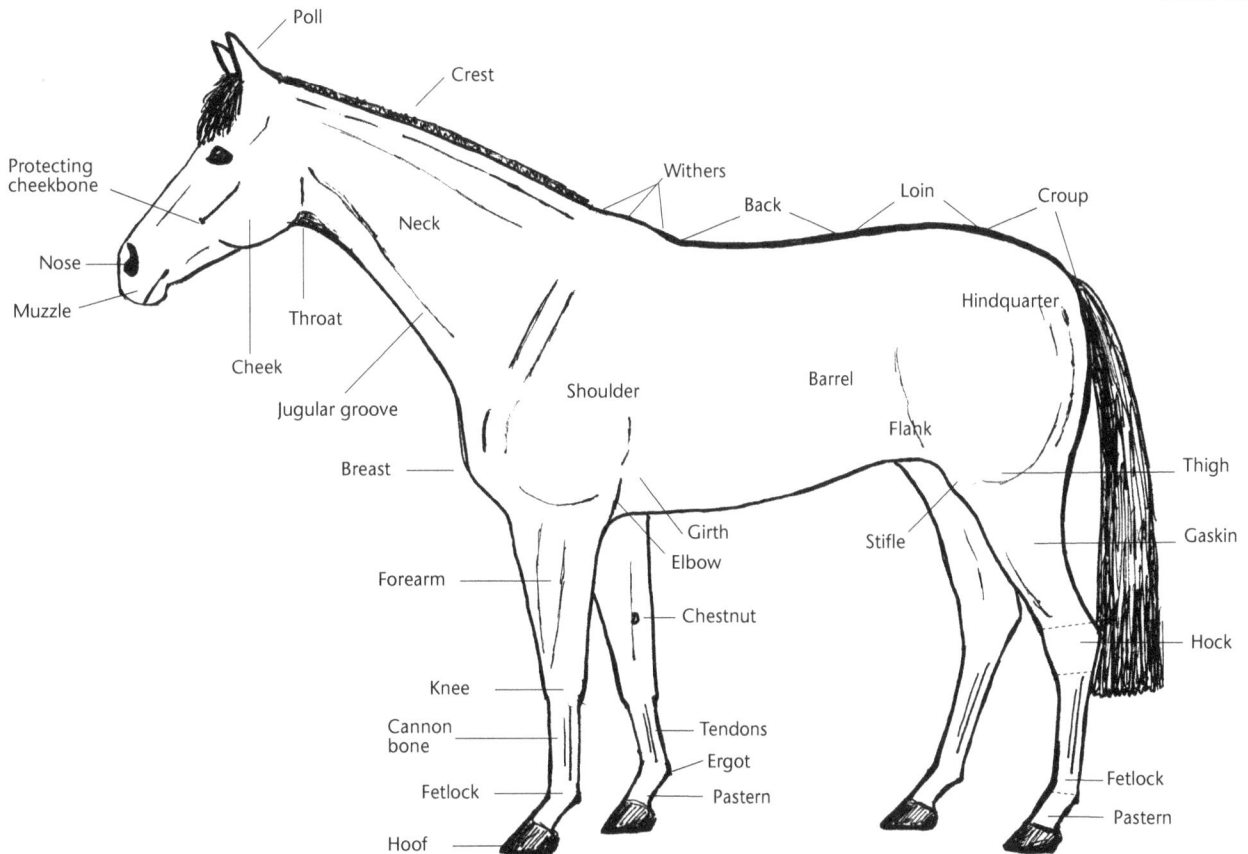

Figure 1. Points of the horse.

Figure 2. Superficial muscles of the horse.

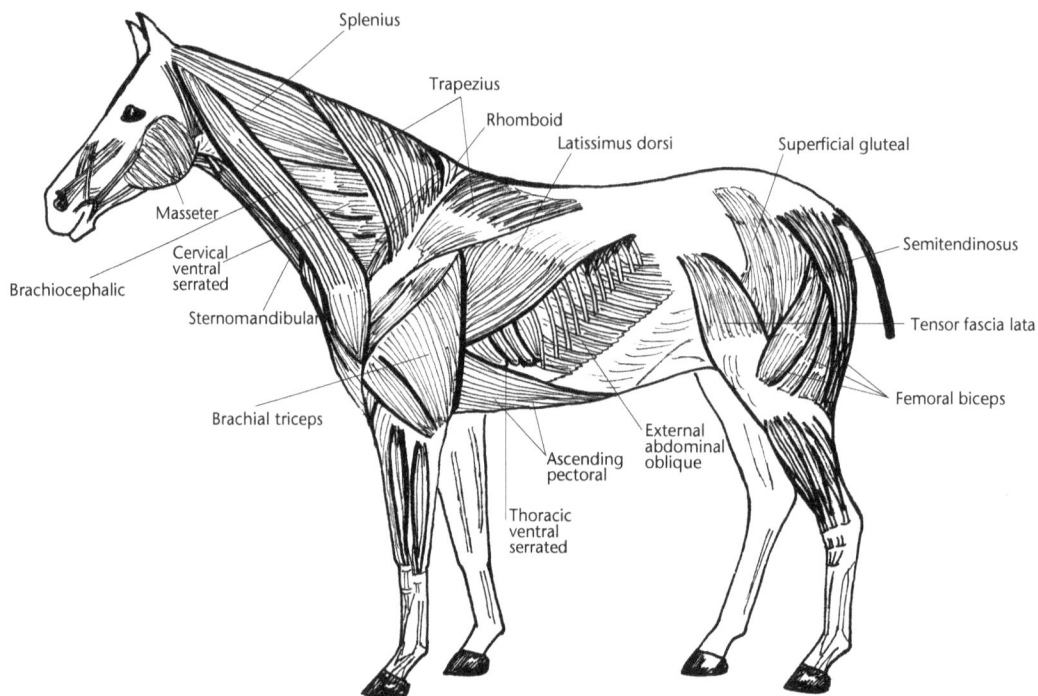

Splenius
Trapezius
Rhomboid
Latissimus dorsi
Superficial gluteal
Semitendinosus
Masseter
Cervical ventral serrated
Brachiocephalic
Sternomandibular
Tensor fascia lata
Femoral biceps
Brachial triceps
Ascending pectoral
External abdominal oblique
Thoracic ventral serrated

Figure 3. Internal organs of the horse.

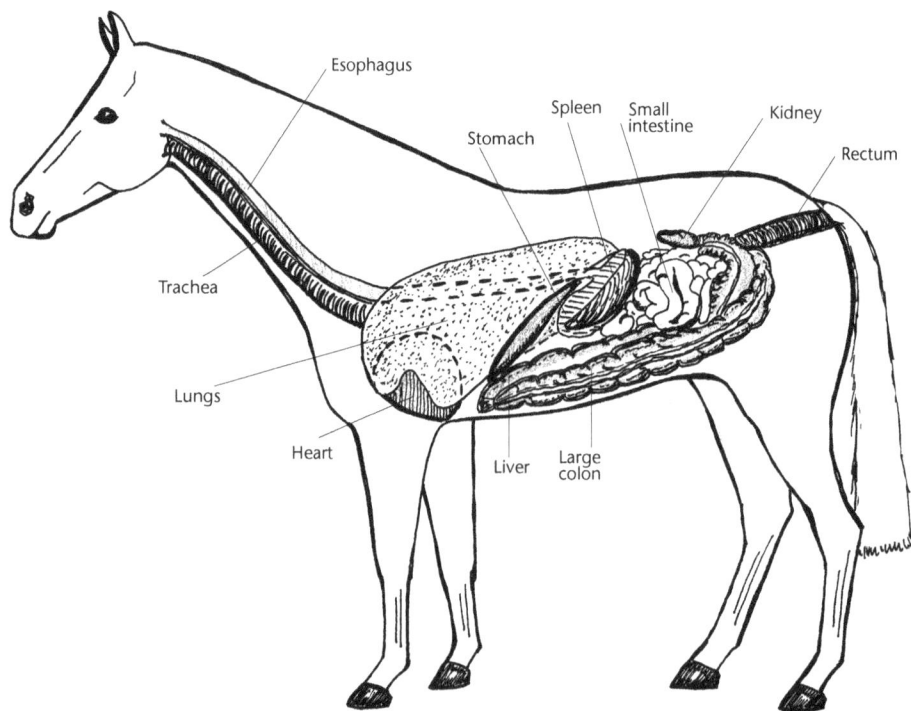

Esophagus
Spleen
Small intestine
Kidney
Stomach
Rectum
Trachea
Lungs
Heart
Liver
Large colon

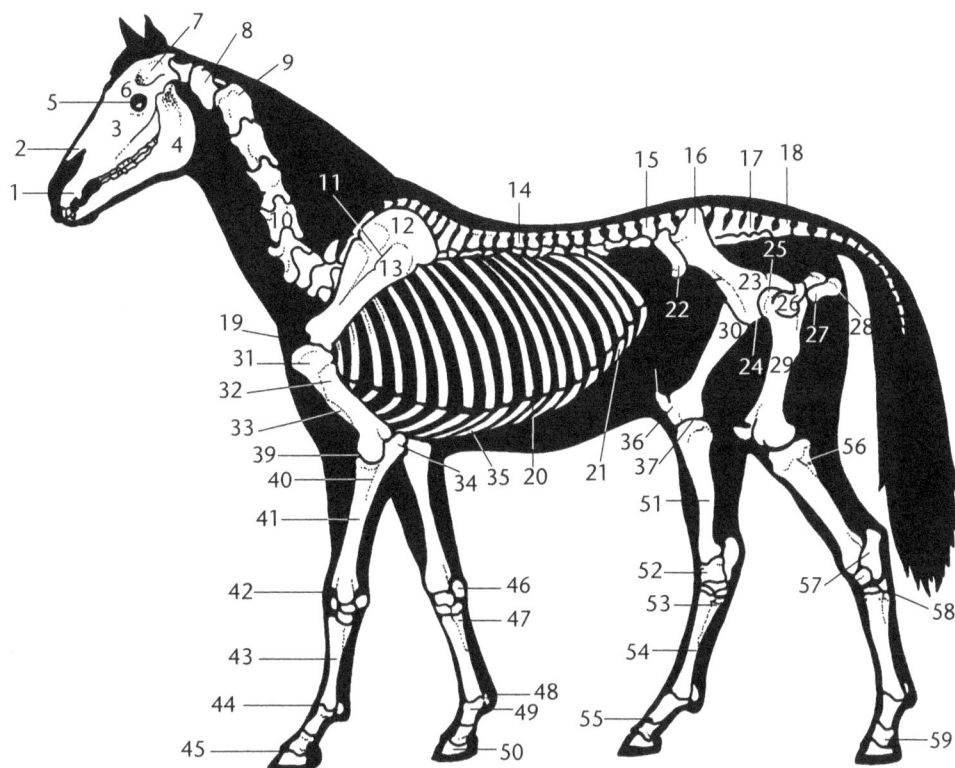

Figure 4. Skeleton of the horse.

1. Incisive bone (premaxillary)
2. Nasal bone
3. Maxillary bone
4. Mandible
5. Orbit
6. Frontal bone
7. Temporal fossa
8. Atlas (first cervical vertebra)
9. Axis (second cervical vertebra)
10. Cervical vertebra (there are 7 of these, including the atlas and axis)
11. Scapular spine
12. Scapular cartilage
13. Scapular
14. Thoracic vertebrae (there are usually 18 of these)
15. Lumbar vertebrae (there are usually 6 of these)
16. Tuber sacrale
17. Sacral vertebrae (there are usually 5 vertebrae fused together)
18. Coccygeal vertebrae
19. Shoulder joint
20. Ribs (forming wall of thorax; there are usually 18 ribs)
21. Costal arch (line of last rib and costal cartilages)
22. Tuber coxae
23. Ilium
24. Pubis
25. Hip joint
26. Femur, greater trochanter
27. Tuber ischii
28. Ischium
29. Femur, third trochanter
30. Femur
31. Humeral tuberosity, lateral
32. Humerus
33. Sternum
34. Olecranon
35. Costal cartilages
36. Femoral trochlea
37. Stifle joint
38. Patella
39. Elbow joint
40. Ulna
41. Radius
42. Carpus
43. Metacarpus
44. Fetlock joint
45. Coffin joint
46. Accessory carpal bone (pisiform)
47. Small metacarpal bone (splint bone)
48. Proximal sesamoid bone
49. First phalanx
50. Distal phalanx (third phalanx; pedal bone)
51. Tibia
52. Talus (tibial tarsal bone; astragalus)
53. Small metatarsal bone (splint bone)
54. Metatarsus
55. Pastern bone
56. Fibula
57. Calcaneus (fibular tarsal bone)
58. Tarsus
59. Middle phalanx (second phalanx)

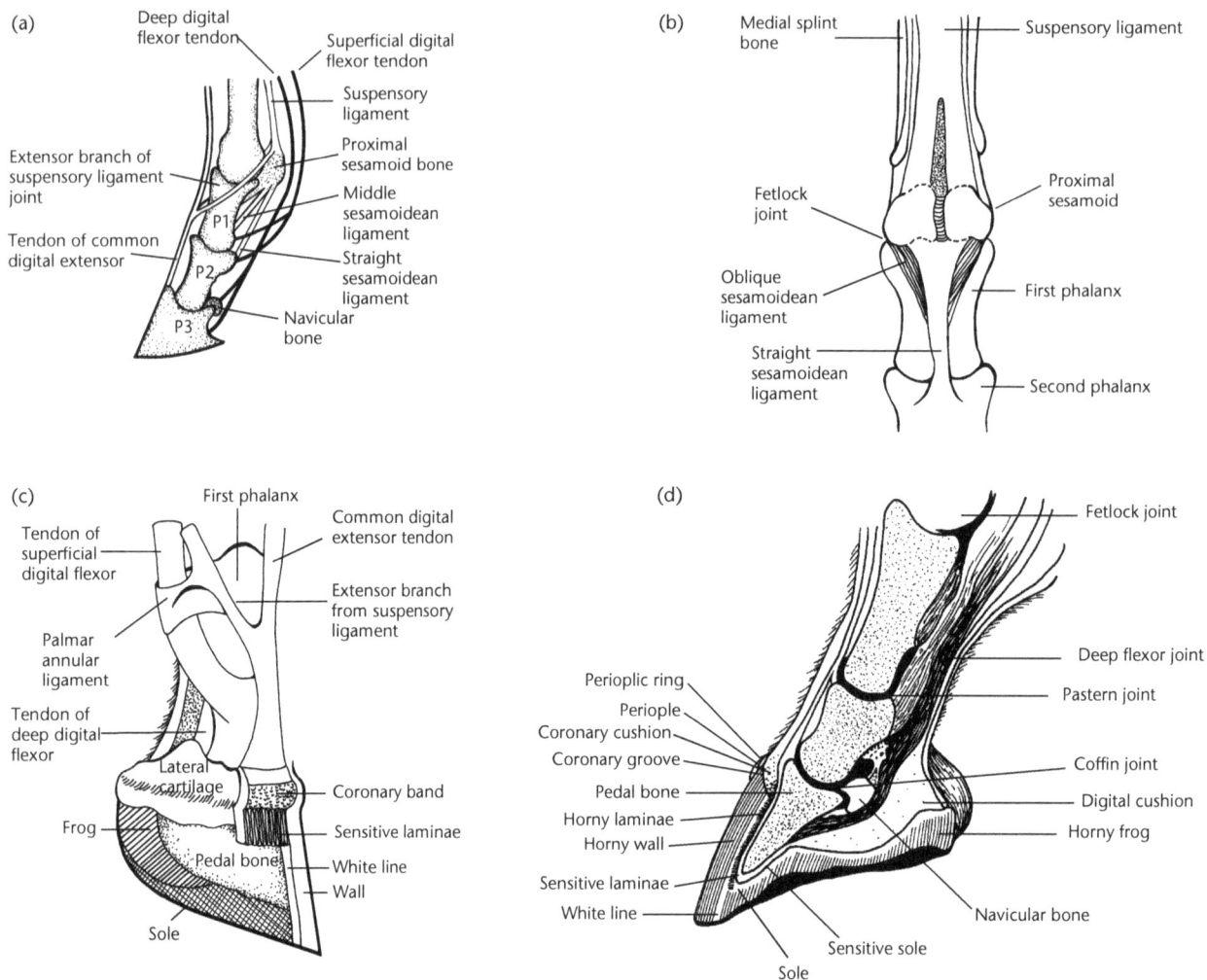

Figure 5. Lower leg anatomy.

The hoof

The old saying 'no foot, no horse' still holds true and in order to correctly care for the horse's hooves, you need to understand their construction. The anatomy of the foot and the hoof are shown in Figures 5 and 6.

Insensitive structures of the hoof
The wall
The wall of the hoof grows from the coronary band at the rate of 6–9 mm/month. As the average hoof is 76–100 mm long at the toe, this means that the horse grows a new hoof almost every year.

The hoof wall is made of a tough material called keratin that has a low moisture content (approx. 25% water), making it very hard and rough. The wall has three layers: the periople, a middle layer and an inner layer.

The periople extends almost 19–25 mm down the wall from the coronet to protect the sensitive coronary band at the junction of the skin and the hoof. It is very tough and of a similar consistency to the frog. After extending nearly 20 mm, the periople changes into a thin layer of material that covers rest of the hoof and gives it the shiny appearance, acting as a protective covering. It helps to control evaporation of moisture from the wall.

The middle layer makes up the bulk of the wall and is responsible for its strength and rigidity. The inner layer forms the insensitive (horny) laminae of the hoof, which mesh with the sensitive laminae that cover the pedal bone (third phalanx) and firmly attach it to the wall of the hoof. The insensitive and sensitive laminae bear most of the weight of the horse.

In general, the main function of the wall is to bear weight, so it is the wall that goes out of shape and needs to be trimmed. When a horse is shod, the nails of the shoe are driven up into and through the wall (see Chapter 2 for details on foot care and shoeing).

The wall grows evenly below the coronary band, so the heel is the youngest part of the hoof and therefore also the most elastic, which helps the heel expand during movement. Because the heel is also softer, care must be taken when trimming it.

The white line
The white line is the junction of the wall and the sole. It is a guide for placing the horseshoe nails to be driven into the hoof. Nails should be placed so that they pass through the hoof wall edge of the white line. They may penetrate the sensitive structures in the hoof if they are placed closer to the sole.

The hoof can sometimes split along the white line, which is called seedy toe, and this can allow an infection to develop in the foot and become an abscess.

Figure 6. The parts of the hoof.

The sole
The sole makes up most of the ground surface of the hoof. It is made up of nearly 33% water, so it is softer than the wall. The structure of the sole is similar to that of the wall, except that it breaks away when it grows to a certain length.

The sole is thickest at its junction with the wall and should be concave in shape, which gives the horse more traction, acting as a suction pump in wet conditions. The sole of the hind foot is normally more concave than that of the front foot.

The function of the sole is to protect the sensitive structures within the hoof. It is designed to bear the horse's weight, not surface weight, and must be protected from constant pressure. Lameness may result from bruising if the sole is hitting the ground.

Figure 7. The action of the frog when in contact with the ground. (a) The foot expands with frog pressure. (b) The foot contracts without frog pressure.

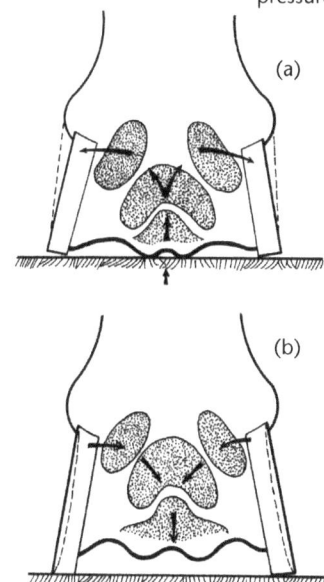

The bars
The bars are the parts of the wall that have turned inward from the heels to surround the frog. Their function is to bear weight.

The frog
The frog is wedge-shaped and made of rubbery and highly elastic material that comprises almost 50% moisture. It has three main functions:

- a shock absorber in its own right, as well as distributing concussion to the internal digital cushion
- a non-slipping mechanism
- an aid to blood circulation and heel expansion because of its position on the foot.

The frog should not be trimmed, except to remove flaking pieces after trimming the rest of the hoof. The frog should be level with the ground surface of the walls of the heels.

Sensitive structures of the hoof
Coronary band
The coronary band is the primary source of growth and nutrition for most of the hoof wall. Injuries to this structure are serious and usually leave a permanent defect in the growth of the wall.

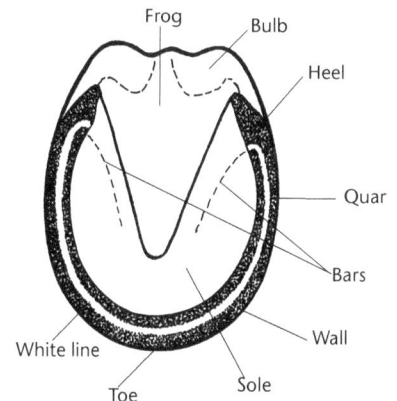

Laminar corium (sensitive laminae)

The laminar corium consists of laminae engorged with blood vessels. The sensitive laminae mesh with the insensitive laminae of the wall on one side, and are firmly attached to the pedal bone on the other side. The bond with the insensitive laminae is disrupted during laminitis.

Solar corium (sensitive sole)

The solar corium consists of hair-like laminae engorged with blood vessels. These laminae mesh with the insensitive laminae of the sole and frog to supply nourishment for growth.

The digital cushion

The digital cushion is a tough, elastic, fatty, yielding structure sitting on top of the frog and behind the pedal bone. It is visible externally as the bulbs of the heels. Its main function is to reduce concussion to the foot, as well as circulating blood back up the foot by acting as a pump when it compresses veins against the lateral cartilages. It also aids heel expansion.

Lateral cartilages

The lateral cartilages slope upward and backward from the wings of the pedal bone and reach the margin of the coronary band, where they can be felt near the heels. The many veins on the internal sides of the lateral cartilages are compressed against the cartilage when weight is placed on the foot and this forces the venous blood back to the heart. When weight is taken off the foot, the compression of the veins is released and the veins fill with blood again. The disease called sidebone occurs if the lateral cartilages turn into bone.

The bones of the hoof

Two bones are completely within the hoof. The pedal bone (also known as the distal phalanx or P3) is the largest and is shaped like the hoof. The significantly smaller, shuttle-shaped navicular bone lies adjacent to the pedal bone and closer to the heel.

CONFORMATION OF THE HORSE

A horse's conformation is its size, shape and proportions of the body and legs. It is the product of genetic and environmental factors including nutrition, disease control, training and foot care.

Conformation affects balance, movement and flexibility, which in turn affect the horse's performance and health. Some horse breeds have evolved (and then been bred) with a conformation that makes them better able to do one type of work than another. Some individual horses, because of conformation faults, are not able to withstand hard work (such as racing or eventing), but can manage light work such as pleasure riding. However, horses with poor conformation should not be used for breeding because the defect may be hereditary and passed on to the offspring.

Conformation is a major factor in the physical soundness of a horse. Poor conformation can lead to problems related to concussion. Concussion is the force that travels up the leg each time the hoof hits the ground and if excessive, leads to injuries and conditions such as ringbone (a disease of the pastern joints). Poor conformation also leads to gait abnormalities (the leg or legs do not travel in a straight line) and gait interference (the legs hit each other in movement).

Poor conformation can even result in behavioural problems if the horse finds it physically difficult to carry out what it is being asked to do. Horses with good conformation are more likely to stay sound through hard work because less stress is being put upon their body.

Very few horses have perfect conformation. Some conformation faults are not serious and others will affect one activity more than another. Just because your horse is not a good jumper, for instance, does not mean that it cannot do well at anything else: it may excel at endurance. Unless you want to compete at the top level, most horses can do most activities within reason. Correct work (i.e. educating the horse) will improve the level of skill. As the horse develops, the stronger muscles will carry the animal better and this can override, to some extent, milder conformation faults. Conformation also varies with breed, within a breed and between individuals.

In addition to affecting athletic ability, a horse with very poor conformation is usually unattractive and does not provide a comfortable ride. However, some horses, despite having poor conformation, will still perform well and sometimes even better than those judged as more superior, because the behaviour and trainability of a horse is just as important as good conformation, but are often harder to judge.

Age and conformation

There are certain conformation characteristics that can appear with advancing age. A dipped back can occur as the horse gets older. Any conformation faults will tend to become more noticeable as the horse ages, which you need to keep in mind when assessing an older horse.

Assessing conformation

To assess conformation stand the horse on a level surface with someone holding the head, but allowing the horse to stand with its head at the natural height. On first appraisal a horse with good conformation should have the following features.

- Be symmetrical from all angles. Walk around the horse to assess this.
- Be able to stand square. With a little pushing backwards and forwards the horse should be able to stand with all four legs coming straight down from the body to the ground.
- Be in proportion. For example, the head should look in proportion to the rest of the horse, not too large or small. The front end of the horse should not be too small or large for the back end.

The head and neck

The head and neck of the horse act as a counter balance for the rest of the body, so it is important that they are in proportion.

Neck

The neck can be too short or too long, too thick or too thin. Minor imperfections can often be improved with work; for example, a thin neck will improve as it develops fat and muscle. A short, thick neck is difficult to change, but as long as the horse is not required to work collected (i.e. the higher levels of dressage) then it may not be a problem.

Ewe-neck: the neck dips rather than curves up at the crest. A horse that is in very poor condition will always have a ewe-neck because of muscle wastage and fat loss. Some horses, however, have a ewe-neck even when they are in good condition, in which case it is a conformation fault. Correct work will improve a ewe-neck, but will not eliminate it. A horse with this problem will find it difficult when being ridden in a collected frame.

Swan neck: the neck is arched at the top and dips before the wither. Again, correct work will improve, but not eliminate the problem.

Parrot mouth/overshot jaw: the lower jaw is set back. In a mild case the horse is usually unaffected, but in a more severe case the horse is unable to graze properly because the front teeth (incisors) do not meet to clip the grass; these horses often ingest dirt as they try to graze and must be hand fed. In addition, the upper and lower incisors do not wear each other down as they would in a normal mouth, and the molars develop hooks because they are out of alignment. Therefore the horse needs more frequent attention by a dentist (see Chapter 2 for dental care).

Sow mouth/undershot jaw: this is the opposite condition to parrot mouth. The lower jaw is set forward from the top jaw. It is not as common as parrot mouth and horses with this condition also require extra care with feeding and dental attention.

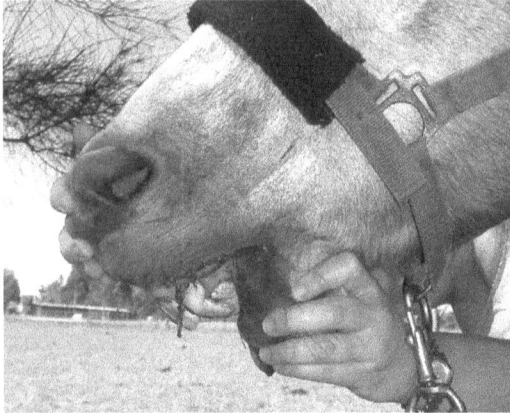

Figure 8. Parrot
mouth
(courtesy of Jane Myers).

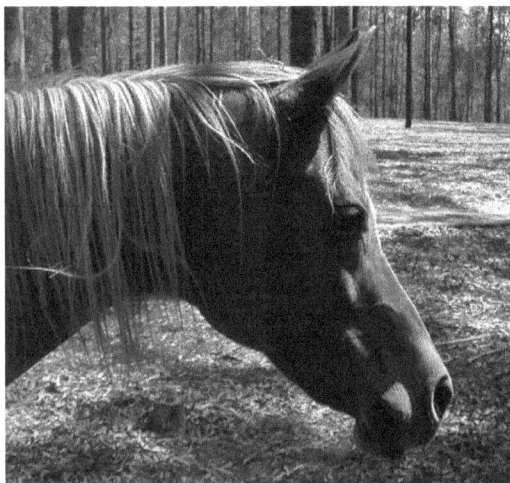

Figure 9. Dished
face
(courtesy of Jane Myers).

Figure 10. Roman
nose
(courtesy of Jane Myers).

Head
Roman nose/dished face: these are only faults in certain breeds; for example, a draught horse is expected to have a roman nose, but an Arab is expected to have a dished face. If a horse's face is too dished it can affect the breathing.

Head set: the head should be 'well set on'. The neck should taper towards the top and join the head in a way that allows the horse to flex the head in and out. Draught horses tend to be thicker through the jowls than other breeds, but should still be able to flex the head to some extent.

Head size: the head can be too big or too small. A large head is more common than an undersized one and results in the horse having problems with balance, particularly when working in a collected frame. Correct work will improve this condition by strengthening the muscles of the neck and back.

The back and thorax
Thorax
Withers: should be neither too low nor too high. Low withers tend to shift the centre of gravity forward and the saddle will also tend to move forward, which can cause soreness. Horses that have very high withers usually require a custom-made saddle.

Shoulder: should be sloping, which is believed to give the horse a longer stride. Heavier horses (draught types) tend to have more upright shoulders. The slope of the shoulder usually corresponds to the slope of the pasterns. If a horse is upright rather than sloping it will tend to suffer more concussion-related injuries and therefore speed work on hard surfaces should be limited.

Girth: should be deep to allow room for the lungs and heart.

Ribs: should be 'well sprung' (i.e. wide as opposed to flat) to give the lungs room to expand.

Chest: should be neither too narrow, (which tends to lead to leg interference) nor too wide. Very wide-chested horses tend to 'roll' as they move, which can be uncomfortable for the rider.

Coupling: where the loins join the croup, at the sacroiliac joint, should be smooth and well muscled. Horses that have old sacroiliac ligament damage may have enlargements on either side of the spine at the highest point of the rump. These are known as 'jumpers bumps' and can be a sign of an injury that may cause lameness and poor performance.

Back
The horse's back should be reasonably short and strong. A short back tends to be stronger than a long back and draught horses are naturally short in the back for power. However, if the back is too short the horse is more likely to hit itself (see over-reaching in the later section on gait abnormalities).

Roach back: the spine curves upwards in the loin area and, unless very pronounced, is not usually a problem. It is certainly preferable to a sway back and is more likely to affect a horse used for riding than one used for driving.

Sway back: the back dips excessively in the loin area, which weakens the back and can cause problems with fitting a saddle. Despite this, some sway backed horses do very well.

Dipped back: the whole back dips down between the wither and the point of the croup (sacroiliac joint). It is usually a problem of old age and should be treated with care.

(a)

(b)

Figure 11. (a) Poor coupling and (b) good coupling (courtesy of Jane Myers).

Straight back: the back is straight from the withers to the point of the croup. Straight-backed horses tend to be uncomfortable to ride because of being rather rigid. It is, however, a strong back and is not as much a fault in a driving horse as in a riding horse.

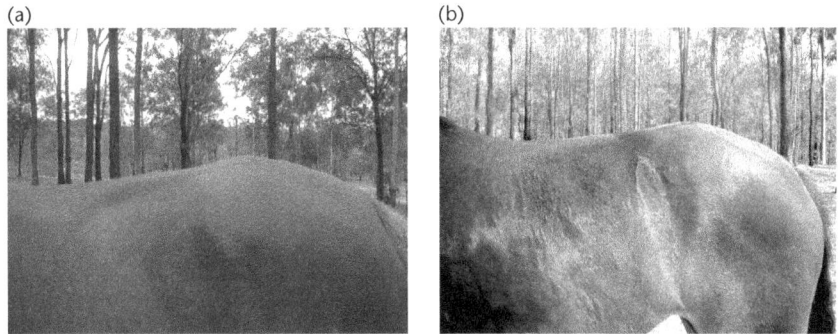

The hindquarters

The hindquarters should be well muscled, powerful and symmetrical when viewed from behind. Lean, long muscles are good for endurance as opposed to bulky muscles that are good for power. If you were to compare a top endurance horse (usually an Arab) with a top Quarterhorse (cutting bred) they would be very different in their hindquarters. The Arab would have long, lean muscles whereas the Quarterhorse would have short, bulky muscles, yet both horses would be in top condition.

Figure 12. Dipped back (courtesy of Jane Myers).

The horse should be as long as possible from the point of the hip to the pin bone (point of the buttock). Most good racehorses (Thoroughbreds) are long in this area. The horse should also be long from the hip to the hock and short from the hock to the floor. In other words, the hocks should be low to the ground. The hindquarters are the powerhouse for the rest of the body, so it is important that they are correct and strong.

The slope of the croup varies between breeds and individual horses. Arabs and Thoroughbreds tend to have a flatter croup whereas driving/draught breeds are more sloping. Flatter croups are supposed to indicate galloping speed in a horse whereas sloping croups are more powerful for pushing/pulling and jumping. Many top-level show jumpers have quite sloping croups.

The forelegs

The forelimbs bear 60–65% of the weight of the horse, and therefore are usually subjected to more concussion and trauma than the hind legs. The ratio varies from horse to horse and discipline to discipline.

Figure 13. Normal conformation: front view. A line dropped from the point of the shoulder bisects the limb.

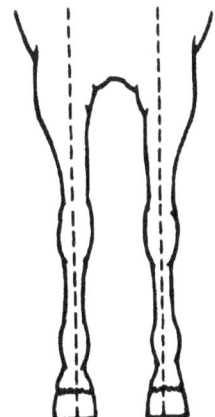

Viewed from the front, you should be able to draw an imaginary line from the point of the shoulder equally bisecting the limb all the way down the leg and the hoof. The knees should look large and of even size.

When viewing the horse from the side an imaginary line should bisect the forearm, knee and fetlock and should drop to the ground behind the heel. The knees should appear flat; however, in a young horse (approx. 1–3 years) the knees will sometimes look bumpy as they are still 'open' (i.e. the joints are still forming).

Faults in the forelimb (viewed from the front)

Base narrow: often accompanied by toe-in or toe-out conformation. The limbs are set close together on the body, which causes the horse to bear more weight on the outside of the foot and leg, subjecting the outside of the limb to more strain.

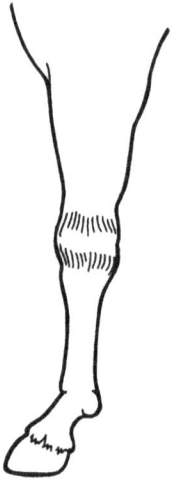

Figure 14. Open knees are recognised as bumps on the knee.

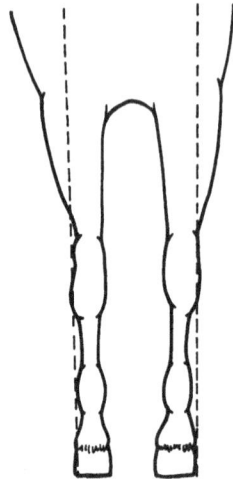

Figure 15. Base narrow conformation.

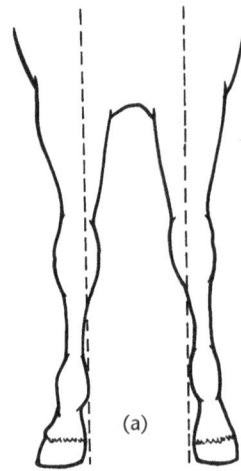

Figure 16. Base wide conformation. (a) Toe out. (b) Toe in.

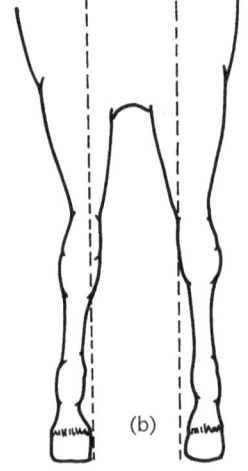

Base wide: often found in narrow-chested horses and can also be associated with toe-in or toe-out conformation. The limbs are set wide apart on the body and closer together at the hooves, which causes the horse to bear more weight on the inside of the foot and leg, subjecting the inside of the limb to more strain.

Toe out/splay footed: the pastern bones and hoof are twisted and pointed outward in a toe-out conformation. Most of the weight is placed on the inside of the foot, causing it to wear more than the outside, which flares out. Horses with this fault tend to 'wing in' when moving and, in bad cases, will actually cause interference when moving.

Toe-in/pigeon toed: the limb may rotate inward from as high as the chest or as low as the fetlock. The horse may 'paddle' (outward deviation) when moving.

Medial deviation of the carpus (knock knees, knee narrow): this causes strain on the inside of the limb.

Figure 17. Knock knees. Lateral deviation of the carpus (bow legs, bandy legs): this causes strain on the outside of the limb.

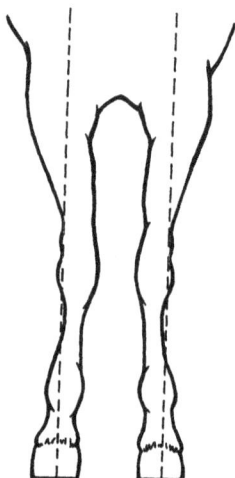

Figure 19. Bench, or offset, knees. Note that the cannons are set too far to the outside of the knee.

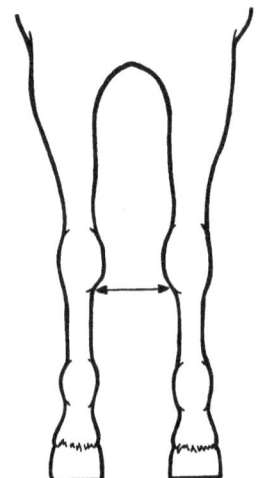

Figure 18. Bow legs.

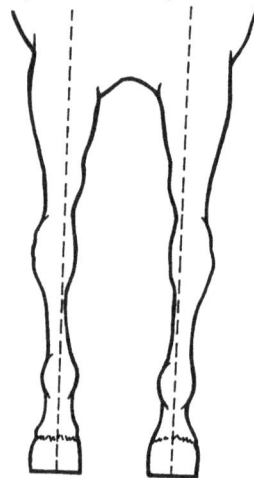

Lateral deviation of metacarpal bones (bench knees/offset knees): the inside splint bone is under more stress than usual, so the horse is prone to splints, which are not usually a problem, but will earn a penalty in the show ring.

Faults found in the forelimb (viewed from the side)
Backward deviation of the carpus (calf knees, back at the knee): this results in strain on the ligaments and tendons that run down the back of the front leg. It can also often result in chip fractures in the bones at the front of the knee (carpus).

Forward deviation of the carpus (buck kneed, knee sprung, over at the knee): this usually causes less trouble than backward deviation; however, it can cause strain on some of the tendons and ligaments. Foals may be born with this condition, but it usually disappears by 3 months.

Figure 20. Normal conformation: side view.

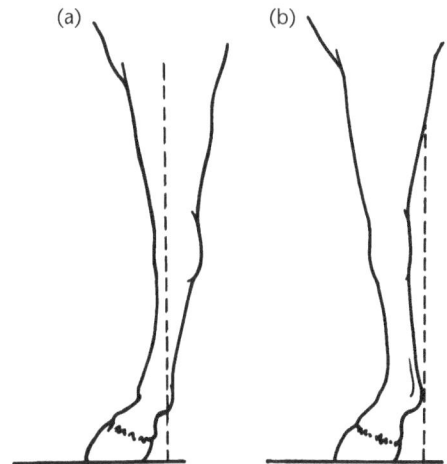

Figure 21.
(a) Back at the knee.
(b) Over at the knee.

Tied-in below the knee: the lower leg is narrower below knee than just above the fetlock, which can cause strain on the tendons just below the knee.

Standing under in front: the entire forelimb leans backward underneath the horse, which leads to causes excessive wear and fatigue of bones, ligaments and tendons down the front of the leg.

Camped out in front: the opposite of 'standing under in front'. It causes strain to the bones, ligaments and tendons down the back of the leg.

The hind limb
Less lameness occurs in the hind limb, but it is still important for the horse to have good conformation.

The horse should have good hocks in particular, because that joint does the most work in the hind limb. Good hocks are large, strong, clean (i.e. no puffiness or bony lumps in the wrong places) and well defined.

From the side view, a line drawn from the point of the buttock should touch the point of the hock, run down the rear aspect of the cannon and touch the ground 7–10 cm behind the heels. The angle of the stifle and hock should be neither too straight nor too acute.

From the rear view, the hindquarter should look square with a well-rounded croup, which indicates good muscling down into the thigh and stifle. An imaginary line drawn from the point of the pelvis should divide the leg into two equal parts.

Figure 22. Pastern conformation. (a) Normal angulation of the hoof and pastern. (b) Short upright pastern. (c) Long upright pastern. (d) Long sloping pastern.

Faults found in the hind limb (viewed from the side)
Sickle hocks: the horse stands under from the hock down, which places stress on the ligament just below the point of the hock and predisposes the horse to curb.

Straight hind legs: the hock is almost straight, which predisposes the horse to bone spavin in the hock and

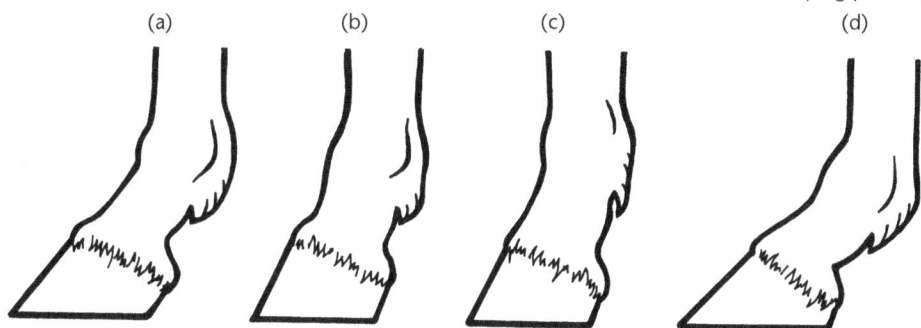

(a)   (b)   (c)   (d)

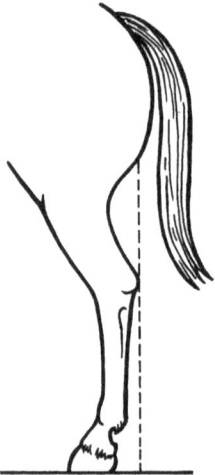

Figure 23. Normal hind leg conformation: side view.

upward fixation of the patella (locked stifle), because the patella (true knee joint) also will be too straight.

Standing under behind: the hind limb is placed too far forward or the horse has sickle hocks.

Camped out behind: the entire limb is placed too far back, often associated with upright pasterns behind. These horses will find it difficult to do collected work.

Faults found in the hind limb (viewed from the back)
Knocked down hip: is an acquired conformation fault; the points of the hipbones are not level because of a past injury.

Base narrow: the legs get closer together towards the bottom of the legs, which causes excessive strain on the outside of the limb and is often accompanied by bowlegs.

Base wide: the legs get wider apart towards the bottom of the legs, which causes strain to the inside of the hind leg, particularly the hock. These horses often also have cow hocks.

Cow hocks: the hocks turn in and the toes turn out, which causes excessive strain on the inside of the hock joint and may cause bone spavin. Many Clydesdale horses are naturally cow-hocked to some extent because the cow hocks give traction in heavy conditions and are therefore not detrimental to the draught horse. It is detrimental to a horse that is expected to do fast work.

The pasterns
The pastern should be in proportion to the rest of the leg length and be at the same angle as the hoof wall (this is called the foot–pastern axis). A short, upright pastern increases the concussion on the joints and can predispose to arthritis or navicular disease. This type of conformation is often associated with the 'pigeon-toed' defect and horses with straight shoulders, short legs and powerful muscles. A long, upright pastern predisposes to fetlock arthritis, but not ringbone. A long, sloping pastern is commonly seen in combination with sloping shoulders in a rangy type of horse. It puts extra strain on flexor tendons, suspensory ligaments and the sesamoid bones.

The foot
Good hooves should have:

- a front surface angle of almost 45° (front) and 55° (rear)
- angle of the toe equalling the angle of the heel
- thick walls with a glossy surface

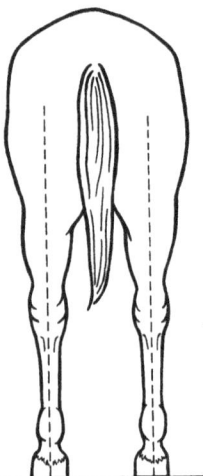

Figure 24. Normal hind leg conformation: rear view.

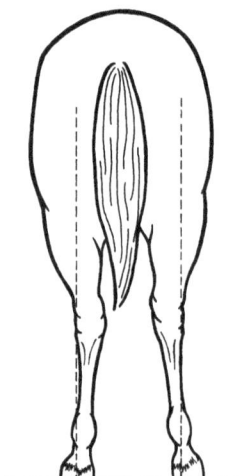

Figure 25. Cow hocks.

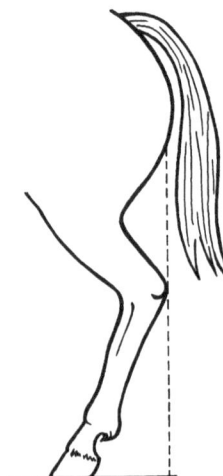

Figure 26. Sickle hocks.

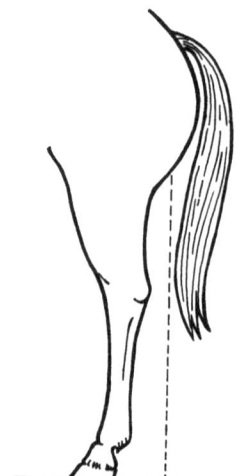

Figure 27. Too straight behind.

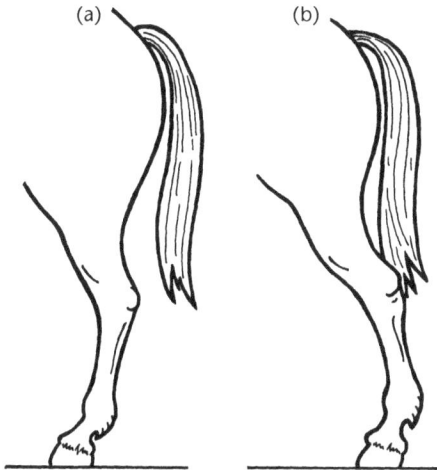

Figure 28. (a) Standing under behind. (b) Camped out behind.

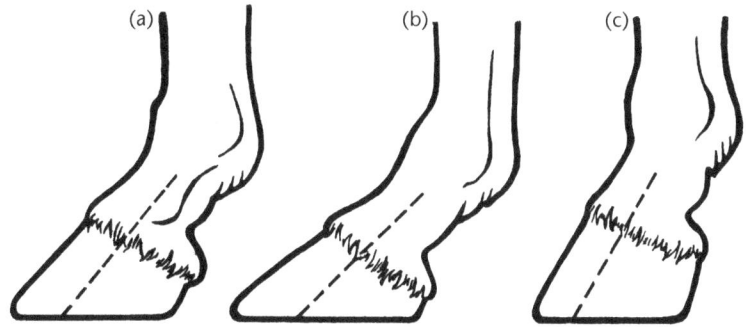

Figure 29. Side view of the foot–pastern axis. (a) Normal foot–pastern axis for the front foot (approx. 47°). (b) Foot–pastern axis that is less than normal (less than 45° in front or less than 50° behind). (c) Foot–pastern axis that is greater than normal (greater than 50° in front or greater than 55° behind).

- concave and thick soles
- large, well-developed, high and open heels
- large, strong frogs
- even length of heels
- correct foot–pastern axis.

In addition they should be in matching pairs, both front and hind, and should be neither too large nor too small for the horse.

The foot–pastern axis is where an imaginary line (viewed from the side) passes through the centre of the pastern and down through the hoof. The foot axis should follow the same axis as the pastern.

In the front feet the foot–pastern axis should be between 45° and 50° to the ground. The hind feet are usually more pointed than the front and slightly more upright (50–55°), so the correct foot–pastern axis will also be in this range. An examination of the foot flight patterns shows what tends to happen with different conformations.

- With a correct foot–pastern axis: the foot breaks over easily and is carried in a rounded arc of moderate height.
- If the pastern and hoof are too sloping: the foot is slower to break over plus there is strain on the tendons at the back of the leg.
- If the pastern and hoof are too upright: the foot breaks over too quickly and lands hard, causing an increase in concussion.
- With a sloping pastern/upright hoof: the foot breaks over too quickly and lands hard (concussion) and the sloping pastern is under more strain along its front surface.
- With an upright pastern/sloping hoof: the foot is slower to break over and there is strain on the back of the pastern and the tendons of the leg.

Faults of the feet

Flat feet: the hoof is large and the sole is close to the ground. Flat feet are more common in heavy breeds, but may also occur in Thoroughbreds. Horses with flat feet are prone to sole bruising and therefore lameness.

Contracted foot/heels: characterised by the back half of the foot appearing to be narrow. Bad shoeing can cause this defect. Because the heels cannot expand on contact with the ground, the foot does not work as well as it should.

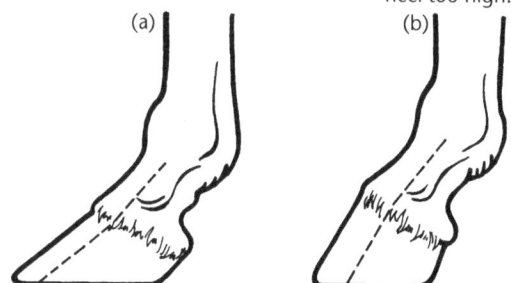

Figure 30. Examples of broken foot–pastern axes. (a) Toe too long and heel too low. (b) Toe too short and heel too high.

Clubfoot/feet: the foot has an axis of 60° or more, which increases concussion. This condition can be inherited or can occur because of a nutritional deficiency. It is possible for the defect to occur in one or more feet.

Other faults:

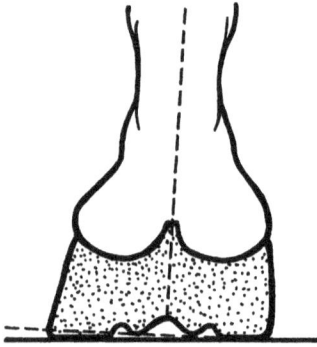

Figure 31. Uneven heels because one side of the foot is being over-worn or has been excessively trimmed.

- Brittle, shelly hoof walls that are prone to cracking.
- Small feet, which increases concussion.
- Poor frog development: the foot is less able to absorb concussion, which may lead to heel contraction.
- Dropped sole: occurs after laminitis.
- Excessive hoof wall rings: these relate to changes in the diet or season, severe illness or laminitis
- Uneven heels: uneven length of heels.
- Low/under run heels: can lead to heel bruising and more strain on the tendons at the back of the leg and on the navicular bone.

Improving your assessment of conformation

Look at as many horses as possible and assess their conformation. Being able to pick conformation faults will enable you to select the most suitable horse for your purposes, saving time, money and heartache in the future.

If a horse has conformation defect that can be improved, concentrate on the work that will strengthen that area. However, if it is more appropriate, the horse should be used for another type of work altogether.

MOVEMENT OF THE HORSE

A horse inherits its way of moving and although it cannot be changed, it can be improved with appropriate work and other factors such as corrective shoeing and training.

The different speeds at which a horse moves are called gaits. When changing from one gait to another the horse is in a transition; for example, when a horse goes from a walk to a trot it is called a walk–trot transition.

Normal movement

Walk

Walking is a symmetrical gait with evenly spaced intervals between footfalls. The walk is a four-beat gait because the individual stroke of each foot can be heard during movement. The sequence in which the legs move is left hind (LH), left fore (LF), right hind (RH), right fore (RF).

Trot

The trot is characterised by the legs moving in a sequence of diagonal pairs: the LH and RF move together, then the RH and LF follow. It is a symmetrical gait and two distinct beats can be heard: the first beat is when pair of legs hit the ground and the second beat is when the other pair hit the ground.

Figure 32. Toe conformation. (a) Normal. (b) Toe-out. (c) Toe-in.

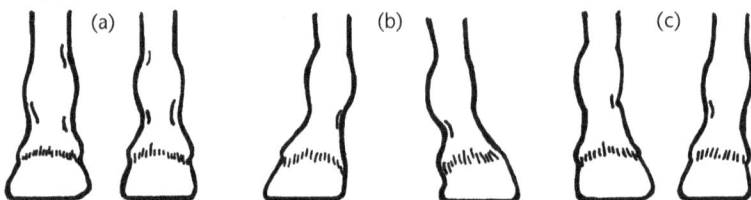

(a)   (b)   (c)

Pace

Pacing is characterised by the legs moving in a sequence of lateral pairs: the LH and LF move together, then the RH and RF move together. It is also a symmetrical gait with two distinct beats.

The difference between pacing and trotting is that the horse moves lateral pairs of legs in the pace rather than diagonal pairs as in the trot. It is commonly seen in Standardbreds, which are raced at the pace, as well as at the trot.

The pace is said to be an unnatural or artificial gait, but to many horses it is instinctive. Some gaits are variations of the pace: the amble, the rack, stepping pace, pacing walk, slow gait, running walk, paso, single foot and tolt. Some breeds of horses that naturally exhibit these more unusual gaits are the Tennessee Walker, American Saddlebred, Paso Fino, Peruvian Paso and the Icelandic Horse.

Canter and gallop

The canter and gallop are very different to the trot and pace because they are not symmetrical. The limbs do not have an even spacing between the intervals of the footfalls. For a canter the movement sequence is RH, then LH and RF together, then LF.

The last foreleg down is called the leading leg. Therefore, the above sequence is left lead canter. If the horse were leading with the right leg the sequence would be LH, then RH and LF together, then RF.

The canter is a three-beat gait. Beat one as the hind leg strikes the ground, two as the other hind leg and diagonal foreleg strike the ground (the 'in phase' legs) and three as the leading foreleg strikes the ground.

'Flying change' describes the horse simultaneously changing the leading leg at the front and the back. Only a trained horse is able to do it on command from a rider. When a horse changes legs of its own accord, when cantering around the paddock for instance, it usually changes one set (e.g. the front) during one stride and then the other set in the next stride.

During the gallop, separation of the 'in phase' pair of limbs occurs (refer to beat two of the canter). Therefore, instead of a three-beat gait, the gallop has four beats. The movement sequence for right lead gallop is LH, RH, LF, then RF.

When the horse accelerates, it increases the length of its stride. Once it has maximised its stride length, the horse increases the stride frequency.

Jumping

The jump occurs in the middle of a stride. For example, cantering through a jump with the right leg leading requires the following gait sequence: LH, RH, jump, LF lands, then RF.

As the horse approaches the jump the front legs prop and the head lifts in order to push the centre of gravity back, thereby preparing to lift/push the horse off the ground. Next, the hocks come under the horse and it pushes off the ground.

As the horse leaves the ground the hind legs straighten and the front legs flex. The hind legs flex mid flight after leaving the ground. Once over the jump the horse begins to straighten the fore legs, but the hind legs remain flexed. The horse lands with the non-leading foreleg, heel first. Then the leading foreleg follows, landing just in front of the other, followed by each hind leg in turn. Before the second hind leg lands, the second fore leg has already pushed off to commence the next stride.

Concussion

Concussion is the force sent vertically up the leg each time the foot strikes the ground. It is a major factor in lameness and is therefore worthy of serious consideration.

Factors that affect the concussive force on the horse's legs are as follows.

- Other than when walking, the camber, surface and hardness of the ground on which the horse is travelling.

- The fitness of a particular animal. The more fit the horse is, the more likely its body will cope with the effects of concussion.
- The conformation of the horse will affect its ability to absorb concussion.

How the horse absorbs concussion:
The horse has several mechanisms to counteract the effects of concussion.

The foot
The high moisture content of the foot, the elasticity of the hoof wall, the collateral cartilages, digital cushion and frog all aid in absorbing concussion.

The legs
When the leg first strikes the ground, the joints above the fetlock are slightly bent (i.e. the knee, shoulder and elbow), which helps to absorb some of the shock. The muscles, tendons and ligaments act like springs, absorbing the concussion of impact by allowing some flexion of the knee, shoulder and elbow, helping to push the weight of the horse back up again.

The bones of the knee
The small bones of the knee (carpus) are spongy with small spaces within the bones, as well as between the bones, to aid absorption of concussion.

Front legs
Instead of a collarbone, the horse has what is termed a 'thoracic sling'. The front legs are not attached to the body by a joint, but instead are connected by tendons and ligaments, which allows the horse's body to dip and spring during motion. Again, this assists absorption of concussion and reduces the shock waves going down the spine.

Hind legs
There is less concussive force on in the hind limbs than the front limbs because they take less of the total weight of the horse. However, the hind limbs are joined to the body by a joint (hip joint), so any shock will travel up the spine. The cartilage disks and spaces between the vertebrae help to absorb this concussion.

The hock assists in the absorption process. As with the knee, it is made up of small bones set on top of each other. However, if the angle of the hock is too straight then the joint will only have a limited ability to reduce the shock.

Fatigue and concussion
In the galloping horse, the leading leg is subject to most concussion. When the horse starts to fatigue, stride frequency reduces and the horse spends longer in the air. The 'out of phase' pair of legs comes into phase again (four beats become three), more stress is place on the leading leg than before (because of the horse bobbing up and down) and the horse begins to change the lead leg every few strides. This is why the leading leg is most susceptible to injury in a racehorse towards the end of the race.

Abnormal movement

A horse should be able to move its legs in a straight line. When the horse is walked and trotted, both away and towards you, the legs should swing straight forwards and backwards without any sideways deviation. Any deviation from this is classified as abnormal movement.

Abnormal movement can lead to interference of the legs. The definition of interference is the horse striking a fore or hind limb with the hoof or shoe of the opposite limb. Sometimes horses that interfere can become unsound because they repeatedly hit themselves. It is therefore important to understand the different forms of interference so that preventative measures can be taken.

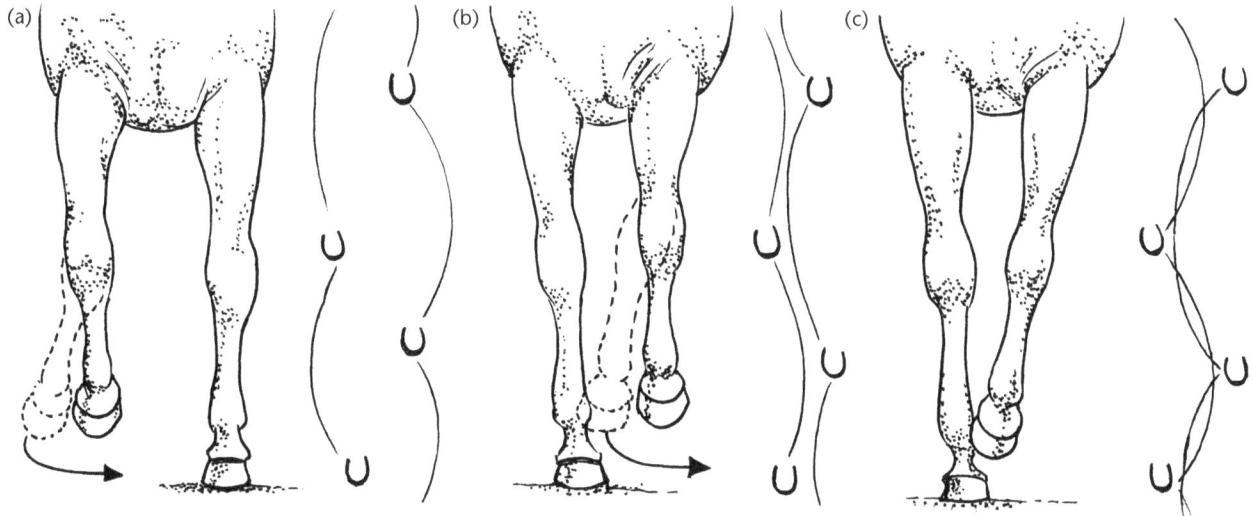

Figure 33. Abnormalities of gait. (a) Winging-out. (b) Winging-in. (c) Plaiting.

Stumbling

A horse will stumble if it catches or digs its front toes into the ground. This is more likely to occur when the animal is tired, has long, overgrown feet, lands on its toe because of lameness, or has reduced flexion of its fetlock or hock joints. A horse may also stumble when recently shod, before the shoes have worn to conform to the horse's action.

Paddling or winging-out

This is a movement of the foot away from and then back toward the median plane at the beginning and end of a stride. It usually accompanies toe-in conformation.

Winging or winging-in

This is a movement of the foot toward and then away from the median plane at the beginning and end of a stride. It usually accompanies toe-out conformation.

Brushing

Either the fore or hind foot strikes the opposite leg between the bulb of the heel, the knee or the hock. This defect occurs in the front legs more often than the hind legs and can happen in the walk or trot. Affected horses often have toe-out conformation (if it occurs in the front) or are cow hocked (if it occurs in the hind).

Plaiting

The front feet are placed directly (or nearly directly) in front of each other. It is usually seen in the walk.

Knee hitting

This is a high interference involving the knee, which is generally seen in Standardbreds when racing.

Elbow hitting

The elbow is struck with the foot of the same limb. It requires extreme flexion of the carpal and digital joints and rarely occurs, except in racing horses and horses with weighted shoes.

Cross-firing

Either the toe or the quarter of a hind foot strikes the inside of the forefoot on the diagonally opposite side. It is common in pacers and gallopers. Gallopers who cross-fire contact the fetlock region of the front leg with the inside of the foot of the diagonally opposite hind leg. The sesamoid bones can be fractured.

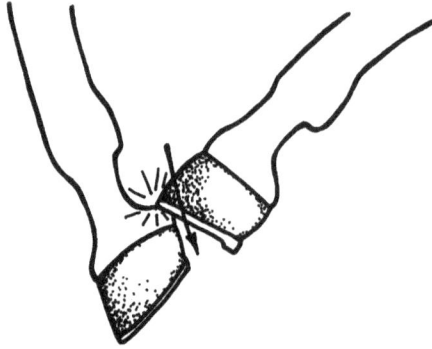

Figure 34. Over-reaching

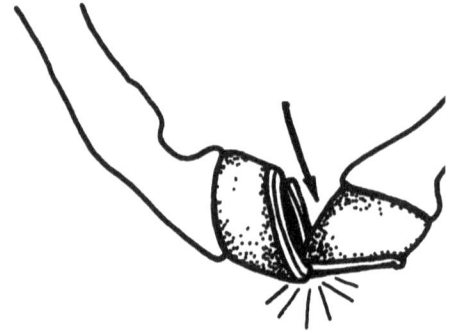

Figure 35. Forging

Over-reaching

The toe of the hind foot strikes the heel or back of the pastern of the forefoot. It can be caused by the horse having a short back and relatively long legs, or short front legs with long hind legs. It is also very common in young horses that have not fully matured and in some the use of over-reach boots will be sufficient until the horse has matured enough to carry itself better.

Forging

The horse makes a clicking noise because the toe of the hind foot strikes the sole of the front foot on the same side; it often occurs at a slow trot. It is very common in young horses that have not yet fully matured. Some horses hit slightly higher up the foot and can pull the front shoe off with the toe of the hind.

Scalping

The toe of the forefoot hits the coronet of the hind foot on the same side in the trot and gallop and occasionally during pacing in which case the horse hits the diagonally opposite leg.

Speedy cutting

This is similar to scalping, but occurs higher up the leg, around the fetlock or pastern or just below the knee (in which case it is called shin hitting). It is usually seen in horses with a high action and can occur during trotting, galloping, jumping etc. Horses that have long backs may be more prone to speedy cutting, which is a term often applied to a number of interferences.

Going down on the bumpers

When horses are galloping and jumping they can graze the skin on the back of the fetlocks, especially if they have long, sloping pasterns. Horses may need leg bandages, protective boots and/or corrective shoeing.

How to reduce interference

Interference is common in young horses, especially when they are shod for the first time. The young horse has to learn to balance itself, which takes time and practice. Also, if the horse still has to mature that will affect its movement to a large extent.

An untrained horse is more likely to interfere because it is not as balanced as an educated horse. An educated horse has what is termed 'self carriage'; that is, the horse is able to carry itself in a balanced frame rather than rely on the rider to hold it together.

If the horse is badly shod and/or has poor conformation it is even more likely to interfere, so use a qualified, experienced farrier (see Chapter 2). The farrier will study the wear of a shoe when it is removed and then defects of conformation and any faults of the foot can be taken into account when fitting a new shoe. Other questions to ask yourself are:

- Does all the gear fit properly?
- Is the horse being asked to work harder than it should?
- If it is a ridden horse does the rider ride well or badly?
- Is the horse too young to do the level of work it is being asked to do?
- Is the horse being fed properly?

All of these factors must be taken into account as they will affect how the horse moves.

BREEDS OF HORSE IN AUSTRALIA AND NEW ZEALAND

The breeds of horse found in Australia and New Zealand are listed here in alphabetical order and some are shown in pages 87–94.

American Saddlebred

The American Saddlebred originated in Kentucky from the crossbreeding of the Narragansett Pacer, the Morgan and the Thoroughbred. A horse that is suited to such disciplines as hacking, dressage, hunting, show jumping, endurance and trail riding, the American Saddlebred stands at 16 hands. It is known for its unusual, five-gaited action.

Andalusian

The Andalusian, an ancient and noble breed, has only recently been permitted by the Spanish Government to be exported. The most common colour is grey; other colours are bay, brown, roan and black. The Andalusian is very strongly boned and stands at between 15 and 16.2 hands. Its 'spirited' and 'arrogant' appearance belies a calm, cooperative temper, steadiness under pressure and a remarkable disposition for training. It possesses strength and endurance and its movements are elevated, agile and smooth. It successfully competes in dressage, harness, in-hand and under-saddle classes.

Anglo-Arabian

Resulting from the crossbreeding of Thoroughbreds and Arabs, Anglo-Arabs are admired for their noble appearance, balanced action and endurance. The Thoroughbred gives the Anglo-Arab its size, shoulders, powerful hindquarters and speed, and the Arabian provides beauty, bone structure, good feet and stamina. The combination usually produces an ideal performance horse.

Appaloosa

The most distinguishing feature of the Appaloosa is its coat pattern, which varies from horse to horse. No two Appaloosas have identical markings and in some horses the markings change with age. Most Appaloosas, however, are white over the loins and hips with dark, round or egg-shaped spots, varying in size from tiny specks to 100 mm in diameter.

Other characteristics of the breed are a white ring around the eye (the sclera), mottled skin and striped hooves. The most valued quality of the Appaloosa is its quiet, sensible disposition. Stock work, sprint racing and show ring competitions, in-halter, dressage, performance and jumping are the main uses of the Appaloosa.

Arabian

The Arabian horse has been bred pure for many centuries and is the basis for many other breeds. It originated in the deserts of Arabia and the Near-East, where a hardy body and tough will was required for survival. Today, because of its versatility, it is used for all types of riding, but it is especially renowned for its abilities as an endurance horse.

Large eyes and a beautifully chiselled triangular face distinguish the Arabian horse. The neck is long and arched, leading to a compact and graceful body. The usual height ranges between 14.1 and 15.1 hands. The luxurious coat is usually bay, grey or chestnut, although there is an occasional black.

Australian Pony

Soon after imported ponies began arriving in Australia, there was a need for definite breeding guidelines. Selected pony sires were mated with mares of known pedigree and with certain foundation mares of a pony type to produce the Australian Pony.

The Australian Pony should not exceed 14 hands in height and all colours are acceptable. It should have good conformation, free of hereditary defects. A good temperament is of the utmost importance. It makes an ideal riding or harness pony.

Australian Riding Pony

The Australian Riding Pony is a smaller edition of the show hack, ranging in height from 12.2 to 14.2 hands. It is a 'quality' animal with superb conformation. Free, straight action from the shoulder and not the knee and an impeccable manner are its other main characteristics.

The Australian Riding Pony was derived from selective crossing of Thoroughbred, Arabian and pony bloodlines and is mainly used as a children's show pony.

Australian Saddle Pony

The Australian Saddle Pony is derived mostly from Australian Pony, Welsh Pony, Thoroughbred and Arabian bloodlines. It is slightly heavier in bone and slightly calmer in temperament than the Australian Riding Pony. The height of these ponies varies from between 11 and 14.2 hands. Colours include black, grey, brown, palomino, buckskin, palouse and pinto. A very versatile pony, the Australian Saddle Pony is particularly suited to pony club competition.

Australian Spotted Pony

The Australian Spotted Pony must be less than 14 hands at maturity and show similar coat colour patterns to the Appaloosa. It must have good conformation, with the head and feet in proportion to the body, which should be well muscled. The action of the pony should make it easy to ride.

The Australian Spotted Pony is generally for young riders, but it is also suitable for many adults. The disposition of the pony should be such that it can be handled and trained by children.

Australian Stock Horse

The Australian Stock Horse has evolved over the past 150 years as Australia's own horse. The harsh conditions and vast open spaces of early Australian settlement demanded a tough durable horse with a mild manner and a comfortable canter. Infusions of Thoroughbred blood have added to the versatility of the breed.

Apart from its main role as a working stock horse, the Australian Stock Horse is now also used for camp-drafting, polo, dressage, show jumping, hacking, pony club, rough riding, endurance riding and tent-pegging. Its ideal height is 15–15.2 hands, but 14–16 hands is acceptable. The coat may be any colour, but solid colours are preferred.

PLATE 1 – DESCRIBING THE HORSE   **87**

Figure 1. American Saddlebred (courtesy of Inspiration Farm).

Figure 2. Andalusian (courtesy of Tracie Sullivan).

Figure 3. Appaloosa (courtesy of Vera Welsh).

Figure 4. Australian Spotted Pony (courtesy of Mrs John Grice).

Figure 5. Australian Miniature Pony (courtesy of Janice Morgan).

Figure 6. Australian Stock Horse
(courtesy of Australian Stock Horse Society).

Figure 7. Arabian (courtesy of Sharon Meyers).

PLATE 1 – DESCRIBING THE HORSE   89

Figure 8. Clydesdale (courtesy of John Knight).

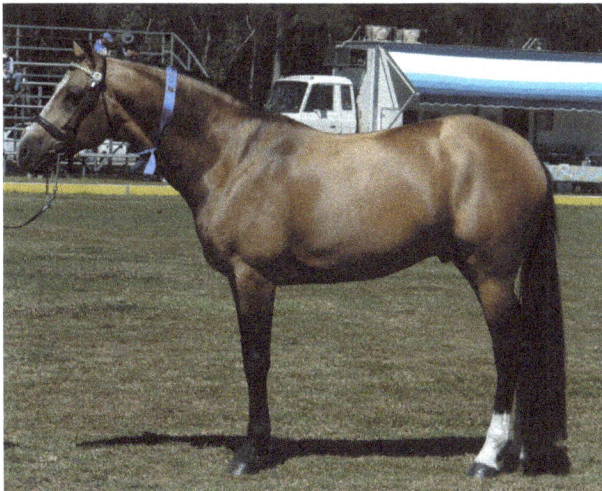

Figure 9. Buckskin (courtesy of Annie Minton).

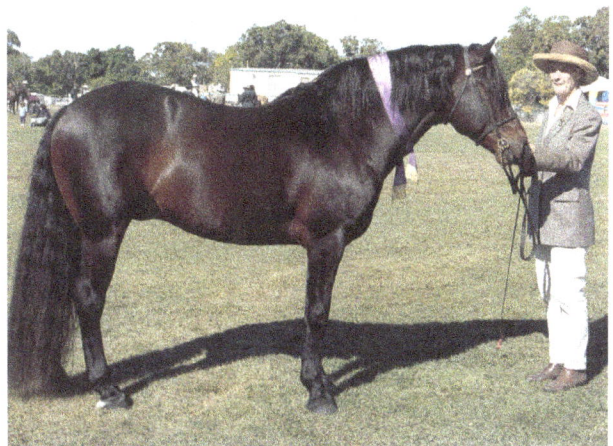

Figure 10. Connemara Pony (courtesy of Narelle Wockner).

Figure 11. Haflinger (courtesy of Arkley Haflinger Stud).

Figure 12. Friesian (courtesy of The Carrock Stud).

Figure 13. Irish Draught (courtesy of Barbara Utech).

PLATE 1 – DESCRIBING THE HORSE   91

Figure 14. Pinto (Tobiano) (courtesy of Margaret Quinlivan).

Figure 15. Palomino (courtesy of Annie Minton).

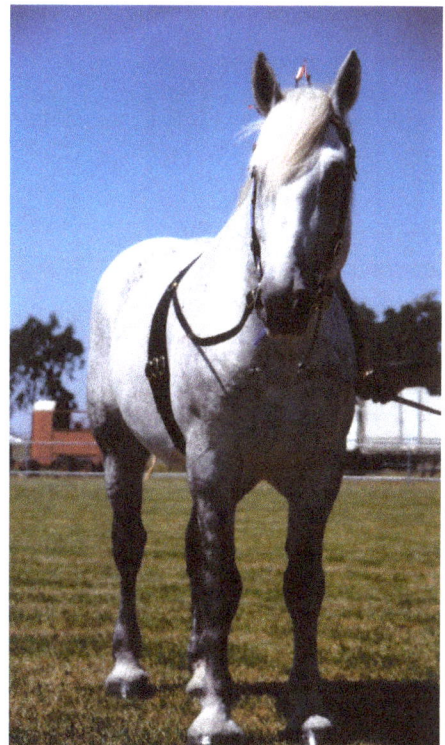

Figure 16. Percheron (courtesy of Trevor Kohler).

Figure 17. Shetland (courtesy of Annie Minton).

Figure 18. Perlino (courtesy of Annie Minton).

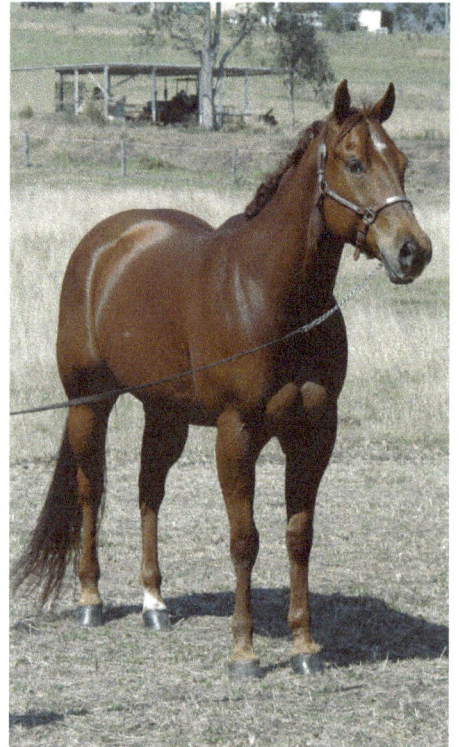

Figure 19. Quarterhorse (courtesy of Annie Minton).

PLATE 1 – DESCRIBING THE HORSE   93

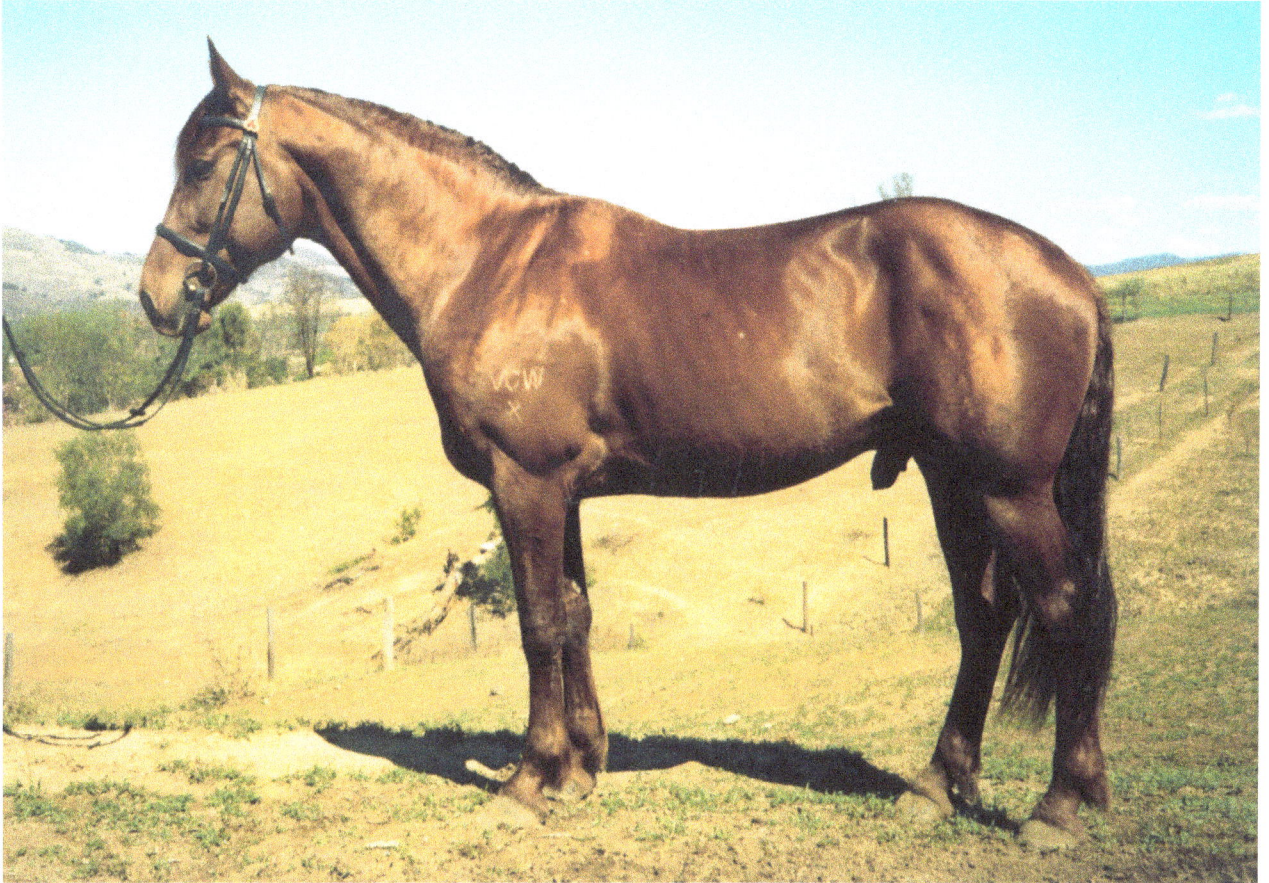

Figure 20. Waler (courtesy of Talara Waler Horse Stud).

Figure 21. Warmblood (courtesy of Annie Minton).

Figure 22. Thoroughbred (courtesy of Annie Minton).

Figure 23. Standardbred (courtesy of Shirley Williams).

Figure 24. Welsh Section A (courtesy of Helen Sloane).

Figure 25. Welsh Section B (courtesy of Janette Murphy).

Figure 26. Welsh Section C (courtesy of Karen Czora).

Figure 27. Welsh Section D (courtesy of Susanne Fritzsche).

Australian Stud Saddle Pony

The Australian Stud Saddle Pony is derived from the mating of any registered pony breed and horses with Arabian blood. It is characterised by its quiet temperament under saddle, its ability to learn and its affectionate and loyal character. It is hardy and healthy, with strong hoofs, and has a maximum height of 14.2 hands. It must be a whole colour. These ponies have achieved outstanding success in all facets of the show ring.

Australian White Horse

Australian White Horses are classified by their colour, rather than their breeding. There are two varieties of White Horse: cremello and perlino. Both are genetically termed pseudo-albino because they differ from the true albino by having blue eyes and limited pigment in their skin.

Belgian Draught

The Belgian is descended from the Flanders Horse and has been developed through selective breeding into an exceptional utility horse of great power. Weighing more than 1200 kg and standing at 17–18 hands, the Belgian Draught has an amply muscled croup and a broad chest.

Buckskin

As with palomino, different breeds can be buckskin because it is a colour not a breed. Therefore, the type can vary. Each horse, however, must be a good representative of its type. Buckskin can range from very dark to very light in colour and is often mistaken for dun, which is a separate colour (see later section on identifying horses).

Caspian

The Caspian horse originated in the Caspian Mountains in Iran. Thought to have been extinct for centuries, the Caspian Miniature horse was rediscovered in 1965 in a village in northern Iran. Since then Caspian studs have been established in several countries, including Australia. Despite the establishment of breeding programs, it is believed that there are less than 300 purebreds in existence.

Although small, with an average height of 11 hands, the Caspian is not regarded as a pony, but as a small horse. They have a slim body with a graceful neck, long sloping shoulders and a deep girth. Limbs are slender, with strong, dense bones and little or no feathering at the fetlock. The breed is light, well balanced, agile and strong. Their jumping ability is extraordinary for their size.

Cleveland Bay

Though once regarded as superb coach horses, today the Cleveland Bay is valued for its ability to produce top quality, crossbred performance horses. Cleveland Bays are sound, clean-legged and active horses with substance and good temperament. When crossed with Thoroughbred mares in particular, they produce good competition horses, combining the speed and ability of the Thoroughbred with the straight movement and hardiness of the Cleveland Bay.

Clydesdale

The Clydesdale was bred to be a very versatile draught animal. It is one of the few, if not the only, flat-boned draught horses in the world. Together with its fine joints, the set and flexibility of its pasterns and its close setting, both fore and hind, enable it to maintain very sound limbs and to perform the job for which it was principally bred – moving large loads quickly.

The Clydesdale is short-coupled and has deep, sloping shoulders. Its very broad feet give it a firm grip on the ground and its large springy heels act as shock-absorbers under the pressure of hauling heavy loads.

The Clydesdale is very docile and has a majestic 'outlook'. It is a great favourite in the show ring. After being ignored as a power source for the past 40 years, the draught breeds are again being bred in the United States, Europe, Japan and Australia as a source of power for smaller farms and short-haulage work.

Connemara Pony

The Connemara Pony is one of the larger pony breeds, ranging in height from 13 to 14.2 hands . The major colours are grey and dun, although some ponies are bay, black, chestnut, palomino or roan.

The Connemara tends to be more sturdy than dainty, having evolved in rugged stony country where food is scarce. It is capable of carrying 75 kg or more over great distances, but is amenable enough for children to enjoy. Its good temperament make it suitable for a wide variety of activities including dressage, stock work, all facets of pony club, show jumping and harness work. A popular cross is the Connemara and Thoroughbred, which usually produces a good event horse.

Dartmoor Pony

The Dartmoor is a good-looking riding pony with a small head set well on a strong, but not heavy, neck. Its shoulders are well laid back, its back is of medium length, and its loin and hindquarters are strong and well-covered with muscle. Its tail is full and set high. The Dartmoor's usual colours are black, brown and grey, and its maximum height is 12.2 hands. Their small size, calm temperament and affectionate nature makes Dartmoor Ponies particularly well suited for harness work and for children.

Fjord

Also known as the Norwegian Fjord Pony, the Fjord is a native breed of Norway with origins dating back to the post-glacial era. It is an exceptionally good natured breed suitable for both harness and riding.

Standing at 13 hands on average, the Fjord is classified as a pony for competition purposes, although its admirers insist it is a horse. It is a dun coloured animal with a clearly defined dorsal stripe, bordered on the mane by lighter hairs. The Fjord's head is well defined with small ears and a broad forehead, it has a short thick neck, a strong and compact body, sloping shoulders and a long rump, a bristly mane, and short, thick, sturdy limbs.

Friesian

A horse especially suited for harness work and high-level dressage, the Friesian originated in Friesland, a province of the Netherlands. These horses were bred there exclusively for a period of 1000 years and have had little influence from other breeds.

The average height of the modern Friesian is between 15 and 16 hands and they are always black with no white markings. They make a flamboyant impression in the show ring because of their extravagant natural knee action and their long wavy mane and tail. They are also often used for film work and equestrian displays.

Hackney

There is a hackney pony and a hackney horse. Each has a bold head, large eye and small ears. When trained, the hackney pony excels at a collected trot with a most energetic display. The hackney horse's action is not quite as dynamic. Size ranges from 11 to 16 hands. Hackneys excel in the show ring.

Haflinger

Originating in Tyrol, Austria, the most distinctive feature of the Haflinger is its flaxen mane and tail, which, combined with its body colour that ranges from pale to dark chestnut, makes it a distinctive breed.

The Haflinger physique is a head that tapers to the muzzle, a broad chest, deep girth, short legs and a long, broad back. Docile, good tempered and adaptable hard workers, Haflingers are frugal, tough, sure-footed and long lived. They are used for trail riding, show jumping, dressage, harness work and as a pack horse in difficult terrain.

Highland

The Highland Pony is native to the highlands of Scotland and certain adjacent islands. Its unique two-layer coat enables it to withstand the bleakest winter. The two recognised types of Highland Pony are the Scottish Mainland and the Western Island. The latter is smaller and lighter in build than the former. The strength and good temperament of these animals is renowned.

The Highland is a true pony of great beauty and strength. It ranges from 13.2 to 14.2 hands in height and is generally dun, grey, black or brown in colour. It has an attractive head with prominent eyes, a strong neck, shoulders that are set well back, a deep chest, and well-sprung ribs. Its quarters are powerful and its thighs strong, but its action is free and straight.

Irish Draught and Irish Sport Horse

The Irish Draught has versatility, endurance, and a pleasing disposition, characteristics thought to have resulted from when it was the selected horse of Irish farmers.

Descended from the Norman heavy horses and infused with the blood of the Andalusian, Barb and Thoroughbred, the Irish Draught is used intensively in show jumping, although its good straight action and willing temperament also make it ideally suited to dressage.

The Irish Draught, and its Thoroughbred cross, known as the Irish Sport Horse, have proved ideal performance horses, combining the best qualities of both breeds.

Lippizaner

Lippizaners are generally grey in colour and stand between 14 and 16 hands. They possess great power and stamina. Their most important feature is their elegant movement with lift and their ability to remain suspended in the air for a long period of time. They are slow to develop and mature.

Miniature Horse

Miniature horses include the Falabellas of Argentina and the American Miniature Horse. Enthusiasts insist that in its true form the Miniature horse is a breed apart from the Miniature pony because it is bred and developed to resemble a true horse in miniature, whereas the Miniature pony generally appears stockier than a horse. Ponies, particularly the Shetland, have, however, played a major role in the development of the Miniature horse.

The Miniature Horse Association of Australia has defined its own breed standards. Miniatures must not measure more than 8.2 hands (34 inches) at the base of the last hairs on the mane and its size must be in proportion to the length of its neck and body. Miniatures must also possess broad foreheads with large, expressive eyes set well apart. The New Zealand Miniature Horse Association has two size categories: Category A horses must not measure more than 34 inches in height, and Category B horses not more than 38 inches.

Miniature Pony

The Australian Miniature Pony is the result of crossbreeding Shetlands with other small breeds such as the American Miniature Horse and the Falabella. Selective crossing with the Welsh Pony, Australian Spotted Pony and Arabians has also given the Miniature Pony greater coat colour variation and a finer, smaller head.

To be eligible for registration as an Australian Miniature Pony, it must not exceed 87 cm in height from the highest point of the wither and must have a certificate of soundness. Any colour is permissible. The whole body should be balanced and in proportion to the pony's height. The body should have well-sprung ribs, a slight natural curve of the back and a strong well-rounded rump.

Morgan

The Morgan stands at about 15 hands and is usually solid black, brown or chestnut. It has a compact physique with a broad, short head and thick neck. It has strong shoulders and back, and powerful hindquarters. Considered a good all-rounder, the Morgan is capable of show driving, weight pulling in collar and working horse harness, as well as pleasure riding. An increasingly popular cross with the Morgan is the Arab, called a Morab.

New Forest Pony

This is one of the nine native pony breeds of Britain. It is a true pony, and renowned for its sensible temperament and hardiness. It has exceptional speed and jumping ability for its size. When crossed with Thoroughbreds and Arabs, a very useful one-day-event type is obtained.

Any colour is acceptable except piebald, skewbald and blue-eyed cream. Height is up to 14.2 hands and there is no lower limit, although New Forest Ponies are seldom less than 12 hands. The New Forest is a riding pony and therefore coarseness is undesirable. However, it must have sufficient strength to be capable of a day's hunting or stock work.

Paint

Paints are being bred in Australia by mating broken-coloured mares (pintos) to Quarter Horse and Thoroughbred stallions, and to registered Paint stallions. Breeders give at least as much importance to conformation, ability and bloodlines as to colour, so the Paint can compete on equal terms with all other light-horse breeds.

The Paint should be short-coupled for a high degree of action and have the ability to move quickly in any stock horse event.

Palomino

Palominos compete in numerous events because they are not a breed but a colour. Any light-horse breed is acceptable, providing the colour and conformation standards are maintained. A Palomino is a horse with a golden coat, and a white, silver or ivory mane and tail (no more than 15% dark or chestnut hairs throughout the mane and tail). The skin is dark, the eyes may be black, brown or hazel, but both must be the same colour. White stockings should not extend above the knees and hocks except in a diminishing spear-shape. White markings on the face are permissible. There should be no white or black spots on the body, nor should there be a dorsal stripe or zebra markings on the legs or body.

Percheron

The Percheron originated in the area of France known as Le Perche where abandoned Arabian horses were crossed with massive Flemish heavy horses, the progeny of which provided the foundation for the Percheron breed.

Percherons stand at 15.2–16.2 hands and can weight up to 900 kg for mares and 960 kg for stal-lions. As a heavy draught horse the Percheron has great muscular development, combined with style and activity. Grey or black are the only colours allowed for entry in the stud book.

Pinto

Pinto is a colour rather than a breed and is a collective term for broken-coloured horses, which incudes tobiano, sabino, overo and splashed white. The colour range is grey/white, black/white, bay/white, chestnut/white, palomino/white, and so on. Broken colour must be visible from a normal standing position, and only white on the body is counted for colour qualifications. Horses and ponies of any height may be registered, but they must be a light-horse breed.

Quarterhorse

The Quarterhorse is big in the haunches, compact, heavily muscled and can run exceptionally fast over short distances, hence the colloquial name 'quarter-miler'. Its gentleness and easy-going disposition make it an ideal family horse. It is sure-footed and can start, stop and turn with ease and balance.

An innate 'cow sense', thriftiness, ease of training and ability to learn are its other attributes. The Quarterhorse is a versatile horse and is particularly suited for Western-style events, including barrel racing, cutting, Western pleasure, Western riding, reining, working cowhorse, English pleasure and trail class.

The most common colours are chestnut, bay and dun. Less common colours are black, palomino, roan and brown.

Shetland

As one of the smallest breeds of pony, the Shetland is known throughout the world, and is remarkable for being probably the strongest breed in relation to its size. It is well suited as a saddle-pony for small children, but is also marvellous in harness, moving along at a brisk pace.

The average height of a Shetland is approximately 9.3 hands and the limit is 10.2 hands. It can be of any colour, including paint, and should have an attractive head with small, well-placed ears. With its deep, hardy little body, set on short and busy legs, it is most lovable.

Standardbred

The Standardbred is so named because in the past each animal had to be capable of performing to a set standard before being eligible for inclusion in the stud books.

Standardbreds may either trot or pace. They vary in conformation because speed is considered more important than beauty, although the modern type is often very attractive. The most common colours are the solid colours; however, some roans and broken colours appear from time to time.

Standardbreds are a popular racing animal because they give the owner an opportunity to both train and drive. Standardbreds are gaining popularity as good riding horses after retirement from racing.

Thoroughbred

The Thoroughbred or blood horse was developed in England nearly 300 years ago for racing purposes, using local and Arabian stock. Today it is noted for its stamina, speed, fineness of conformation and long, well-muscled legs. It has a temperament most suited to competitive events.

Common coat colours are bay, brown, black and chestnut, although there are also some greys and roans. White markings on the face and legs are common.

The Thoroughbred has been successfully used to improve the quality of many other breeds. Its major use is racing, but its blood is also found in most horses used for hunting, show jumping, dressage, polo and stock work.

Waler

Walers originated in Australia during the 1700s and 1800s. They were originally sired by Thoroughbreds, Arabians and Anglo-Arabs out of local mares of undetermined breeding. Many of these mares would have had heavy-horse bloodlines.

Regarded as excellent saddle horses, Walers were bred and exported in large numbers, chiefly as cavalry horses. The legendary World War I Australian Light Horse Brigade, numbering some 40,000 men, was largely mounted on Walers. After the war, interest in the declined and, largely uncontrolled, they were left to breed in a wild and natural state where, in the process of natural selection, only the strongest survived.

Interest in the breed revived during the 1970s and 80s. A society and a program to identify and locate as many horses of the Waler type as possible were begun and a stud book was established. The society believes that the Waler population in Australia is difficult to estimate as many people do not recognise the type. Many are still believed to be running wild.

Ranging from 14.2 to 16 hands Walers have a general character suited to a strong Warmblood type with a kind temperament. Three categories of Waler have been established: light, medium and heavy.

Warmblood

Grouped under the name German Warmblood are the Hanovarians, Trakehners, Holsteiners, Oldenburgers, Westphalens, Rheinlanders, and other similar breeds. Cleveland Bays are not classed as Warmbloods by the Australian Warmblood Horse Association.

A Warmblood has a strong but attractive conformation, quiet temperament, long forward active movement and straight action. The usual height of the Warmblood ranges from 16 to 17 hands. It is a horse that, through generations, has been bred for dressage, show jumping, cross-country and leisure riding. It is suited for mating with Thoroughbred and other selected types of mares to produce successful dressage and show jumping horses.

Welsh Ponies and Cobs

This group of ponies is divided into four sections: the Welsh Mountain Pony, which must not exceed 12 hands (section A), the Welsh Pony; which stands at up to 13.2 hands (section B), the Welsh Pony of Cob type, which also stands at up to 13.2 hands (section C) and the Welsh Cob, which exceeds 13.2 hands (section D). Pones and Cobs may be of any colour except piebald and skewbald.

The Welsh Mountain Pony was bred in the mountains and wild regions of Wales. This harsh environment has ensured that the breed has a hardy constitution, sound limbs, and good temperament. It makes an ideal child's pony that can be both ridden and driven.

The Welsh Pony is similar to the Welsh Mountain Pony, but greater emphasis is placed on its riding qualities. It is hardy, fast and has a natural jumping ability that, combined with its good temperament, makes it suitable for performance competitions and the show ring.

The Welsh Pony of Cob type is a stronger counterpart of the Welsh Pony with Cob blood. It is suitable for both adults and children for trail riding, jumping and harness work.

The Welsh Cob has been bred for its courage, tractability and power of endurance. It is a good hunter, suitable for driving and is a most competent performer in all competitive sports.

IDENTIFICATION OF THE HORSE

Identification of horses is one aspect of horsemanship that is often overlooked. A thorough description of a horse requires the gathering of a lot of information and familiarity with a lot of terminology.

Individual identification can aid in:

- tracing bloodlines
- effective change of ownership
- preventing fraud at shows, races and competitions
- obtaining loans and certifying insurance claims
- tracing stolen horses and acting as a deterrent to theft
- health programs, health certificates and quarantine regulations.

Why is identification necessary?

On the stud farm

Identification is an important aspect of managing a stud. Visiting mares have to be identified to ensure the correct stallion serves the correct mare. Foals should be identified as soon as they are born because mares have been known to swap foals. Young stock also has to be identified for breed registration.

In the sale ring

Adequate identification is needed in the sale ring to protect the buyer from mistaken identity, whether intentional or unintentional. When buying a horse privately the identification should also be checked.

For competition

In order to compete (other than at small, local shows) the horse will need to be registered with the relevant authority, which requires identifying the horse. At the racetrack, in the showing ring or in any other competition, adequate proof of the identity of the horse should be available.

The identification document

A certificate of identification should be complete, precise and yet easily understood by racing officials, owners and breeders, airport and shipping personnel, customs officers and quarantine veterinarians.

When filling in the identification document it is recommended that:

- a black ballpoint pen is used to permit the production of clear photocopies
- the position of whorls be indicated by an X
- white head markings be indicated by drawing the outline and shading in lightly
- sparse white hairs be indicated by a few lines
- bordering be indicated by drawing a double outline
- flecking and ticking be indicated by small, light lines scattered over the area
- scars be indicated by an arrow.

A standard identification form will usually require information on the horse's registered name (if any), the names of its sire and dam (if known), height, colour, sex, age, and sketched and/or written descriptions of its natural and acquired markings and congenital abnormalities or special

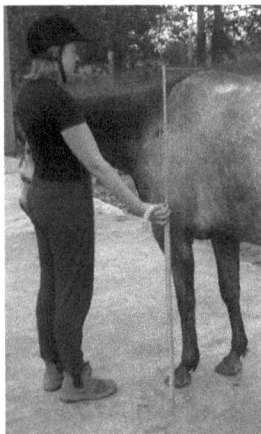

Figure 36.
Measuring height
(courtesy of Jane Myers).

peculiarities. Many stud books now also require verification of parentage via DNA analysis, especially when imported frozen semen has been used.

Height

The height of the horse is measured in 'hands'. A hand equals 4 inches or 10 centimetres. The term 'hand' originated from ancient times when the average width of a man's palm was 4 inches. For example, an animal measuring 15.1 hh (hands high) is fifteen hands and one inch.

The horse is measured from the ground to the top of the withers. To be accurate the horse should be standing square on a flat level surface and a measuring stick with a spirit level incorporated into the cross-bar should be used.

Generally, a horse that measures up to and including 14.2 hh is termed a pony and above that is a horse. However, this is not a set rule; some people regard up to 14 hh as a pony, between 14 hh and 15 hh as a Galloway and above 15 hh as a horse. Some breeds, such as Arabs, are always classed as a horse even if smaller than 14.2 hh.

Name

The registered name of the horse is often different to its 'stable' or pet name and should be used for identification.

Sex

The following terms must be used when referring to the sex of the horse.

- Stallion: an ungelded male horse, aged four years and over.
- Colt: an ungelded male horse up to four years old.
- Mare: a female horse aged four years and over.
- Filly: a female horse up to four years old.
- Gelding: a castrated male horse of any age.

Age

On registration, the date of birth will be required. The following terms are used when referring to the age of a horse.

- Foal: young horse up to one year old. After weaning (which is usually at six months of age) the foal is described as a 'weanling'.
- Yearling: one year old and less than two.
- Aged (varies between older than seven in the racing industry to 13 years or more elsewhere).
- Rising: the horse is about to turn one year older. A horse may be describes as five rising six when it is nearer six years than five.
- Off: the horse has just turned a year older e.g. 'six off'.

Horses that are older than one year are described according to their age in years; for example, 'a two year old'. See the later section 'Ageing a horse by its teeth' for identifying a horse that is not branded.

Breed

The correct name of the breed should be used (see earlier section on breeds of horse in Australia and New Zealand).

Colours and markings

Horses come in many colours and have a variety of markings. Both are important in the identification of individual horses and for this reason, there are certain terms that must be used when referring to coat colour and markings.

Genetics determine the colour of a horse. The genetic inheritance of a horse is called its genotype, but what we see (i.e. the end result of the genes) is called the phenotype. For example, some Appaloosas do not have spots and to the inexperienced, look like an ordinary bay or chestnut horse. Regardless of their external appearance, they are still an 'Appaloosa' and can be registered as such because the horse has the genes of the Appaloosa breed. Some horses can only be accurately be identified by knowing the colour of the parents; this is because certain colours will only produce offspring of a particular colour.

Base coat colours

The basic coat colours are brown, bay, black and chestnut. They are referred to as solid colours.

Brown: the body coat is uniformly brown and the mane, tail and lower parts of the legs range from brown to black. Some registries do not recognise brown as a colour and the horse is then referred to as bay.

Bay: the body coat is brown, but can range from very light, yellowish brown to very dark reddish brown. The mane and tail are black and there is black on the lower parts of the legs and tips of the ears.

Black: the black pigment is evenly spread throughout the body coat, limbs, mane and tail. There is no pattern other than white markings. There is no evidence of brown on the flanks and muzzle or the horse must then be classified as brown. Some blacks do 'fade' to brown in the summer sun.

Chestnut: the body colour ranges from a light, washy yellow, through golden and reddish shades to a dark liver colour. The mane and tail are not black, but are darker or lighter than the body coat. Chestnuts may have a flaxen mane and tail and are sometimes confused with palominos.

Patterns

There are many patterns that can be superimposed on the base colours, but the most common is grey or roan.

Grey: the body colour is an uneven mixture of coloured and white hairs. The skin of a grey horse is dark. The foal is born with one of the basic coat colours, but with increasing age, white hair gradually develops and the horse goes through a stage of being dappled and/or flecked (flea bitten) and eventually the whole coat may appear white. Some foals are born already showing grey and their coat will tend to lighten quickly. Greys can have a mane and tail that is darker or lighter than the body.

Roan: the body coat is a fairly even mixture of coloured and white hairs. A foal is born roan and remains roan throughout life because the basic coat colour is permanent and does not change with increasing age as in a grey. The white hairs are not usually present to any extent on the face, lower parts of the leg or in the mane and tail. The colour of the mane, tail and legs is associated with the basic coat colour. The horse should be described as black roan, bay roan, chestnut roan etc. Roans can appear to be quite a different colour during winter because their coats sometimes lighten. Summer sees the loss of the lighter, thicker coat to that of a thinner, slightly darker, summer coat. The terms blue roan, red roan and strawberry roan are no longer used.

Broken colours

Pinto is the collective word for broken-coloured horses (the terms 'piebald' and 'skewbald' are no longer used). Pintos are further described as follows.

Tobiano: white patches from the legs up and the spine down; in some respects, it is an overo in reverse (Plate 1, Figure 14). There can either be a large or small amount of white. It has normal face markings or even none at all. It may have a bi-coloured tail and can have none, one or two wall (blue) eyes.

Sabino: white patches going up the legs and onto the underbelly. There can be slashes, spots and ticking on the body also, which can be confused with 'roan'. Sabinos have a blaze and chin spot on the head and none, one or two wall eyes. Clydesdales are often sabino. It is also a common coat colour in Arabians and Welsh Ponies.

Overo: white patches that appear to spread sideways along the body. The top of the neck and back and the legs are coloured. The horse has none, one or two wall eyes. Overos carry a gene that produces white-foal syndrome in which the foal is born dead or dies only hours after birth. This can especially occur if overo is bred to overo, so an overo should only be bred to a solid colour horse, which then gives a 50% chance of an overo foal.

Splashed white: white on the lower parts of the body and a bald face with wall eyes. It can be as little as white socks and a bald face to extensive white with just a small amount of colour along the back. It can be confused with sabino; however, the edges of the markings are more defined and the splashed white does not usually have ticking. All 'splashed white' horses are thought to be deaf.

Spotted horses
Spotting occurs in many breeds, the most well known being the Appaloosa (Plate 1, Figure 3). All true spotted horses (including solid Appaloosas) have other distinctive characteristics: stripes on the hooves, a white ring around the eye (the sclera) and mottled skin, which can be clearly seen around the mouth and genitalia. The mottled skin is not always present at birth. Once, many spotted horses had a thin, short, mane and tail, but this feature has been bred out of many lines. Spotted horses often have 'texturing' in the coat, which can be recognised by the different thickness of hair (courser or finer) within the spotting when compared with the base colour.

Dilute colours
The presence of 'diluting genes' produces a variety of colours, examples of which are given.

Dun: there are four classifications of dun: red, yellow, mouse and blue. Duns have a distinctive dorsal stripe, a 'mask' of darker hair on the front of the face, and primitive, transverse markings on the legs (see later description of zebra stripes). The coat of a dun does not shine to the same extent as a buckskin. There are several breeds that have duns, including Highland Ponies.

Taffy: variations are red taffy, blue taffy and silver dapple. The dilution gene affects the black pigmentation of the horse, and to a much lesser extent the red, so that instead of being black, the coat is lightened. Taffy occurs in many breeds, including some of those common to Australia.

Buckskin: the body colour that ranges from cream to near black (these horses are often mistaken for black) (Plate 1, Figure 9). The mane, tail and points of buckskins are darker than the body, usually black, but they can be various shades of brown. Buckskin is a colour, not a breed. Buckskins are no more likely than any other horse to have the dorsal stripe and primitive leg markings that duns always have.

Palomino: a true palomino is gold with a white mane and tail, but can range from a dark gold to a much lighter shade of gold (Plate 1, Figure 15). The best palominos do not have black hairs in the mane or tail or anywhere else on the body. Palomino is not a breed, so, for example, if two palominos are bred together there is only a 50% chance of a palomino foal.

Cremello and perlino: these light coloured horses are often mistakenly referred to as albinos. It is believed that albino does not occur in horses. A cremello is usually born of palomino parents and has a completely pink skin, light blue eyes, white feet, off-white to cream body hair and a mane and tail that are lighter in colour than the body hair. Perlino horses, on the other hand, are usually born of Buckskin parents. The have blue eyes, pink skin, cream body hair and a mane and tail darker than their body hair. It can be difficult to tell cremello and perlino apart unless the parentage is known. The light blue eyes and pink skin make cremellos and perlinos susceptible to sunburn and they need additional protection with rugs, nose flaps on the headcollar and sunscreen.

White markings

White markings must be accurately defined, with their extent and position stated precisely. White markings are present in many types and breeds of horse; however, some breeds are penalised for their existence (e.g. a Friesian (Plate 1, Figure 12) cannot have any white markings). The presence of white markings does not change the general colour description of a horse; a bay is still a bay with or without the white. It is recommended, for the purposes of identification, that the following terms be used to describe white markings of the face.

Star: any solid white marking on the forehead is described as a star. The position, size, shape and intensity of the star should be described, as should any coloured markings in the white. A star may be qualified by one or more descriptive terms such as large, small, faint, mixed, bordered, crescent-shaped (right or left open), and conjoined. Any marking on the forehead that only consists of a few white hairs or a patch of mixed hairs should be described as such and not referred to as a star.

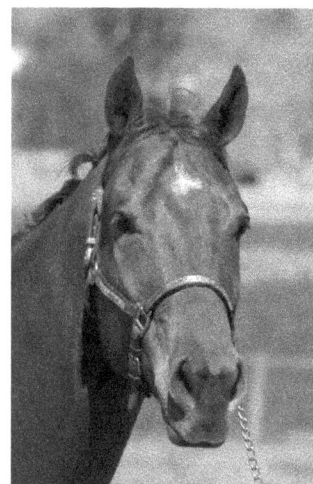

Figure 37. Star
(courtesy of Annie Minton).

Stripe: a stripe is a solid white marking down the face, but no wider than the flat anterior surface of the nasal bones. The stripe may be continuous with the star or separated from it (interrupted). A stripe may be qualified by one or more of the following descriptive terms: conjoined, interrupted, broken, mixed, narrow, broad, broadening, irregular, bordered, faint, inclined or curved to left or right. Any markings on the site of the stripe that consist of a few white hairs or a patch of mixed hairs should be described as such and not as a stripe.

Figure 38. Blaze
(courtesy of Annie Minton).

Blaze: a solid white marking covering almost all of the forehead between the eyes, extending down the front of the face, (usually to the muzzle) and involving the whole width of the nasal bones. Any variation in direction, and the point of termination, should be described. The presence of any coloured markings in the white should be noted.

Bald face: solid white marking covering the forehead and front of the face, enclosing at least one eye, which will be a wall eye. The mark extends laterally over the sides of the cheeks and towards the mouth. The extension may be unilateral or bilateral and should be described accordingly. The presence of any coloured marking in the white should be noted.

Snip: an isolated white marking, independent of those already described, and situated between the nostrils. Its size, shape, position and intensity should be specified.

Chin spot: a white spot located on the lower lip or chin.

White markings on the legs are described as follows.

Coronet: solid white marking immediately above the hoof. The extent and location of the white marking should be described with special reference to any variation in the height of the marking or on the various aspects of the leg. The presence of coloured markings in the white should be noted. Any marking on the coronet that consists of a few white hairs or a patch of mixed hairs should be described as such and not as a coronet.

Figure 39. Bald face
with a blue (wall)
eye
(courtesy of Jane Myers).

Heel: a white marking in the area at the back of the pastern extending from the bulbs (of the heels) to the ergot. It should be noted if the white is confined to one or both bulbs of the heel. Conversely if coloured marking is present within the white it should also be noted.

Pastern: refers to the area extending from the top of the hoof to immediately below the fetlock joint. The extent of the white should be described 'half pastern', 'three-quarter pastern' etc. Special note should be made of variations in the height of the white marking and on the various aspects of the leg. The presence of coloured markings in the white should be noted.

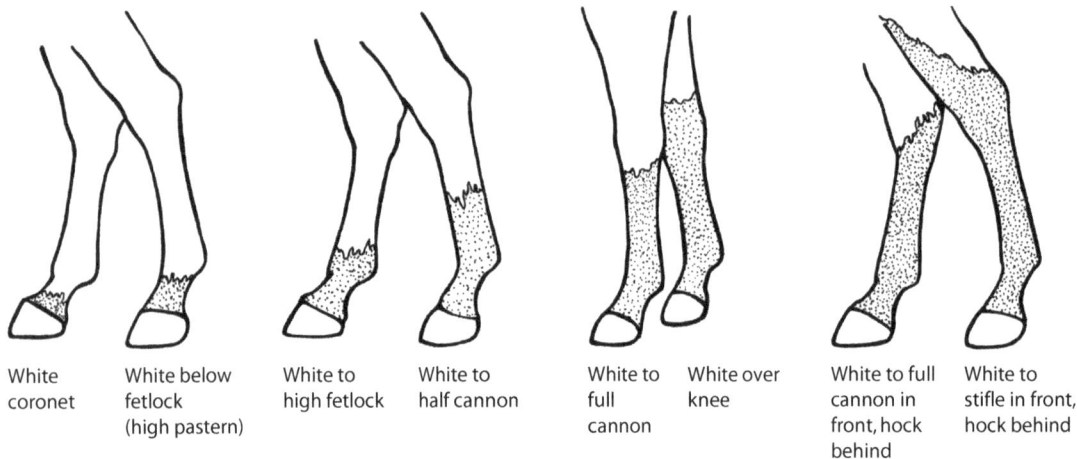

White coronet

White below fetlock (high pastern)

White to high fetlock

White to half cannon

White to full cannon

White over knee

White to full cannon in front, hock behind

White to stifle in front, hock behind

Figure 40. Leg markings.

Fetlock: refers to the area extending from the top of the hoof to immediately above the fetlock joint. The extent of the white should be described using such terms as 'to half fetlock', 'to three-quarters fetlock' and variations in the height of the marking on the various aspects of the leg should be specified. The presence of coloured marking in the white should be noted.

Cannon: refers to the area extending from the top of the hoof to immediately below the knee or hock. The extent of the white marking should be described using such terms as 'to half-cannon', 'to three-quarters cannon' and variations in the height of the marking on the various aspects of the leg specified. The presence of coloured marking in the white should be noted.

Knee or hock: A knee or hock marking is from the top of the hoof to the knee or hock. Variations in the height of the white marking on the various aspects of the leg should be described and the presence of coloured markings in the white noted.

Other markings

Flesh marks: patches where the pigment of the skin is absent, often seen around the nostrils.

Ticking or flecking: ticking/flecking is the presence of isolated white hairs distributed throughout the coat in any part of the body and it can vary from light to heavy.

Ermine marks: black or dark marks on white leg markings, often just above the coronet.

Blood markings: darker hair patches, such as a red patch on a grey horse, often seen in the shoulder area where it is referred to as a 'bloody shoulder'.

Smuts: small, circular collections of hairs (not white) in any part of the body that are a different colour to the general body colour. The colour of the smuts should be stated.

Zebra stripes: transverse, dark stripes across the back of the forelegs and front of the hind legs, which are regarded as primitive markings. Duns always have them and they sometimes occur on other colours.

Dorsal stripe: prominent dark stripe running from the withers down the centre of the back. It is always seen in dun horses and sometimes in other colours (e.g. a bay). Thoroughbreds can have a thin, black, dorsal stripe.

Acquired markings

Acquired markings are usually white or bald and include scars and the white patches that occur in the regions of the body where gear has created a pressure sore (e.g. from bandages on the legs or saddle pressure sores).

Congenital abnormalities or individual peculiarities

Any congenital marking or individual irregularity that has not be included in the description must be clearly described and shown on the sketch if it is a permanent characteristic of the horse.

Those commonly encountered include wall eye, partly coloured eye, showing the white (sclera) of the eye, partly coloured hoof, deformities of the hoof and limbs, abnormal dentition such as parrot mouth, muscle indentations (Prophet's thumb marks) etc.

Brands

Completely secure identification of an individual horse combines the natural identification features with a manufactured system of identification. Branding is the most common system and most horses are branded when they are weaned at 5–8 months old. The brand contains a lot of information that can be easily understood with a little practise.

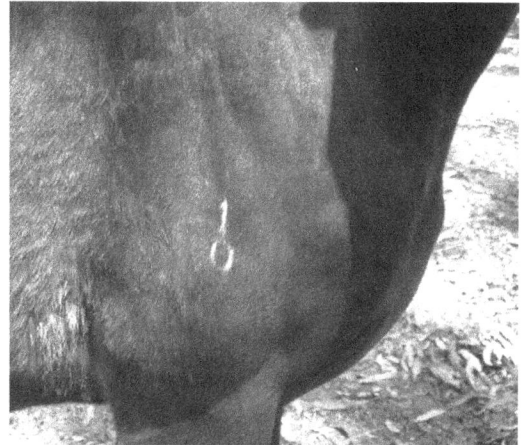

Figure 41.
Individual number and age brand. This horse was the first foal born at this particular stud in 2000
(courtesy of Jane Myers).

A horse is usually branded with a combination of an owner/breeder brand, individual number brand and an age brand. Some horses also have a breed society or association brand. The owner or breeder brand distinguishes the stock belonging to different owners or breeders. Most States have a brand register to prevent duplication of owner brands, but Victoria and Tasmania do not register brands. The owner/breeder brand is traditionally applied to the near shoulder. The individual number brand differentiates individual horses born at the same stud in the same year. The first foal is branded with a 1, the second with a 2 and so on. This brand is placed on the offside shoulder above the age brand, except in Queensland, where it goes on the near shoulder. The age brand denotes the year of foaling. The last number of the year is used, 1990=0 and 1992=2, for example. Horses with the same age brand may be 10 or 20 or even 30 years apart in age, but the horse's teeth and other signs will allow differentiation of the real age. The age brand is placed under the individual brand. The breed society or association brand is usually applied to the horse by an official representative of the society/association with which the horse is registered. Different societies have differing regulations and owners should refer to the various societies for more information. Andalusians, Hanoverians and Trakehners are branded with a society brand that can be either a hot or freeze brand.

The alpha angle coding system brand is applied to Standardbreds on the right side of the neck under the mane. It is a coded brand, with each symbol corresponding to a number, so that the brand can be generated by a computer and stored electronically. The alpha angle coding system is practically unalterable. It relies on only two symbols: two parallel bars and two bars set at right angles. These symbols are rotated to produce a complete set of ten numerals. The brand always starts with S for Standardbred, followed by a symbol for the postcode prefix of the State where the horse was born. The next two symbols are the last two digits of the year in which the horse was born and the four symbols underneath are the registration number of the horse. It is always applied as a freeze brand.

To accurately read the alpha angle brand, the horse may need to be clipped over the brand. Occasionally, one or more symbols may be difficult to read. It is important in all documentation, that the symbols are recorded as they actually appear. Even with missing or indistinct symbols the freeze brand is a unique mark enabling rapid identification of individual animals. Contacting the Registrar of the local Harness Racing Authority and quoting the registration number of the horse from its freeze brand can quickly ascertain the identity of the horse and its owner.

Freeze branding

Freeze branding is now the most common method of identifying horses. It uses intense cold to destroy the pigment-producing cells in the skin and hair of the horse, resulting in white hair that permanently outlines the brand.

The advantages of freeze branding are that it is relatively painless, the brands can usually be read from a distance, brand quality is less variable compared with fire brands and series of numbers suitable for computerised recording can be used. The disadvantages of freeze branding are that it is more expensive than fire branding and some owners do not like its appearance.

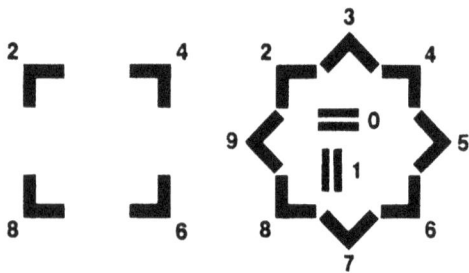

Figure 42. Alpha angle coding system symbols.

The brand should be between 30 and 80 mm high. The brand will grow with the horse, so young horses should be branded with a small iron approximately 30 mm high. The width of the iron should not be more than 5 mm.

In order to apply a freeze brand the hair is clipped from the branding area to allows complete contact with the skin. The area is thoroughly wetted with methylated spirits just before applying the brand. The horse will need to be restrained, but does not usually require twitching.

White hair will appear after a few months. Do not fill in the brand details on a registration certificate until the white hair can be seen.

The horse may need to be clipped over the brand in order to read it accurately.

Fire branding

Fire branding used to be the most frequently used method of identifying horses, but is now less common. A hot branding iron is used to cauterise the skin to produces a permanent scar. Hair then grows on the raised surface of the scar and the brand can usually be read when the horse is turned towards the light. Smoothing the hair down by hand or dampening the hair will aid identification, but in some cases the hair will have to be clipped to read the brand.

The brand should be between 30 and 80 mm high. The brand will grow with the horse, so foals or young stock should be branded with a small iron (i.e. approx. 30 mm high). The brand on a foal will increase in size by almost one-third by maturity. Large brands are unsightly and 50 mm is usually big enough for a brand on a fully-grown horse. On average, the width of the burning edge of the iron should not exceed 4–5 mm.

The area to be branded is clipped to ensure that the branding iron is in complete contact with the skin. The area needs to be clean and dry. The horse is positioned against a fence or wall, preferably in a corner to restrict its movements, and may be twitched. Any movement of the horse during application will result in a poor quality brand. If the skin is broken by an overheated iron, apply a dressing to prevent infection.

Fire branding has the advantages of being cheap, quick and easy to apply, and is not unsightly when done correctly. Its disadvantages are that it will cause pain to the horse when applied and it can often be hard to read. It should only be done by an experienced person.

Other methods of identification

Lip tattoos

North American horses are identified with a numbered tattoo on the inside of the lip. Those who feel that brands disfigure the horse may prefer lip tattoos; however, the numbers can be difficult to read in a restless horse.

Figure 43. The brand S7874325, as expressed in alpha angle coding system symbols.

Whorls

Whorls are permanent, irregular settings of coat hairs found on the forehead, nose, throat, neck, shoulder, chest, flanks, and/or buttocks. Whorls are unique to each horse, rather like a finger print, and so are a good means of identification, especially on solid-coloured horses. When describing a single whorl, its exact location and type should be stated. When describing multiple whorls, their exact location, type (diagonal, horizontal, vertical, triangular) and their relationship to one another (conjoined, separate, superimposed) should be stated. When a whorl and a white marking are associated, their relationship to one another also should be stated. For the purposes of identification it is recommended that 3–5 whorls should be located and described.

DNA identification

A relatively new method of checking parentage that provides positive identification is DNA identification. Blood, semen, hair or even tissue from a dead horse can be used. The cost of DNA identification is reducing as the technique becomes used more widely. DNA identification can be used either instead of blood typing or to resolve cases that blood typing cannot answer.

Blood typing

A horse's blood type is a permanent, unalterable record of identification. Theoretically, there are millions of possible blood types in horses, many more than all the horses on earth, so the likelihood of two horses having the same blood type is remote. All horses registered with the Australian Stud Book have their blood type recorded.

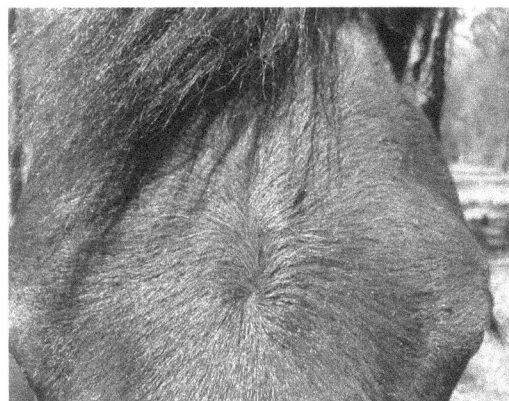

Figure 44. Whorl
(courtesy of Jane Myers).

The main application of blood typing is verification of parentage. Blood types are controlled by genetics, so any factor found in the blood of a foal must be present in either one or both parents. Unlike DNA identification, blood typing cannot prove that a foal is by a certain sire or out of a certain dam. However, it is possible to show that a certain animal is not the parent. In the case of a double covering in which two sires have served the same mare, blood typing can prove which of those stallions is not the sire.

Electronic identification

Microchips can now be used to identify a horse with a unique number. The microchip is implanted under the skin of the horse using a special implanting gun and a recording device, when placed adjacent to the microchip, reads the number. Microchip implants provide a permanent form of identification that does not mark the horse. Microchips are gaining in popularity and are already being used extensively in some disciplines such as endurance.

Ageing a horse by its teeth

It is useful to be able to tell the age of a horse before buying it and for checking the identification of a horse. With practise, anyone can accurately determine the age of young horses, but even experienced people may have to make a 'guestimate' once the horse is older than 10 years of age.

The incisor (front) teeth of the lower jaw are used to determine the age. There are six incisors on each jaw (12 in total) and they are referred to in pairs. The middle pair are called the centrals, the next pair the laterals, or intermediates, and the two outermost teeth are referred to as the corners.

The horse also has 12 molars (back teeth) in the lower jaw and 12 in the upper jaw (24 in total). Most, but not all horses, also have wolf teeth, which are small premolars (1st premolars) that sit in front of the 2nd premolars.

All male horses and some female horses also have canine teeth (sometimes also called tushes or bridle teeth). These erupt at about five years of age.

Horse's teeth continue to erupt throughout life, which is why a horse's age can be estimated from them. The teeth have a different appearance at the different stages of the horse's life. The tables (surface) of the teeth wear down as the horse chews, which can lead to some variation between horses depending upon what they eat. If the horse grazes very coarse grass all its life its teeth will wear sooner than a horse that only grazes fresh green grass. Therefore, these two horses will appear to be different ages when they are actually the same age, unless the diet is taken into account.

(a)

(b)

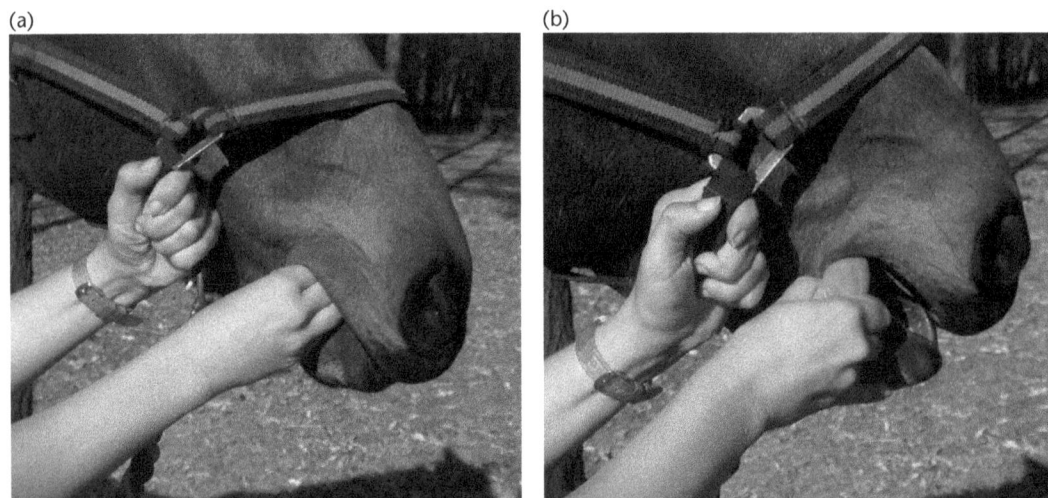

Figure 45. Opening the horse's mouth to check its age (courtesy of Stuart Myers).

If the teeth do not meet correctly (as in parrot mouth) it will affect the tables of the teeth and these horses generally cannot be aged this way. Another factor that affects the tables of the teeth and makes it difficult to estimate the age is crib-biting, which causes abnormal wear of the incisors.

How to open the mouth of the horse
You should watch someone who is experienced at this before you attempt it yourself. The horse must be untied, but wearing a headcollar. Make sure that the noseband strap is loose enough to allow the horse to open its mouth.

To open the mouth, slip three fingers into the side of the mouth between the incisors and the molars (i.e. the bars). When you have put your fingers in the mouth, grasp hold of the tongue and gently pull it out of the mouth. You will then be able to see the tables of the lower incisors, as most horses will not close their mouth when their tongue is pulled out to the side. When you have finished looking, gently let go of the tongue and the horse will pull it back into the mouth.

The four main features of the teeth
The four major features of the teeth used in estimating the age of a horse are:

- the eruption of the teeth
- the disappearance of the cups and the appearance of the stars
- the angle of incidence between the upper and lower incisors
- the shape of the surface (table) of the teeth.

In the young horse, age estimation is quite straightforward; you can tell the age by just determining the eruption of the teeth. In the older horse, age estimation is more difficult and so you need to combine information about the disappearance of the cups and appearance of the stars, the shape of the tables, the angle of incidence between the upper and lower teeth and other factors such Galvayne's groove and hooks on the corners.

The eruption of the teeth
Horses have two sets of teeth during their life. The temporary, milk or deciduous teeth are the first set, which are replaced by the permanent teeth. By the time the horse is 4.5 years old, all of the permanent teeth have erupted and replaced the temporary teeth. By 5 years all teeth are 'in wear'. The eruption and casting of temporary teeth and their subsequent replacement by permanent teeth rarely varies by more than a couple of months in most horses, which means that it is quite easy to accurately determine the age of a horse up to 5 years old. Horses that are older than 5 years are harder to age.

The permanent teeth continue to erupt throughout the horse's life and as they erupt they are worn down on the tables. Consequently, younger horses have long, deep roots to the permanent teeth and very old horses have shallow roots. These large roots can be seen as bumps in the lower jaw line of a young horse (tooth bumps) (see Figure 14, page 26). Bumps are also sometimes seen on either side of the nasal bone on the front of the face. In both cases, they disappear as the horse gets older.

When foals are born the six upper and lower deciduous incisors are generally visible through the gums. They erupt in pairs: the centrals first, at approximately eight days, the intermediates at approximately eight weeks and the corners at approximately eight months. You can look at the teeth of a young horse by simply pushing back the lips with your fingers. Depending on the age of the horse you may be able to see a difference between the temporary and permanent teeth from the side and the front. Temporary teeth are smoother, whiter, smaller and relatively shorter with narrower necks than permanent teeth. Permanent teeth have a yellow tinge caused by the cement that surrounds them and are larger, longer and darker in colour without a well-defined neck.

If a horse does not have any permanent teeth at all then it is less than 2.5 years old. The four central permanent incisor teeth (upper and lower jaws) appear at 2.5 years of age, the laterals at 3.5 years, and the corners at 4.5 years of age. If, therefore, a horse has all its permanent teeth it will be at least 4.5 years of age. It takes almost six months after the eruption of teeth for them to be 'in wear', which means that the upper and lower teeth are touching each other. So at three years of age the centrals are in wear, at four years the laterals are in wear and at five years the corners are in wear and the horse is said to have a 'full mouth'.

The disappearance of the cups and the appearance of the stars
In a newly erupted, unworn tooth the enamel is deeply folded into a cavity (infundibulum) or cup (Figure 46), which becomes darkened by food and is termed the 'mark'. As the tooth wears the cup disappears, which can be used to assess the age of young horses.

The cups disappear in the centrals at 5.5 years of age, in the laterals at 6.5, and in the corners at 7.5 years of age. The cups in the upper teeth are deeper and there is greater variation in the timing of their disappearance, which is why the cups in the lower jaw are used to estimate age. It therefore takes approximately three years of wear for the cups to disappear after the permanent teeth have erupted (Figures 53–57).

The dental star (secondary deposit) starts to appear in the centrals at 8–10 years, in the laterals at 9–11 years and in the corners at 10–12 years (Figures 55–59).

Stars are softer than marks and therefore do not stand out as much from the surface of the tooth.

The angle of incidence between the upper and lower incisors
As the horse ages, the angle made at the meeting of the upper and lower incisors changes and can be used as an indication of age. When viewed from the side, the angle is between 160° and 180° in young horses, and decreases to almost 90° in a very aged animal (Figure 61).

As the angle of incidence decreases, the upper and lower corner incisors do not meet properly and a hook on the upper corner incisor appears at about seven years of age (the seven-year hook; Figure 54). It is quite noticeable for about two years and then disappears. In some horses a hook may re-appear for a short time at 13–14 years of age and then disappear again.

The shape of the surface (table) of the teeth.
As the horse gets older the surface of the tables of the teeth changes. In a horse up to seven years the tables are oval, from 9 to 13 years the tables are triangular and in a horse older than 13 the tables are rounded (Figures 48–61).

Other aspects of the teeth used in estimating age
Another factor commonly used in estimating the age of aged horses is Galvayne's groove, which is a dark groove on the outer surface of the upper corner incisor. Not all horses have it, so it is not

routinely used in age estimation. It will appear on the side of the corner incisors at about 10 years of age and reaches the bottom of the tooth at about 20 years of age. Therefore, if a horse has a Galvayne's grove that extends more than halfway down the tooth, then you can be sure that the horse is at least 15 years old.

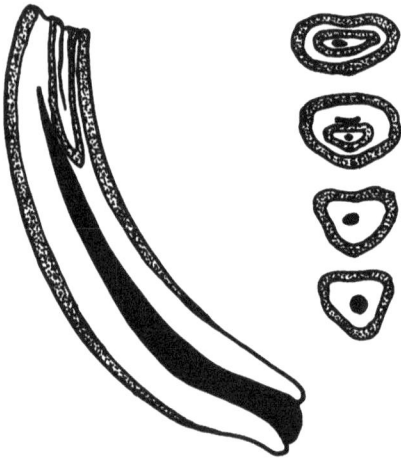

Figure 46. A permanent middle incisor tooth at different ages and stages of wear. The reasons why wear brings about a change in angle of incidence, shape of the table surfaces and disappearance of cups with appearance of dental stars are evident in this figure. As wear progresses, surface enamel is worn away, leaving two enamel rings, one around the margin of the table surface and the other around the cup. As wear continues, the cup almost disappears and the dental star or pulp cavity makes its appearance (at about 8–10 years). Tip of the dental star first appears surrounding the cup. In old age dental star appears dark, round and centred in the tooth. Changes in shape from oval to angular are shown in the cross-sectional views as wear progresses toward the root.

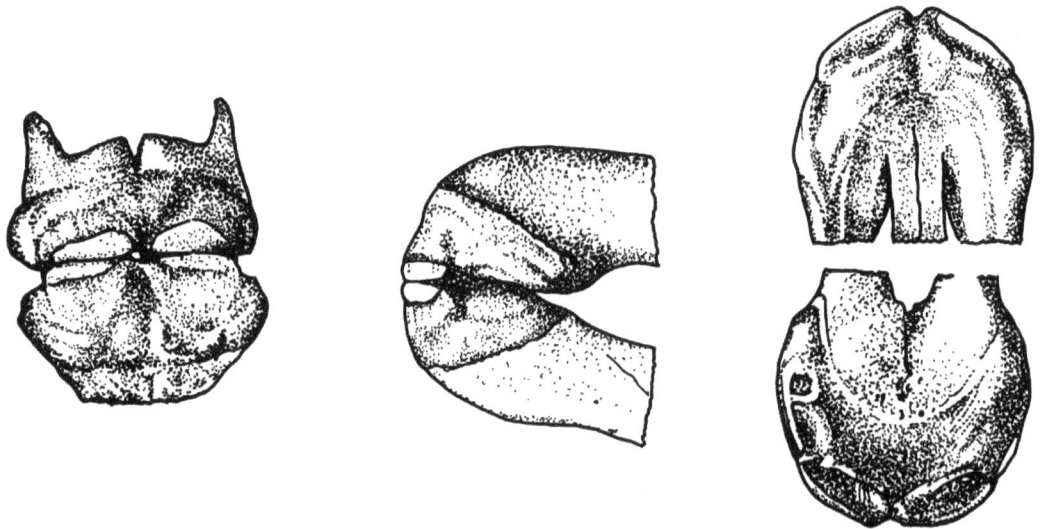

Figure 47. Birth to 2 weeks. The teeth have either not penetrated the gums or just appeared.

Figure 48. One year of age. All temporary teeth are present. The corners are not yet in wear.

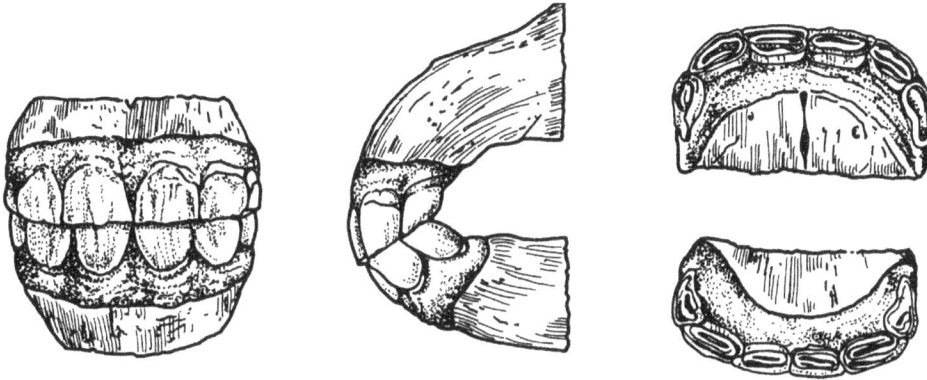

Figure 49. The 2-year-old mouth showing corners in wear. The temporary teeth are identified by the well-defined neck joining the root and gum, lighter colour and smaller size than permanent teeth.

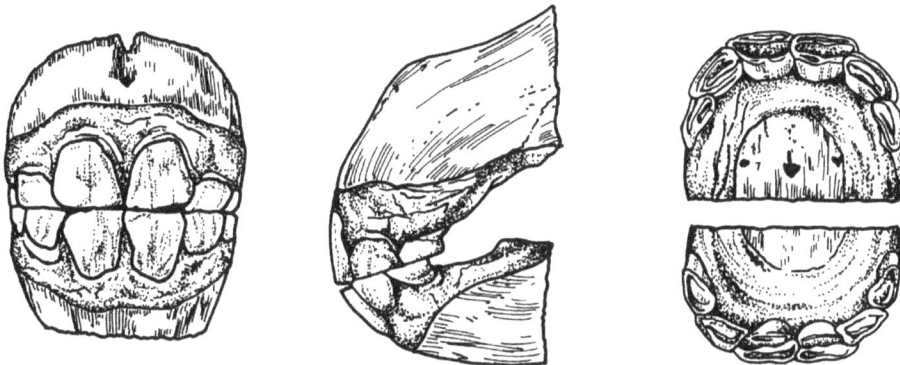

Figure 50. A typical 3-year-old mouth, showing the large permanent central teeth, both upper and lower. Contrast these with the adjacent small, light-coloured temporary teeth.

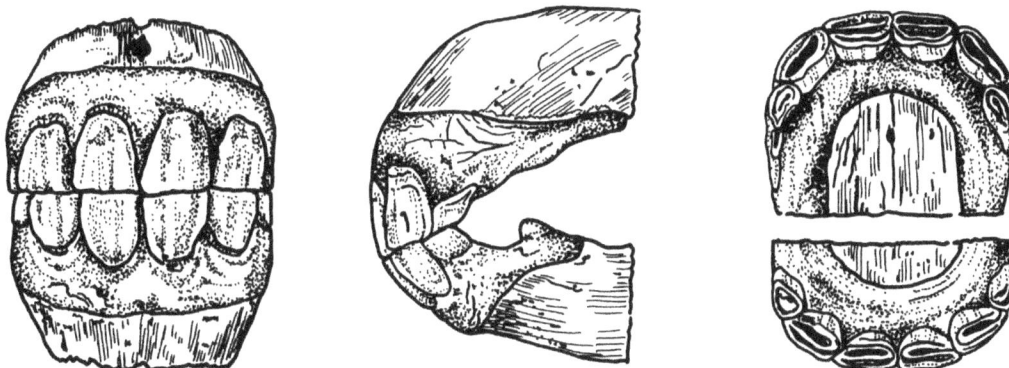

Figure 51. Note the well-developed permanent centrals, immature intermediates, and milk teeth at the corners in this 4-year-old mouth. Bridle teeth have appeared.

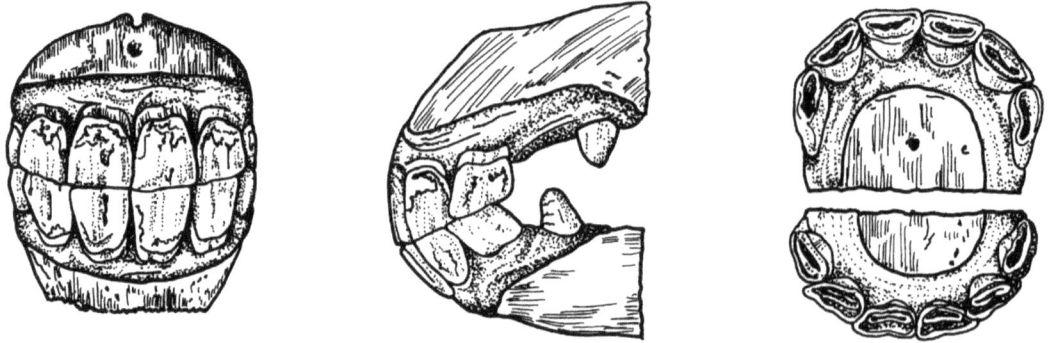

Figure 52. At 5 years, all the temporary teeth have been replaced by permanent teeth. This is called a 'full mouth.' Although the corner teeth are well matched from a profile view, they show very little wear in the view of the upper jaw. The upper centrals are beginning to appear round on the inside back surface. Cups are very obvious, both above and below, with little wear.

Figure 53. This 6-year-old mouth shows some wear on the corner teeth as viewed from the side. Cups in the centrals of the lower jaw should be worn reasonably smooth at this age. Bridle teeth have reached their full length. The angle of incidence shows little change.

Figure 54. At seven years the 'hook' has usually developed to its maximum. The rounded corner shown in this illustration does not give it the appearance that is ordinarily seen in a 7-year-old mouth. The angle of incidence is not very sharp, perhaps typical for this age. Cups are out of the lower centrals and intermediates, but very prominent cups still show in the corners and in all of the upper teeth. Dental stars have not appeared.

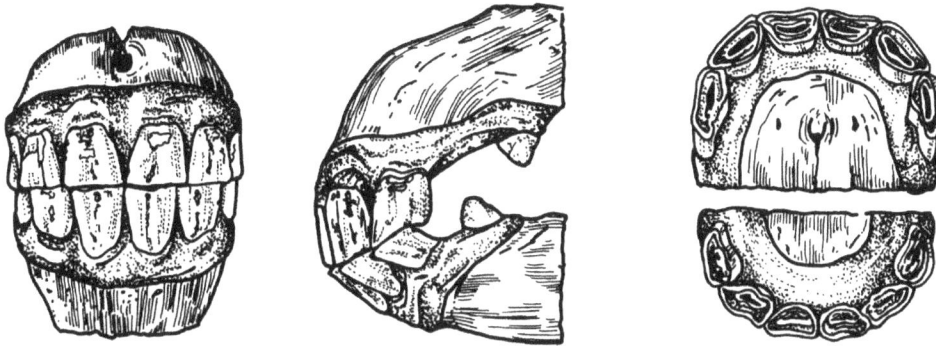

Figure 55. This 8-year-old mouth shows the hook almost completely gone from the upper corner tooth. Cups have all disappeared from the lower jaw. The teeth are showing a much more oval shape on the back surfaces than at younger ages, and the angle of incidence is becoming sharper. Dental stars have appeared in four lower and two upper incisors.

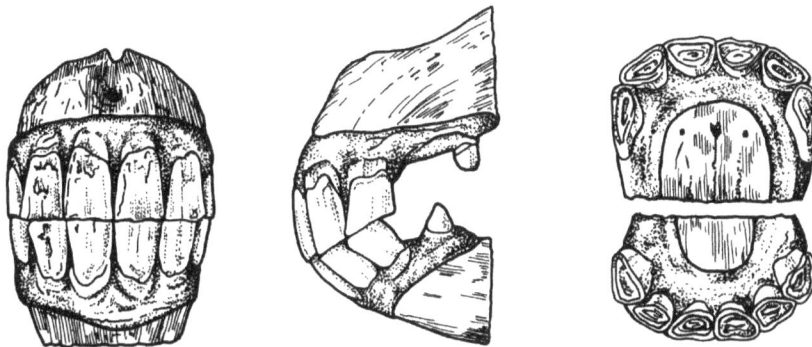

Figure 56. The front view of this 9-year-old mouth shows a tendency toward less width of the teeth and more length. The profile view shows more steepness to the angle of incidence; however, this illustration does not show the acuteness of a typical 9-year-old mouth. Cups are gone from the lower jaw. All teeth are tending toward a more oval shape, except the upper corners. Dental stars are merging with the central enamel rings, which are becoming small and round.

Figure 57. This 10-year-old mouth shows a typical angle of incidence with reappearance of the hook on the upper corner. Cups have gone from the teeth. Back surfaces of upper centrals are changing from oval to angular in shape.

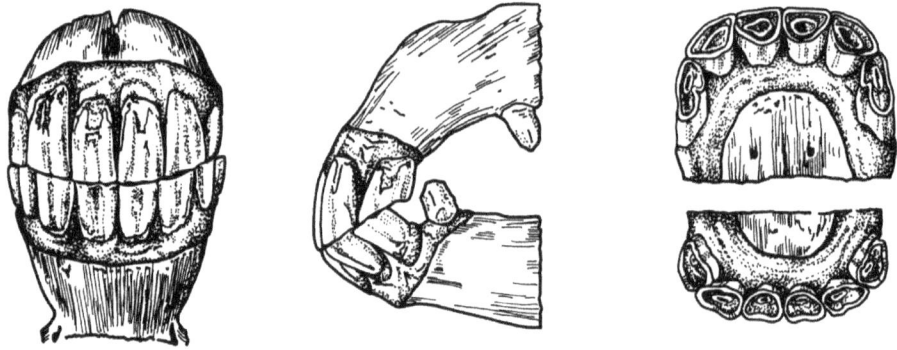

Figure 58. Much length of teeth relative to width can be seen in the front view of this 11-year-old mouth. A profile view shows considerable angle, with the upper corners almost missing the lower ones. Cups have all gone.

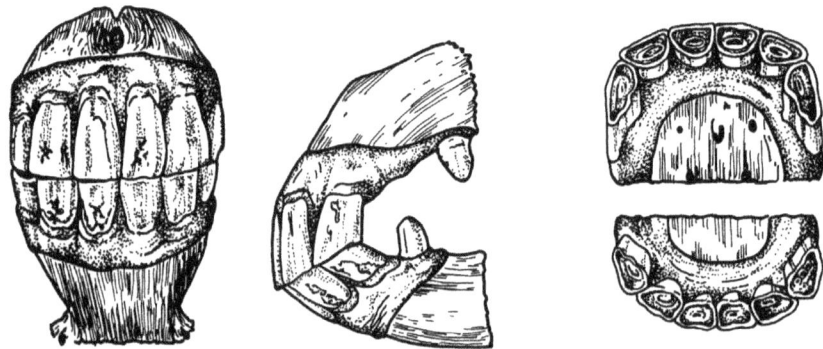

Figure 59. This 12-year-old mouth cannot be differentiated from the 11-year-old mouth, except for the cups in the upper corners and a decrease in the size of the central enamel rings.

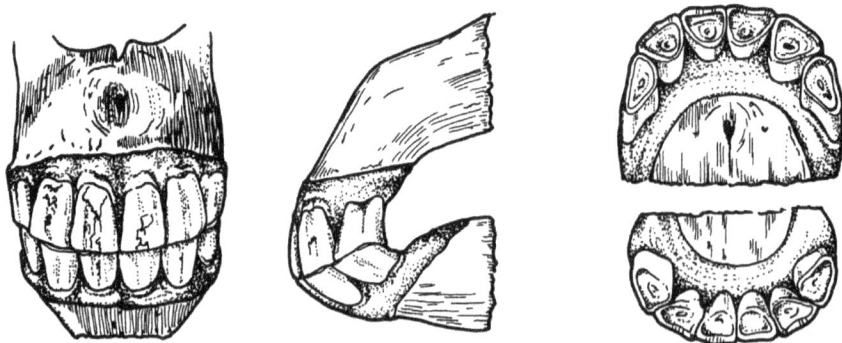

Figure 60. A 15-year-old mouth showing that all cups have gone. The central enamel rings are prominent, but are very small and round. All teeth have become angular.

Figure 61. Length of teeth, acute angle of incidence, and triangular surfaces characterise this 21-year-old mouth. Spaces have appeared between the teeth.

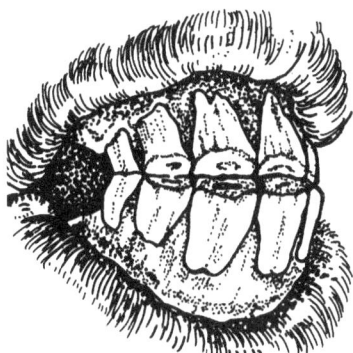

Figure 62. Crib-biter showing typical signs of wear on the incisors.

WHERE TO GO FOR MORE INFORMATION

Jeanette Gower has a website (http://gowerphotos.tripod.com/chalani), which has some information about colour genetics and Australian Stock horses.

Go to www.horsecouncil.org.au/breedsoc.htm for a list of breed societies in Australia and their contact details. The Virtually Horses website (www.worldzone.net/recreation/virtuallyhorses/breeds.html) includes a list of breed societies in New Zealand.

www.haynet.net is a valuable search engine for all topics relevant to horses (US based).

CHAPTER 5

FEEDING THE HORSE

HORSE DIGESTION

Correct feeding of the horse is based on an understanding of its digestive system. Unlike ruminants such as cattle and sheep, which have four stomachs, the horse has evolved with only one, small stomach and the main process of digestion takes place in the large hindgut (caecum and colon) where billions of micro-organisms (bacteria and protozoa) produce the enzymes that ferment plant fibre, enabling the horse to survive on a high-fibre diet. The caecum and colon (the hindgut) together hold approximately 96–109 litres (21–24 gallons) of liquid. The microbes need a constant environment and any sudden changes to the diet that alter the conditions in the gut will destroy them and affect digestion. Therefore, the horse has to eat small amounts often and chew the feed well before swallowing (see also Chapter 2). Fresh water supply must always be available.

Although horses cannot digest poor-quality feed as efficiently as cattle or sheep, their digestive system is capable of utilising a wide variety and quality of feeds. Factors such as breed, age, weight and level of work greatly influence the horse's nutritional requirements. Most horses at rest can be maintained on pasture, but because it can vary in nutritive value, particularly in late summer and winter, supplementary feeding may be needed. Your guide to the need for extra feed is the body condition of the horse (see later section in this chapter). Horses doing heavy work, young growing horses and lactating mares may also need supplementary feeding.

BASIC FEEDING PRINCIPLES

Horses must be fed well before they can perform well. Correct feeding is both a science and an art – the science is in knowing what to feed to satisfy the horse's requirements, and the art is in knowing how to feed to get the best performance.

The feed ration

A well-balanced ration has the correct amounts of fibre, energy, protein, minerals and vitamins for the horse's particular requirements. The ration must also be palatable, economical and practical.

Many horse owners underestimate the contribution pasture can make to the overall ration. If you use lots of supplements in the ration your horse may be unnecessarily overfed at great cost to you. Good pasture or roughage is often a cheaper and preferable alternative to expensive supplements that your horse may not need. Remember that the horse evolved as a grazing animal with a high-fibre diet and the high grain/low roughage diets fed to many horses

Figure 1. The digestive system of the horse
(P. Huntington et al. *Control Colic Through Management*. Kentucky Equine Research).

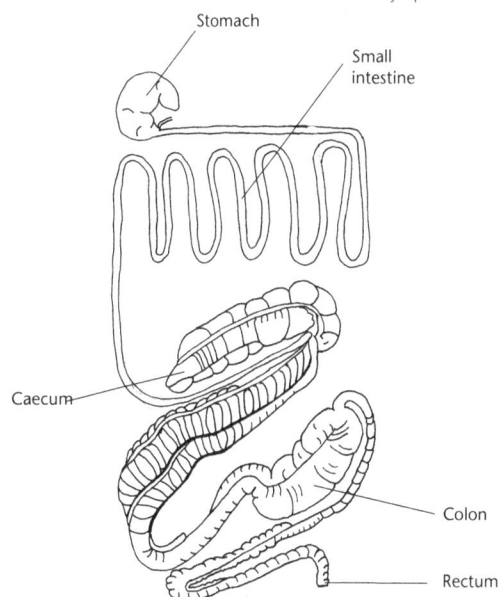

Stomach

Small intestine

Caecum

Colon

Rectum

are unnatural and can lead to many problems, including colic, gastric ulcers, diarrhoea, laminitis, tying up and behaviour problems.

Rations need not be complicated mixtures and usually can be made from locally available ingredients. The most common roughages for horses are made from pasture, legumes or cereals, the most commonly used concentrates are grains such as oats, barley, sorghum, maize and rice, and the protein supplements (in order of preference based on amino acid balance and palatability) are soybean meal, lucerne, lupins, cottonseed meal, sunflower seeds, linseed meal and peanut meal.

It is possible to calculate the best-balanced ration scientifically, but adjustments will probably have to be made to suit the individual horse's taste. Good feeding must be complemented by good management, with careful attention being paid to general health, dental care, parasite control and regular exercise.

Good feeding practices

Pasture and roughage
The horse evolved to eat one type of feed – grass. Good quality pasture or roughage should be the prime ingredient in the diet. Roughage contains fibre, which aids digestion, satisfies hunger and energy requirements, and decreases the incidence of wood chewing. If supplementary feeding is necessary, use a minimum number of good-quality feeds (this will also help reduce feeding errors if other people have to do the feeding for you at any time). Do not introduce too much variety to the diet unless it is essential for balancing the ration.

Body condition scoring
The horse's body condition or weight is your guide to the adequacy of both the quantity of feed and its energy supply. Record the condition scores or weights for later comparison and make changes to the diet as required to maintain good condition (see later section 'Monitoring weight and body condition').

Feeding and exercise
Avoid long-term confinement in stables or yards without regular exercise. Exercise periods should coincide with times of minimal food intake. Do not work a horse until at least two hours after its last feed.

Working horses should have the energy level of the ration decreased on the days when they are not being worked (see later section Feeding Adult Horses).

How much and when to feed
Horses require a minimum of 1% of bodyweight in forage every day to maintain correct digestive function. The amount of supplementary feed will depend on the amount and quality of pasture available and is calculated on a 'dry matter' basis (see later section). Horses eat by weight, not by volume, so always weigh the feed and remember that different types of feeds have different densities. The proportion of the diet that is roughage (including pasture) must always be greater than the concentrate proportion (on a weight basis). Record the quantities that are fed.

The horse must naturally eat often and in small amounts. Working horses and stabled horses need at least three feeds each day and would benefit from four. When fully hand-feeding (three feeds a day), give one-quarter of the concentrate in the morning and noon feeds and the remaining half at night.

Feed all grains other than oats in a mixture containing at least 50% oats. Replacement must be made on an equivalent basis, using the following guide.

5 kg oats = 4.2 kg barley
= 3.5 kg maize
= 3.5 kg wheat

Feeding should be at the same time and in the same place every day so that the horse's digestive system is not upset by erratic feeding. Mix the ration in amounts sufficient for only one day's feeding to prevent it souring, turning rancid or attracting flies. Clean any left-over feed out of the trough or bin before each feeding.

Using feed bins and hay racks and nets, rather than feeding on the ground, will reduce wastage, feed contamination and worm infestation. In stables, racks can be placed over the feed bin to reduce wastage by the horse nosing out the feed. Place hay nets up high enough so that horses can not get their legs caught in them.

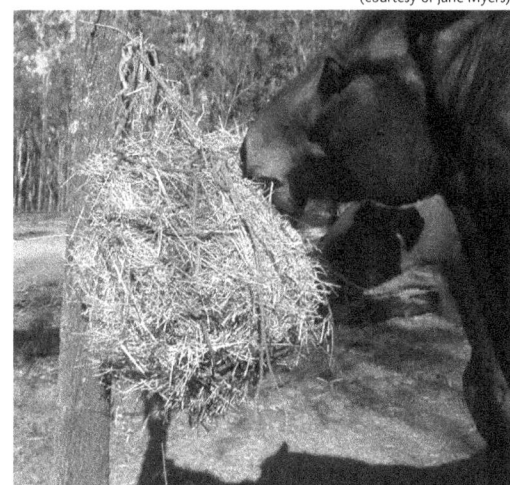

Figure 2. Feed bin made from a tyre (courtesy of Jane Myers).

Use good quality feed
Poor-quality, unpalatable, dirty or contaminated feeds can cause more problems than savings and should not be used unless there are no alternatives. Never give mouldy feed to horses because it can cause digestive upsets and possibly death. Dusty feed should be avoided because the dust affects the horse's respiratory system. If there is no alternative, the feed should be damped down before it is fed to the horse.

Always change the diet gradually
Sudden feed changes, especially in grain feeding, can lead to digestive upsets, colic or diarrhoea. A slow replacement over 7–10 days or longer is required, whether you are increasing the amount fed, changing from one grain to another or even just changing to grain from a new source.

Commercial mixes
Commercial mixes that contain grain, protein, minerals and vitamins can be more convenient to feed than mixing your own ration. The contents, but not the amount of energy, of prepared mixed feeds must, by law, be listed on the container, so compare the protein and mineral contents of the feeds you wish to use. Always supply plenty of good-quality roughage.

Water
Clean, fresh water, free of organic matter and sediment, should be available to horses at all times. A horse can drink up to 70 litres of water each day.

After heavy work, the horse's water intake should be limited to between 2 and 4 litres, given over a period of 15–30 minutes, until the horse has completely cooled down. In the stable, place the water trough on the opposite side to the feed trough because horses tend to splash water when they drink, which can wet the feed and make it messy or spoil.

Figure 3. Hay net placed at the correct height (courtesy of Jane Myers).

Good management
Regular dental care is necessary to prevent feeding and digestive problems (see Chapter 2) and a regular program of manure management and drenching will minimise the build-up of worms in the environment and heavy burdens in the horse (see Chapters 6 and 8), with the consequent effects on health and digestion.

Shelter from cold or hot weather will reduce stress on the horse and thus the amount of feed needed to maintain body condition.

When feeding groups of horses, observe the social hierarchy (pecking order) and ensure that the dominant (bossy) horses are not preventing the lower order horses from accessing the feed supply. Spread the feed bins out and place one or two extra feed bins in each paddock. Segregate horses by age, type and body condition to minimise pecking order effects and ensure that you can give special attention to those horses that require extra feed.

Table 1. Daily nutrient requirements (500 kg mature weight)

	Body weight (kg)	Daily gain (kg)	Daily feed (kg)*	Digestible energy (MJ)	Crude protein (g)	Lysine (g)	Calcium (g)	Phosphorus (g)
Maintenance	500	–	7.5	69	697	24	30	20
Pregnancy (last 90 days)	–	0.65	8.75	88	922	32	47	31
Lactation (early)	–	–	15	118	1414	49	60	40
Lactation (late)	–	–	12.5	102	1215	42	47	31
Foal 3 month old, suckling	155	1.0	4.75	69	825	35	43	28
Requirements above milk	–	–	2.75	41	485	20	28	21
Weanling, 6 months old	250	0.9	6.25	75	895	38	46	31
Yearling, 12 months old	350	0.55	7.0	87	935	39	49	33
Yearling, 18 months old	425	0.35	8.5	83	890	33	48	32
2 year old	475	0.2	8.3	69	835	33	47	32
Light work	500	–	8.75	93	766	27	33	22
Moderate work	500	–	11.25	103	836	29	36	24
Heavy work	500	–	12.5	137	906	32	39	26

*Dry matter (DM) basis: (DM value × 10 = weight of feed) (DM weight = weight feed × 0.9).
*From Kentucky Equine Research Recommendations.

SUPPLEMENTARY FEEDING

When pasture is not enough to maintain a horse's body condition, supplementary feeding is required to provide the energy, protein, minerals and vitamins that may be deficient. There are many supplements that purport to make horses run faster, jump higher, reproduce more readily and grow better feet and hair. Are these supplements really necessary?

In order to make informed decisions about the use of supplementary feeds, you must evaluate the current nutritive value of the ration and whether it meets the requirements of your horse. Feed requirements are calculated on a 'dry-matter' basis, which means after any water has been evaporated from the feed. Most supplementary feeds are 90% dry matter, but pasture varies from 30% to 60% dry matter. To calculate the dry-matter weight, multiply the raw weight of the feed by 0.9, except in the case of pasture.

Mature horses will generally eat 1.5–3% (in dry matter) of their body weight per day, depending on the type of feed and individual variation. Growing foals may eat up to 3% of their body weight in a day. The daily nutrient requirements of a mature, 500 kg horse are shown in Table 1. As noted earlier, horses require a minimum of 1% bodyweight in forage every day to maintain correct digestive function and the rest of the diet should be designed around that minimal forage requirement.

Supplementary feeds are either concentrates or roughages. Concentrates are grains and meals, which contain a lot of energy, and roughages are hay and chaff, which are low in energy, but high in fibre. The nutritive value of each of the common supplementary feeds is shown in Table 2.

Energy supplements

The horse's energy requirements determine the amount and type of feed. Horses at rest or only lightly exercised can meet their energy requirement from pasture alone, but young growing horses, pregnant or lactating mares and performance horses usually need more energy than can be supplied by roughage alone. The most common way of supplying energy in the ration is by feeding grain. In some cases, particularly the horse in heavy training, even cereal grains will not provide enough energy because the horse cannot eat enough to meet its requirements.

Table 2. Average nutritive value of feeds commonly used in rations for horses on an as-fed basis

	Digestible energy (MJ/kg)	Crude protein (%)	Lysine (%)	Calcium (%)	Phosphorus (%)
Concentrates					
Oats	12.0	9	0.35	0.1	0.33
Barley	13.0	10	0.4	0.7	3.0
Maize (corn)	14.2	9	0.25	0.02	0.26
Triticale	13.0	11	0.4	0.04	0.25
Rice bran	16.0	14	0.7	0.06	1.75
Sorghum	14	10	0.2	0.04	0.3
Lupins	14.5	30	1.4	0.2	0.3
Tick beans	13.0	25	1.7	0.14	0.5
Peas	14.0	23	1.6	0.2	0.4
Soybean meal	14.0	48	2.75	0.25	0.78
Canola meal	12.0	36	2.1	0.6	1.2
Sunflower seeds	17.5	20	0.6	0.2	0.5
Cottonseed meal	11.5	38	1.5	0.2	1.0
Wheat bran	11.0	14	0.5	0.12	1.0
Wheat pollard	12.0	16	0.6	0.15	0.7
Molasses (70% DM)	11	4	0	1.0	0.1
Vegetable oil (100% DM)	37.5	0	0	0	0
Copra meal	12	20	0.5	0.1	0.55
Roughages					
Oaten hay/chaff	7.5	8	0.25	0.2	0.2
Wheaten chaff	7.0	7	0.21	0.15	0.15
Lucerne hay/chaff	9	16	0.8	1.2	0.2
Grass hay	7.5	8	0.25	0.4	0.2
Clover hay	8.5	13	0.5	0.8	0.2
Pasture (100% DM)					
Improved, green	9.5	22	0.8	0.6	0.4
Improved, dry	8	10	0.33	0.4	0.2
Native, green	8.5	15	0.5	0.6	0.3
Native, dry	7	8	0.25	0.3	0.15
Minerals					
Limestone	–	–	–	36	0
Dicalcium phosphate	–	–	–	22	18

Values are 90% dry matter (DM) unless stated. (Data from 'Nutrient Requirements of Horses' (1989), Department of Food and Agriculture, Vic., CSIRO, KER database.)
Note: the values stated are averages. The composition of different feeds will vary and thus these figures are a guide only. For precise information on feed composition you should have your feed analysed at an appropriate laboratory.

The amount of energy in feeds is expressed in units of megajoules (MJ) per kilogram (kg). A megajoule is a unit of energy and is the metric equivalent of the calorie. Digestible energy (DE) is provided by four different dietary energy sources: starch, fat, protein and fibre.

Starch

Utilising energy from starch is very efficient because of the simple enzymatic process involved in its digestion. It takes fewer kilograms of grain to get the same amount of energy that is in a much larger amount of roughage. Grains are an excellent source of starch, which comprises 50–70% of the grain's dry matter, but they can be hazardous if fed to excess.

The starch molecules in grains are complex polysaccharides that are digested in the small intestine by the enzyme amylase, which breaks up the polysaccharides into simple sugars (glucose). These sugars are easily absorbed into the bloodstream and once in the blood, the glucose units are used for a number of different purposes.

1. Oxidised directly to produce ATP.
2. Used to make muscle glycogen, liver glycogen or body fat.

Muscle glycogen is an important source of energy during exercise. The glycogen stored in the liver is available for the production and release of glucose into the blood during exercise. Maintaining blood glucose levels during exercise is very important because glucose is the only energy source available to the central nervous system. Hypoglycaemia (low blood glucose) is another potential cause of fatigue in exercised horses.

Starch is the dietary energy source of choice for glycogen synthesis. The digestion of starch directly increases blood glucose and insulin, two of the most important factors involved in glycogen synthesis

The limiting factor of starch digestion is the production of amylase in the intestinal tract, which can vary between horses. If amylase production is too low or the amount of starch is too high, undigested starch passes through to the large intestine, but in the hindgut the amount of energy extracted from the grain by the microbial population is less. The fermentation of large amounts of grain in the hindgut creates lactic acid, which lowers the pH of the hindgut and upsets the microbial balance by killing other bacteria, leading to the release of endotoxins into the blood and possibly laminitis or colic.

Starch molecules come in different sizes, some of which are more easily digested than others. The starch molecules contained in oats for example are smaller than those in corn and barley and are more easily digested by amylase. However, if corn and barley are treated with heat, the nature of the starch molecule is altered and it becomes more easily digested. Therefore, for some grains processing increases their nutritional value.

Fat

If a concentrated source of energy is required, then a fat supplement may be added to the diet. Fat contains three-fold more digestible energy (DE) than oats and 2.5-fold as much DE as corn (maize) (see Table 2).

Fat is a less versatile energy source than starch because it must be oxidized aerobically to produce energy or be stored as body fat. Fatty acids cannot be converted to glucose or synthesized to glycogen. Fat is, however, an extremely useful dietary energy source for horses for a number of reasons. Performance horses in heavy training have a very high daily requirement for DE and may not be able to eat enough grain and roughage to meet it. Adding fat will increase the energy density of the diet so that less feed is required.

Fat supplements are either vegetable oil (e.g. corn oil, canola oil and soybean oil) or high fat content feedstuffs such as sunflower seeds, full-fat soy or rice bran. The 'dry fat' sources are easier to feed than vegetable oil. Never use recycled oil or oil that smells rancid.

How does the horse utilise dietary fat? There is a small percentage of fat in the basic diet of grass (1–4%) and in the diet modified by the addition of oats (4–5%) and hay (1–3%), so fat in the diet of the horse is not unnatural. The horse is better able to digest fat than cattle or sheep and can tolerate up to 20% of the diet as fat, although in practice that quantity is rarely ever fed.

Fat must be digested completely in the small intestine. If too much fat passes into the hindgut, the microbial balance will be upset, leading to digestive disorders and interference with the absorption of some nutrients. Signs of too much fat in the diet include diarrhoea or loose droppings that have a soapy appearance. As with all changes to the horse's diet, the introduction of fat must be done gradually to allow the digestive processes to adapt; add 50 mL of oil every three days, taking two weeks to reach 250 mL/day or if adding high-fat feeds, change the diet slowly over 7–10 days.

Where does the fat go after it has been absorbed from the small intestine? It is taken in the bloodstream to the liver where its future use is decided. If the body needs energy for muscle contraction, then the fat is sent by the bloodstream to the muscle cells where it is further broken down into two carbon units that are used as an energy source for the muscle cell. If the body does not need the energy from the fat at that particular time, then it is stored in the adipose tissue (body fat) distributed throughout the body.

Horses need time to adapt their digestive and metabolic processes to a higher fat diet. The digestive system will take 14–30 days to adapt and a minimum of 30 days is needed before the metabolic processes in the muscles switch to using fat as an energy source in preference to glucose. It takes 3–4 months for the optimum development of the energy production in the muscle, so you will not get the benefit of a high fat diet overnight.

Protein

If the protein intake of a horse exceeds its requirement, the excess can be used as a source of energy. The amino acids are broken down by the liver, the resulting nitrogen is excreted as ammonia and the carbon units can be used to make glucose or fat. However, it is not recommended to increase the protein concentration of the diet to supply energy because protein metabolism results in an increase in the levels of ammonia, which can harm the horse. Simply increasing the feed intake to meet the energy requirements of the horse results in an increase in daily protein intake that will more than adequately meet any additional requirements over and above maintenance levels.

Fibre

Fibre is an energy source that is often overlooked in horse nutrition. The fibre portion of the plant is made up primarily of cellulose, hemicellulose (digestible fibre) and lignin (indigestible fibre). Cellulose and hemicellulose are broken down by the microbial population in the horse's hindgut to volatile fatty acids (VFAs) that are used by the horse as a form of energy. The VFAs are readily absorbed into the bloodstream where they are either transported to sites that need energy or are stored in the form of fat or glycogen. Lignin comprises the structural support of the plant and is not digestible by the microbial flora. As a plant matures, the proportion of lignin increases and consequently the digestibility of the forage decreases, therefore the longer and more mature the grass, the less digestible it is. Rigid stalked plants such as grass have a higher proportion of lignin than leafy plants such as lucerne. As a rule, if there is more leaf than stem in a forage source and if the stems are not stiff and inflexible, the digestibility will be higher. Fresh, green, spring grass will have more digestible fibre per kilogram than dry summer grass, and hay or chaff cut prior to maturity will have more nutritional benefit than hay or chaff that is allowed to mature before cutting. Pasture is also more digestible than hay or chaff because the drying process results in a loss of digestible fibre.

Protein supplements

Protein is important for rebuilding damaged and growing tissues, transporting nutrients in the blood, making blood-clotting factors and a host of other functions. Growing horses and brood mares usually require more protein in the diet than is provided by roughage. Young, growing horses need the additional protein to produce muscle and bone, whereas brood mares need it to either nourish the growing foetus or to produce protein-rich milk during lactation (see later section on 'Feeding specific classes of horses').

Lucerne, soybean meal and canola meal are natural sources of quality protein because they contain the necessary amounts of particular amino acids (especially lysine and methionine). Synthetic sources of lysine and methionine are also available.

Mineral supplements

Minerals are required in different amounts depending on the age and function of the horse.

Calcium and phosphorus

Calcium and phosphorus comprise most of the mineral matter in the horse's body; approximately 80% of the phosphorus and 99% of the calcium are located in the bones and teeth, so they are vital in the formulation of the diet. The horse must receive adequate quantities of calcium and phosphorus and in the correct ratio (see Tables 1 and 2 for the calcium and phosphorus requirements of horses and the amounts contained in various feeds).

In growing horses, the ratio of calcium to phosphorus in the diet must be 1:1. In mature horses, a calcium-to-phosphorus ratio ranging from 1:1 to 6:1 is not harmful if there is adequate phosphorus in the ration. If the amount of phosphorus exceeds that of calcium, the excess phosphorus will interfere with the absorption of the calcium, resulting in serious effects on the skeleton.

During the development of the skeleton, a cartilage base forms first and is then replaced by bone in a process known as ossification. This process requires:

- sufficient amounts of calcium and phosphorus in a digestible form
- the correct ratio between the two minerals
- a small amount of vitamin D.

Milk is the perfect food for building bone, and for the first three months of a horse's life, little improvement can be made to that source of calcium and phosphorus (assuming the mare is providing normal quantities to her foal). What happens after the foal is weaned and throughout its adult life?

Fortunately, many feeds and forages also contain abundant calcium and phosphorus. Legumes, such as alfalfa and clover, are rich in calcium, and grass hays, such as timothy and orchard grass, also contain calcium, but at lower levels than in legume hays. The phosphorus in hay is more readily available to the horse than that found in cereal grains.

Problems arise if the horse is fed large volumes (more than 50% of the total ration) of unfortified cereal grains that are high in phosphorus without the addition of a calcium supplement. A calcium–phosphorus imbalance can lead to conditions such as secondary hyperparathyroidism in which bone is replaced by fibrous connective tissue (e.g. 'big head' disease; see Nutritional diseases).

Deficiency of calcium causes skeletal deformities (rickets) in young horses. In addition, a deficiency or imbalance of calcium or phosphorus can result in osteomalacia (adult rickets) or osteoporosis (thinning of the bone) in older horses. Strenuous exercise (more than 16 km/day) increases the need for these minerals in the diets of mature horses. The bones of all horses are being constantly remodelled to some extent, so the need for some calcium and phosphorus is lifelong.

Although the effect of excessive calcium intake in the horse has not been well established, it seems that high levels in the diet may interfere with the absorption of trace minerals such as iron, zinc, and copper. Additionally, an extremely high calcium intake has been implicated in developmental orthopaedic disorders (DODs), including osteochondrosis (OCD) and epiphysitis (see later section in this chapter).

Providing the correct calcium–phosphorus ratio

1. Weigh each of the constituents of the ration given to the horse each day.
2. Calculate from Table 2 how much calcium and phosphorus is given with each constituent; or
3. Add up the total calcium and phosphorus fed each day.

Finely ground limestone is given as a calcium supplement, and dicalcium phosphate is given when both calcium and phosphate supplements are required: 30 grams (1 ounce) of finely ground limestone (calcium carbonate) provides at least 10 g of calcium; 30 g of dicalcium phosphate provides approximately 7 g of calcium and 6 g of phosphorus.

If the total amount of calcium is less than the amount recommended then add enough finely ground limestone to make up the calcium to the amount required.

If the total amount of phosphorus is less than the amount required, and add enough dicalcium phosphate to make up the phosphorus to the amount required.

Recheck that the ration will meet the recommended requirements for calcium, and if necessary add more finely ground limestone.

As a general rule, if you are feeding mostly grain and oaten chaff then additional calcium, but not phosphorus, will be required. Approximately 60 g of limestone each day is required. Thoroughbred mares in late pregnancy and during lactation should be given 90 g of limestone daily. Horses being fed on dry pasture with hay supplement may require phosphorus to improve the palatability and availability of the calcium in the feed.

Trace minerals

Trace minerals can be supplied in either a mineral supplement or a premixed feed. Always consult the label to determine the contents before supplementing and compare products or ask a nutritionist for advise.

Iron

Iron is the trace mineral most often associated with exercise, even though its actual relevance is questionable. The concern with iron stems from its well-known function as part of the heme molecule that carries oxygen in the blood. The first symptom associated with iron deficiency is anaemia, specifically a hypochromic, microcytic anaemia. In practice, there are few instances when the diet would result in iron-deficiency anaemia.

Clinically significant anaemia in the horse is rare. Exceptions are horses with severe burdens of intestinal parasites, horses with gastric ulcers that lead to blood loss and perhaps in horses suffering from severe exercise induced pulmonary hemorrhage (EIPH). The suggested iron requirement of a 500 kg horse is 500, 600 and 1200 mg/day for light, moderate and heavy exercise, respectively.

Manganese

Manganese is involved in bone formation, oxidative phosphorylation in the mitochondria of the cell, and in fatty acid synthesis and amino acid metabolism either as a coenzyme or as an activator of enzymes. There is very little (<0.2 mg/L) manganese in horse sweat and horses at low levels of exercise require approximately 350 mg/day whereas moderate to high work intensities would require 500 mg/day in the diet. Manganese deficiency is most likely when excess calcium and phosphorus in the diet interfere with its absorption.

Selenium

Strenuous exercise is known to induce oxidative stress, leading to the generation of free radicals, which may induce lipid peroxidation and tissue damage in both the respiratory system and working muscle tissue, particularly if the horse is deficient in vitamin E and/or selenium (antioxidants and scavengers of free radicals).

Selenium deficiency is related to the soil content of selenium and the subsequent effect on pasture, hay and grain. Selenium-deficient soils occur in certain parts of Australia and New Zealand and this can be identified by blood testing the horses on your property.

The selenium requirement for mature horses at rest has been estimated by the National Research Council to be 0.1 mg/kg (NRC, 1989), based on studies that evaluated the relationship between selenium intake and blood selenium. Other authorities have suggested that the appropriate concentration of selenium in the total diet of a horse is 0.3 mg/kg. Therefore, if a concentrate mix comprises 50% of the diet and the forage component of the diet supplies 0.06 mg/kg of selenium, the grain mix would need to supply approximately 0.6 mg/kg. Before supplementing selenium, contact your local agricultural consultant to check the selenium content of the soils in your local area.

Iodine
Iodine is part of the thyroid hormones, thyroxin and triiodothyronine, and a deficiency or toxicity of iodine may result in goitre (an enlarged thyroid gland). In the performance horse, the most important role of thyroxin is controlling metabolism. The thyroid hormones stimulates the mitochondria in the cell to increase their oxygen consumption and thus their energy production. Performance horses in light work require 1.75 mg/day of iodine, in moderate work 2.5 mg/day and in intense training 2.75–3.0 mg/day. Iodine toxicity from overfeeding seaweed meal is more common than iodine deficiency.

Chromium
Chromium is a component of glucose tolerance factor, which is thought to potentiate the action of insulin in chromium-deficient tissue. Insulin promotes glucose uptake by the cell, stimulates amino acid synthesis and inhibits tissue lipase. In athletic humans, chromium excretion is increased and the chromium requirement is increased by physical activity; therefore, a similar situation probably occurs in performance horses.

Copper
Copper is essential for several enzymes involved in the synthesis and maintenance of elastic tissue, mobilization of iron stores, preservation of the integrity of the mitochondria, and detoxification of superoxides. There is a small loss of copper in the sweat of horses (approx. 4 mg/L) and performance horses in light, moderate and heavy work require approximately 130, 170 and 190 mg/day of copper, respectively. Approximately 50–75 mg/day will be supplied by pasture or alternative forage sources, depending on their quality. (See also later section on developmental orthopaedic disease.)

Zinc
Zinc is involved as a cofactor in many enzyme systems. A small amount of zinc lost in sweat (20–21 mg/L). Performance horses require about 400 mg/day during light work and 500 mg/day during moderate and heavy exercise. (See also later section on developmental orthopaedic disease.)

Vitamins and electrolytes
Performance horses require vitamins and electrolytes in the diet to achieve optimum athletic results.

Vitamins
Vitamins are required by the horse in small amounts and are either fat soluble or water soluble.

Fat-soluble vitamins
The fat-soluble vitamins (A, D, E and K) are absorbed with dietary fat and stored in body fat, so they can accumulate in the body if high levels are fed, which in the case of vitamins A and D can cause toxicity. Most horses that are fed good-quality feed or spend some time outside in the open will not require extra A, D, or K supplementation. Check the label of your horse feed and if these vitamins are listed, be careful about adding additional vitamin supplements. Never add more than one vitamin A, D, and K supplement to the feed without the advice of a veterinarian or nutritionist. Fat-soluble vitamins are excreted in the faeces via the bile.

Vitamin A: Carotenes, which are abundant in green forage, are the natural source of vitamin A for the horse. Unfortunately, much of the carotene content is destroyed by oxidation in the process of sun curing, so horses are not able to absorb sufficient quantities of vitamin A from hay to meet their requirement, except possibly from freshly cut good-quality, early bloom alfalfa hay. Performance horses require 43 000–63 000 IU/day of vitamin A.

Vitamin D: Vitamin D is actually a hormone and adequate sunlight results in the production of sufficient vitamin D from 7-dehydrocholesterol in the skin. Hence, vitamin D is not required in the diet if sufficient amounts of sunlight are received. Sufficient vitamin D must be present for calcium and phosphorus to be absorbed and its deficiency markedly reduces absorption of both minerals. Performance horses housed indoors require 4300–6300 IU/day of vitamin D.

Vitamin E: Vitamin E is an antioxidant, protecting cell membranes from peroxidative damage. It differs from the other fat-soluble vitamins by not being toxic when fed in large quantities. High levels of vitamin E supplementation (800–1200 IU/day) may be warranted in horses undergoing strenuous training to help the horse's immune system and prevent some of the muscle damage that occurs in the high-level performance horse. Sources of vitamin E include specific vitamin E supplements, wheatgerm and lucerne. It can also be injected.

Vitamin K: The major function of vitamin K is in blood coagulation. Although no requirement has been established, 20 mg/day of vitamin K can be safely added to the ration of a performance horse.

Water-soluble vitamins

Water-soluble vitamins, such as the B-complex vitamins and ascorbic acid (vitamin C), are not stored in the body to any great extent. These vitamins are either synthesized in the body (vitamin C in the liver from glucose) or produced by the microflora in the large intestine (B vitamins). Because supplemental vitamin C is not absorbed well by the intestines of the horse, high levels have to be fed in order to make an impact on the blood level. Supplementation of vitamin B is probably still warranted for performance horses because they are often given high grain rations that can compromise the microbial fermentation in the hindgut.

The recommended levels of water-soluble vitamins for performance horses are:

- vitamin C: 1000 mg/day
- thiamine: 24 mg/day
- riboflavin: 40 mg/day
- niacin: 120 mg/day
- pyridoxine: 12 mg/day
- pantothenic acid: 24 mg/day
- choline: 600 mg/day
- vitamin B12: 120 ug/d
- folic acid: 12 mg/day.

Brewer's yeast is an excellent source of vitamin B. Biotin is a B vitamin that is given as a supplement to improve hoof growth, usually in combination with zinc and the amino acid methionine (see later section on hoof problems).

Electrolytes

Electrolytes are body salts that are lost in sweat. Horse sweat contains a large quantity of chloride, followed by sodium, potassium and smaller amounts of magnesium and calcium. An electrolyte replacement product must contain electrolytes in similar proportions to those lost in sweat, but should not contain bicarbonate as this is not lost in sweat.

The amount of sweat that is produced by a horse will depend on the duration and intensity of exercise, temperature and humidity. Horses working for one hour at a trot and a canter, or for short periods at high speed can lose more than10 litres of sweat in that time. Therefore, a considerable volume of electrolyte supplementation is needed to replace what has been lost in sweat.

Unfortunately, most of the traditional horse feeds are low in sodium and chloride, so normal feeding will not replace them. The basic electrolyte replacement program is to give your horse

Table 3. Total daily electrolyte requirements (g/day) as a function of sweat loss

	Resting	Sweat loss (litres/day)			
		5 litres	10 litres	25 litres	40 litres
Sodium (Na+)	10.0	26.5	43.0	92.5	142.0
Potassium (K+)	25.0	34.0	43.0	70.0	97.0
Chloride (Cl−)	10.0	40.5	71.0	162.5	254.0

access to a salt block. However, not all horses use a salt block effectively, so you may also need to add some electrolytes to the feed. It is vital to adjust the amount of electrolytes fed according to the sweat loss of the horse. Feed more electrolytes on days when the horse works hard. The amount of electrolyte supplementation recommended by some manufacturers is not sufficient; you need at least 90 grams of a well-formulated electrolyte replacer to adequately replace the losses in a hard-working horse.

Table 3 shows the amount of sodium, chloride and potassium required per day by a horse at rest, and after exercising hard enough to lose either 5, 10, 25, or 40 litres of sweat. The best way to determine this loss is to weigh the horse before and after exercise (before the horse is allowed to drink).

The consequences of inadequate electrolyte replacement are dehydration, overwhelming, failure to sweat (anhydrosis), poor recovery after work and failure to perform to expectations, as well as tying-up and thumps, both of which are serious medical conditions.

Herbal supplements

Many alternative therapies are being used to improve the quality of life and health of horses. Massage, acupuncture, physiotherapy and herbal supplements are all being routinely employed by riders to complement (and sometimes replace) traditional methods of injury management and prevention. The degree of success achieved by any of these methods depends as much on the experience and training of the practitioner as it does on the level of expectation of the horse owners. Many horse owners are looking for a 'quick fix', whereas often the answer lies in a change in the mindset of the owner and a long-term commitment to better training and management of the horse.

Many horse owners erroneously believe that because a product is derived from a plant it is 'natural' and not a drug, and that anything sold as a feed supplement (usually in a bucket or as a liquid), is automatically 'unnatural' and therefore inferior. In truth, most modern drugs are derived from plants; for example, willow bark produces salicylic acid, the base product of aspirin. And of course, green does not automatically imply safety. Oleander is highly toxic to horses and humans alike, and comfrey can be used externally, but is known to produce liver damage in humans if be taken internally.

Another common misconception is that herbal supplements will not be detected in a swab. This is not true. Valerian, commonly used as a calming agent in many herbal preparations, is tested for and will produce a positive drug test on urine analysis. As the testing methods improve, and the use of herbs increases, so the number compounds derived from herbs will be tested for and detected.

Care must be taken when applying the nutritional principles of one species to another. Many of the recommendations made by herbalists for horses are based on research and experience in another species, humans. Such practice is unprofessional and at worst is potentially dangerous. It is always better to purchase products from large, reputable companies that conduct the research into their products on the species for which it is being sold, in this case horses, rather than to trust an individual, no matter how well meaning, who sells herbs in bags bearing neither label nor product details.

Be cautious and diligent about everything that goes into your horse's feed bin, whether a herbal or proprietary supplement. You should only purchase products that meet the following criteria.

- Clearly labelled with the active ingredients or amount of each herb, per daily dose. (Be wary of products that hide the composition details INSIDE the container i.e. you have to purchase the product before you know what is in it!)
- Indication of expected mode of action such as 'assists in reducing mental stress in horses'. Where a direct claim is made, the product should be registered with the appropriate authority.
- Show batch number and or date of manufacture, and expiry date.
- Provide contact details for the manufacturer.
- Contain an accurate measuring device.

It is also important to recognise the form of the product that you are feeding and how that affects its efficacy. Fresh crushed garlic has some wonderful properties, from an immune booster to an in-feed fly repellent, but heating and drying garlic to powder or granules destroys much of its effectiveness.

The following is a very brief summary of some of the herbal products commonly fed to horses. Additional reading of texts on herbs for horses is recommended.

Apple cider vinegar
Not a herb but frequently added to feed in the hope of alleviating joint pain. It is high in potassium and may improve feed palatability.

Dandelion
Claims include action as a mild diuretic

Devil's claw
Reputed to be useful as a painkiller and anti-inflammatory.

Echinacea
Several different types of this plant exist, which may or may not influence its efficacy. Supposedly stimulates the immune system.

Licorice root
Used as an expectorant to loosen mucus.

Valerian
Will be swabbed for in performance horses because of its effectiveness as a 'natural tranquilliser'.

Willow
Used as a painkiller; natural source of salicylic acid – will be swabbed for in performance horses.

Yucca
Believed to have anti-inflammatory properties, although very little research has been done in horses to verify its safety and effectiveness.

SUPPLEMENTARY FEEDS
There is a bewildering array of feeds available to the horse owner and it is often difficult to determine which is best to feed your horse. The following is a guide to the various feedstuffs to help you choose the most appropriate diet for your horse.

Concentrates

Oats

Oats are the most popular and safest grain for horses and can be fed whole except for foals, very old horses and those with poor teeth. They are very palatable and are least likely to cause digestive upsets. Oats are lower in energy and less dense than other grains such as maize or corn, so over-feeding is less likely. They are high in fibre, even higher than bran, which is thought of as a high-fibre feed, and less likely to be contaminated by moulds and mycotoxins. Thus feeding oats has a reduced risk of colic, laminitis and weight gain. However, oats can predispose some horses to tying up and there is a widespread perception that they induce nervousness in some horses. However, oats should be the first grain to be included in the feed ration. Good-quality oats are free from dust and are not dull coloured.

The starch in oats is very digestible and crushing will only marginally increase its availability for horses with normal teeth. As with all grains, oats are low in calcium, but have reasonable levels of phosphorus. They are a poor source of vitamins, and of protein because they are deficient in some essential amino acids.

Maize (or corn)

Maize is becoming more popular as a concentrate feed for horses because it has the highest energy level of all the grains and good quantities of vitamin A. However, maize is deficient in some amino acids and is very low in fibre, so large amounts can cause digestive upsets; it should be no more than 25% of any grain mix and should be omitted or reduced on rest days. Do not directly exchange oats for maize unless you want to substantially increase the energy content of the ration (see earlier guide to substitution for oats).

Unlike oats, the starch in maize is not easily digested and the grain requires processing (e.g. extrusion, expansion or micronising). Cracked maize is susceptible to damage from moisture and will deteriorate very quickly. Do not store any cracked grain for more than 4 weeks to reduce the risk of it becoming contaminated by mycotoxins. Do not feed maize that has a mouldy odour.

Barley

Barley is not as palatable as oats or maize. It has a similar protein value to oats and its energy value is between that of oats and maize. Barley is relatively low in fibre and can cause digestive upsets if not mixed with sufficient roughage. It is a hard grain and should be processed (e.g. steam flaking, micronising, expanding, extruding or boiling) before feeding to increase palatability and starch availability. Boiled barley does not have the magical properties some people attribute to it, but it is palatable.

Wheat

Wheat is not commonly fed to horses because it is high in starch and very low in fibre, which can readily cause digestive upsets and lead to laminitis. It can be used as a high-energy feed if it is cracked or soaked and introduced to the diet gradually. A wheat/rye blend, known as triticale, is more useful and is fed soaked or steam flaked. Bran and pollard, the by-products of flour production, are more generally fed to horses, especially in pellets.

Bran

Bran, a by-product of flour production, is the outer covering of the wheat grain. It is very palatable, but does not contain as much energy as wheat pollard or some grains, has a poor amino acid balance and is low in calcium. Bran should not comprise more than 10% of the feed and contrary to popular opinion, it is not especially high in fibre and is not a good laxative. The very low calcium/phosphorus ratio of bran can affect bone growth if it is fed in large quantities without extra calcium to balance the phosphorus. Bran is not an essential horse feed, but is useful for mixing unpalatable supplements or powders into a feed with molasses. Bran should not be fed if it has a musty odour.

Rice pollard/Rice bran

Rice pollard is high in fat (20%), and therefore energy, and is a useful feed for improving both body and coat condition. It is a much more valuable feed than wheat pollard, but it needs care with storage to prevent it becoming rancid from heat or moisture. Stabilised rice bran is an extruded form that stores well. Rice bran has highly digestible fibre and contains the antioxidant gamma-oryzanol.

Sorghum

Sorghum has a similar feed value to maize, but contains less vitamin A and E and is not as palatable. It is also low in fibre and can cause constipation when fed in large amounts. Sorghum can be fed with roughage and should be heat treated.

Lupins, beans and peas

Peas, beans, and lupins are sources of moderate quality protein, but need to be processed to increase digestibility. They should not be fed as the sole source of protein. However, you do not have any problems with 'fines' that the horse may leave in the feed bin when fed a protein meal. Some types of peas and beans are toxic for horses.

Lupins are a good energy source and can provide up to 25% of the total grain mix for performance horses. They are low in starch, but high in digestible fibre. Lupins are digested predominantly in the hindgut and absorbed as volatile fatty acids, which also reduces the starch/glucose load and the amount of undigested starch in the hindgut, thus contributing to the maintenance of caecal homeostasis. Lupins can be used as a protein supplement for young horses and mares.

Soybean meal

Soybean meal is the best protein supplement available because it has the essential amino acids, including lysine, that are present in low levels only in other grains or roughage. However, soybeans are not particularly palatable and must be processed to remove their natural protein inhibitors. Soybean hulls are a source of very digestible fibre. An adult horse will eat up to 500 g/day, but may not require the quality or quantity of the protein supplied.

Full-fat soybean meal contains 20% oil, so it is higher in energy and better for coat conditioning than normal soybean meal. It is the preferred form to use in performance horses.

Canola meal

Canola meal contains nearly 35% protein, with an adequate level of lysine, and can replace soybean meal as the protein and fat supplement. The protein is less concentrated than in soybeans, so a greater volume must be fed. The usual quantity would be 300–400 g/day.

Cottonseed meal

Although the amount of the protein in cottonseed meal (38%) is greater than in soybean meal, it is low in lysine, so it is not a suitable protein supplement for young or breeding horses. Cottonseed meal also has more fibre. Always check that the toxic compound gossypol has been removed from cottonseed meal during processing, but it will not be a problem if cottonseed meal comprises less than 10% of the ration. Cottonseed meal also contains an inhibitor of vitamin E, so do not mix it with Vitamin E supplements. It is best fed boiled.

Sunflower seeds/meal

Sunflower seeds are very palatable and high in fat and energy, so they are a good coat conditioner. Although they contain approximately 20% protein, they are relatively low in lysine, so should be considered as a fat and energy supplement rather than a protein supplement.

Linseed meal

Linseed meal is very popular as a protein supplement for horses because it contains approximately 35% protein; however, it has a poor lysine content. The main reason for its popularity is its coat conditioning ability, which comes from its fatty acid content. Linseed may also have a laxative effect because it is high in fibre. Linseed is traditionally fed boiled as a mash and must not be fed raw because large amounts will be toxic.

Mixed feeds

Mixed feeds contain grains, protein, chaff, minerals, vitamins and molasses. The aim is to provide a complete balanced feed and make the feeding process very simple and convenient. Some mixed feeds contain very low levels of minerals and vitamins, so you may still need to add these, but well-formulated feeds take the guesswork out of feeding. It is important to read the label and select a feed that is appropriate for your horse. Choose a mixed feed from a reputable company that lists the energy level, as well as the nutritive contents, on the label.

Pellets

Pellets are another simple feeding method that reduces wastage and the time spent mixing feeds. They do not contain the 'sweepings off the floor' and they can be fed to racehorses. Pelleted feeds incorporate all the proteins, minerals and vitamins needed to create a balanced ration and most are designed to be fed with hay or chaff. Some pellets can crumble and be dusty when handled in bulk, and you must check that they have sufficient fibre if you are not feeding roughage. Always check the ingredient composition. In particular, be aware of 'cool' pellets that may only contain bran and pollard.

Roughage

Hay and chaff are roughage. Compared with the concentrates, they are low in energy and high in fibre; that is, they add bulk to the diet without adding calories. The fibre is not easily digested and its main purpose is to maintain the normal function of the digestive tract when the horse is being fed a lot of grain. Chaff is chopped hay, so both feeds have a similar nutritional value although there is great variation in the quality and nutritional value of a particular type of roughage. Feeding chaff is popular in England, New Zealand, South Africa, South-east Asia and Australia, but is not a common feed elsewhere. Chaff can be mixed with the other feeds, which prevents a horse from 'bolting' the grain portion of the ration. Molasses is often added to minimise the dust in chaff and some chaff may also have oil added to increase the energy value. There is less wastage with chaff, but it is easy to overestimate the amount being fed because it is very light. A big bucket of chaff may contain less nutrition than a 'biscuit' of hay. Chaff is more expensive than hay, so it is cheaper and easier to feed hay to most horses. The long-stem fibre in hay is also useful for maintaining the proper digestive function of the horse. However, a study done by Kentucky Equine Research found that when long-stem hay and grain were fed together, the rate of passage of the feed increased so that the grain did not stay long enough in the small intestine to be adequately digested. It is not yet known whether the same is true of chaff, but the 'diluting' of grain with forage rather than the horse eating the grain and then the hay probably makes a big difference.

Hay has a lower nutritive value than the pasture from which it is made, but a horse has to eat more of the pasture to get the same amount of nutrients because pasture contains more water. The nutritional value of hay varies greatly according to its quality. Good-quality hay is not excessively weathered, is leafy and has minimal stems. It contains vitamins A, D and E, although these are destroyed after prolonged storage. Mouldy hay can cause colic and respiratory problems.

Lucerne hay/chaff

Lucerne hay or chaff is the most nutritious roughage available, containing the highest amounts of energy and protein, in fact, it contains more protein than any grain and it is high quality. Lucerne hay or chaff is also particularly high in calcium, so it helps to balance a high grain diet.

Lucerne may cause some horses to have loose droppings and will change the characteristics of the urine, but it does not cause most of the problems that are attributed to it.

Wheaten or oaten hay/chaff
Wheaten or oaten hays and chaffs have similar feed values. They do not contain as much protein or calcium as lucerne, but are higher in fibre and just as palatable. Their energy value depends on the amount of grain they contain. A good sample will have a higher energy value than lucerne. As with lucerne, you can mix the chaff with grain to slow down the intake of the concentrates.

Grass and meadow hay
Grass hay varies considerably in quality, particularly in relation to such factors as the pasture species and the time of cutting. The more clover a pasture contains, the greater the nutritive value of the hay. The earlier the hay is cut, the higher its levels of protein, energy and minerals. Good-quality clover hay has a similar nutrient value to lucerne hay and is much more palatable, but it is often hard to obtain.

Other supplements
Molasses
Molasses is a high-energy feed that is usually fed to improve palatability of the ration or incorporate powder supplements into the feed.

Salt
Salt can be added as coarse salt, rock salt or iodised salt. Alternatively, a salt lick can be placed in the feed bin or paddock.

FEED PROCESSING
There is an incorrect impression among horse people that any type of processing of feed is unwarranted; however, feed processing methods have been researched and developed to help horses, not harm them. The practice of baling and preserving hay is a form of processing. Understanding why feeds are processed and the methods used to process the different types of feed will ensure that you can make an informed decision about supplementary feeding.

Why feeds are processed
There is not one simple reason for processing feeds, but there is one common goal – to make the feed better for either the horse or the owner. Processing will improve the digestibility of a feedstuff, extend its shelf life or enable by-products of the human food industry, such as wheat middlings (bran, germ, and all that is left of the wheat grain after removal of the flour), soybean meal and hulls, rice bran and sugar beet pulp, to be used as horse feed.

Processing may improve the convenience of a product; for example, processing concentrates enables the manufacturer to make a consistent product that is safer, easier to chew, and more palatable. Processed feeds may also be easier for the horse owner; for example, hay cubes or pellets make the most efficient use of limited storage space, are easier to carry when travelling, guarantee a consistent intake of nutrients, simplify the balancing of the feed ration and can help the owner feed horses with problems such as poor teeth or respiratory tract disorders.

Grains are processed to increase the availability of the starches to the horse as an energy source. Crushing, cracking and grinding, particularly of corn and lupins, break down the hard outer shell of the grains and reduce the size of the seeds, which aids the horse in chewing the feed. Crimping, flaking, and rolling of oats and barley achieve the same purpose for these grains. Further processing entails adding heat to these processes to produce steam-rolled barley, steam-flaked or super-flaked corn, and micronising of any grain.

Sun curing and dehydrating (quick drying) of hays are processing methods designed to reduce the moisture content of the roughage so it will not develop mould. Chopping the forage to shorten the fibre length (i.e. chaff) is another form of processing. Although not commonly used in horse feed manufacturing in Australia, ensiling or anaerobic fermentation of forage to preserve its nutrients is another form of processing that produces silage and haylage.

Types of processing

Feed processing is least complex when grains are simply mixed and coated with molasses to create a 'sweet-feed'. In more elaborate processing, such as pelleting or extruding, the actual form of a mixture of feed ingredients is changed. The ingredients are ground to improve digestion, decrease segregation and mixing problems, and facilitate the pelleting or extruding process.

Pelleting

In the pelleting process, the feed ingredients are ground to the same particle size, mixed with a binder and then steam heated to 82–87°C for approximately 20 seconds. Next, the mash is pushed through a pellet die of the desired size, cooled, and dried to prevent mould growth. With forage pellets, the forage is dehydrated, ground, mixed with a binder, and then pressed through the die. The size of the die ranges from 2 mm to 200 mm or greater, depending on the purpose of the pellet.

Pelleting causes gelatinisation of the grain starches, making them more available for enzymatic digestion in the horse's gut. This is particularly beneficial for pellets made with substantial amounts of ground corn. If only partial gelatinisation occurs, the outside of the pellet will look very shiny, which is undesirable and it should be rejected.

Pelleting increases feed digestibility less than 5% for horses with normal teeth, but there are other advantages that outweigh the lack of digestibility benefit. Because all the ingredients of the feed are ground to the same size and each pellet is the same size, it removes the problem of the horse sorting through the feed mix and leaving the less palatable ingredients in the trough. Pellets are convenient to handle in any weather compared with sweet-feeds, which will solidify in cold weather.

It costs more to manufacture a high-quality sweet-feed than a high-quality pellet because pellets can be made from by-products. The whole grains used in sweet-feeds can be expensive, particularly oats, and the price fluctuates from crop to crop.

Some horse people question the effects of heat treatment on the viability of vitamins and minerals added to the pellet. Generally, heat will not affect the minerals because they are inorganic and the vitamins have special gel coatings to protect them from oxidation and short bursts of heat. However, if the pelleting process is not done correctly and the pellets become too hot, the availability of vitamins and possibly chelated minerals may decrease.

Fine ingredients are usually combined into a mixing pellet and then used in sweet-feeds. The mixing pellet usually contains the protein sources and the vitamin and mineral premix. The mix blends better if all the ingredients are a similar size. Manufacturers of grain mixes sometimes have problems with the fine ingredients sifting to the bottom of feed batches, so the mix may differ in each bag. Using molasses, in combination with the mixing pellets, has resolved all problems associated with fines in sweet-feeds

A good-quality pellet is relatively long and hard (except for pellets used in special senior concentrates) with good colour, uniformity, a dull surface, and few to no fines. When a pellet can be handled repeatedly without falling apart, it is considered durable, which is an important factor.

Extruding

Extrusion is the process used to make dry foods for cats and dogs, and is now used by some manufacturers of popular horse feeds. Extruded nuggets are made from many of the same ingredients as pelleted feeds. First, the grains are ground and mixed with all other ingredients, and then they are cooked with moist heat at approximately 127°C. The gelatinisation of starches is more extensive during extrusion than during pelleting. The extrusion occurs when the mash is exposed to cooler air, and it begins to expand and pop, like a kernel of corn. The nuggets are dried to a moisture content of approximately 10% before being bagged.

Many of the advantages of pelleting also hold true for extruded feeds, with one additional advantage of being able to add high levels of fat. If there is oil in the mash and oil is sprayed on the outside of the nugget, fat levels can be as high as 20%. Because of the high temperatures used during processing, preservatives are not always necessary (except with high-fat nuggets), which is another benefit. The high heat can affect the vitamin levels, but the manufacturers add extra to compensate. Extruded feeds are more digestible than whole grains, sweet-feeds or pelleted feeds. The physical form of the feed does slow down the intake of the feed, which can be advantageous for a horse that eats its feed too quickly, but it can be a problem if a horse needs a large quantity of feed. Extruded feeds are very low in fines and dust, which makes them ideal for horses with respiratory problems.

The disadvantages of extruded feeds are that they are more expensive to manufacture and are bulky to store. Palatability can be an issue because of the unfamiliar form. Introduction of extruded feeds either at an early age or very gradually to the diet will encourage horses to eat this form of food.

Steam rolling, steam flaking and micronisation

These processing methods for grains such as oats, barley and corn involve the application of heat by steam or infra red heat (micronisation), followed by rolling or flaking. The aim is to improve gelatinisation of the starches and hence palatability, promoting starch digestion in the small intestine and thus make grain feeding safer and more effective. The degree of gelatinisation will vary according to the temperature and time involved. The processing will also remove dust from the feed.

Processed forages

Haymaking

Haymaking is a way of processing forage so that it can be preserved for longer periods of time and be fed when fresh forage is unavailable. Drying and baling hay can be considered processing because the original product is not the same as the resulting product. When fresh forage is cut and dried, protein, carbohydrate and vitamins are irreplaceably lost; for example, if the grass has a crude protein value of 22% (dry matter basis), it will typically only have 12% crude protein after harvesting as hay. Within the first 24 hours after cutting, as much as 80% of the vitamin A (in the form of carotenes) can be lost. If the drying process is extended or the cut grass is rained on, the nutrient losses are much greater. Regardless, hay is still the number one choice of roughage for the horse when fresh forage is not available.

Chaff and chopped forage

Chaff is fed to horses in England, New Zealand, South Africa, South-East Asia and Australia but it is not a common feed elsewhere in the world. Chaff is simply chopped up forage, often with added molasses to minimise dust or oil to increase the energy value.

Traditionally, chaff was a method of feeding poor-quality forage such as straw in a form that was appealing to the horse. The length of the chop varies from 1 cm to 3 cm, depending on where in the world it is made.

Chaff adds bulk to a diet without providing too many calories, and keeps the stabled horse busy eating. Chaff can be mixed into the ration to slow down the feed intake by horses that chronically choke as a result of bolting their food; however, it may not be a good idea for the average horse. Kentucky Equine Research found that when long stem hay and grain were fed at the same time, the rate of feed passage increased, and the horse did not have adequate time to digest the grain in the small intestine. It is not yet known whether the same is true of a diet mixed with chaff, but the 'diluting' of grain with forage rather than the situation where the horse eats the grain and then dives into the hay probably makes a big difference.

Cubes
Another method of processing fresh forage is to compress it into cubes. The hay is dried and chopped before being pressed into a cube with a binder to hold it together. The binders are usually natural materials such as bentonite clay, which is not harmful to horses. Cubes are convenient to handle and there is less waste than with conventional hay bales. Cubes usually contain less dust, so are suitable for horses with respiratory disorders. However, it is very difficult to assess the quality of the forage used in the cube and whether there are any contaminants, so always purchase from a reputable manufacturer. Also, because the forage is partially broken down and therefore takes less chewing, a horse may finish a meal more quickly than usual and without eating to occupy its time, may develop inappropriate behaviours to combat boredom (see Chapter 2). A general recommendation is to feed cubes or pellets in conjunction with pasture or hay, except when none is available.

Haylage
In some countries it is common for horses to be fed hay that has been ensiled and marketed as 'haylage'. The forage is sealed in airtight containers at a high moisture content and allowed to ferment, which retains the nutrients better than sun curing, particularly protein, carbohydrates and many vitamins. Moreover, the higher moisture content keeps the forage practically dust-free, making it ideal for horses with respiratory problems or horses that are sensitive to dusts and moulds. However, there is a danger in feeding haylage. Because of the moist, airtight environment, there is a chance that the botulism bacteria (*Clostridium botulinum*) can grow if the haylage is incorrectly baled or stored. Any horse receiving haylage in the diet should be vaccinated against botulism. Discard any sour-smelling, burnt-looking or slimy silage.

Beet pulp
Beet pulp is a novel source of forage that is commonly fed to horses in Europe and the USA. It is a by-product of sugar production from sugar beets and makes an excellent addition to the horse's diet when other forage is poor quality or the horse will not enough hay. After the sugar is extracted from the sugar beet, the remaining fibrous portion is dehydrated for packaging and mould prevention. The fibre is very digestible, making it an excellent energy source. There is some concern that the amount of sugar in beet pulp is a problem for the horse that is sensitive to sugar (i.e. needs a low sugar/low starch diet), but in fact there is very little residual sugar after the extraction process, although molasses may be sprayed on after dehydration to reduce the dust. Any residual sugar can be removed by rinsing the beet pulp a few times before feeding. Beet pulp is not currently available in Australia as a pure or pre-mixed feed ingredient, but it is available in New Zealand.

PROBLEMS WITH WEIGHT
Monitoring weight and body condition
Accurate estimation of a horse's weight is necessary for dietary management, monitoring growth, and correct dosage of drugs. Scales suitable for weighing horses should be used on a regular basis or you can use methods of weight estimation based on the horse's body measurements and condition score.

Condition scoring
Scoring the body condition ('fatness') gives you an indication of the suitability of the horse's diet, enabling you to make adjustments before problems arise (Table 4, Figure 4).

1. Assess visually, and by feel, the horse's pelvis, rump, back, ribs and neck.
2. Give those areas individual scores. Intermediate assessments can be given half scores.
3. Using the pelvic and rump assessments as the base, adjust that score by a half point if it differs by one or more points from the neck or ribs score.

You can use half scores for intermediate grades.

Weight estimation

Using the height and condition score

A horse's weight can be estimated from its height (in hands) and condition score (Table 5), but is more accurate if the measurements are used with a nomogram (Figure 5.) Height measurement should be performed on level ground when the horse is relaxed and standing squarely. Use the highest point of the withers as the measuring point (see Chapter 4, Figure 36).

Using the girth and length measurements

The most accurate method of weight estimation uses the measurements (in centimetres) for girth and length (Figure 6) with the nomogram shown in Figure 7.

Alternatively, the weight can be calculated from the girth (G) and length (L) measurements using the formula:

$$\text{weight (kg)} = G^2 \times L \text{ (cm)}/12\,000$$

Weight loss

Insufficient intake of calories (energy) is the usual cause of weight loss in horses. Many of the reasons for caloric deficiency are easy to diagnose and correct (e.g. diet, parasite burden, teeth problems), but others may be impossible to diagnose while the horse is still alive (e.g. anatomical problems with the digestive tract). Other reasons that should not be overlooked include psychological and environmental problems.

Causes of weight loss

Teeth

The first thing that should be checked in a horse that fails to maintain weight on an appropriate diet is the condition of its teeth (see Chapter 2).

Parasites

A heavy burden of internal parasites can contribute to weight loss or an inability to put on weight. However, severe cases of parasitism are now rare with modern worming treatments. All horse should be on a regular program of parasite control (see Chapter 6)

Digestive tract problems

Weight loss may be caused by a physiological problem or a disease that prevents food from reaching the intestines for digestion. Teeth problems, obstruction of the throat from strangles or abscesses, muscle weakness because of hyperkalemic periodic paralysis (in Quarterhorses) or botulism, or blockage of the oesophagus, such as with abnormal growths, scar tissue from past episodes of choking or by a foreign body, can seriously alter a horse's eating habits and reduce appetite and feed intake. The diagnosis of the cause of the difficulty in swallowing is made by examination with an endoscope or an X-ray. If the obstruction cannot be removed, changes to the diet should be made to allow the horse to swallow more easily.

Gastric ulcers can reduce the appetite because of the pain and discomfort in the stomach (see later).

The physical conditions in the small intestine, large intestine and caecum can affect the horse's ability to digest and utilise food. If nutrients move too quickly through the digestive system, as

Table 4. Body condition scoring system

	Neck	Back and ribs	Pelvis and rump
0 Very poor	Marked ewe neck	Skin tight over ribs	Angular pelvis – skin tight
	Narrow and slack at base	Spinour processes sharp and easily seen	Deep cavity under tail and either side of croup
1 Poor	Ewe neck	Ribs easily visible	Rump sunken, but skin supple
	Narrow and slack at base	Skin sunken either side of backbone	Pelvis and croup well defined
		Spinour processes well defined	Deep depression under tail
2 Moderate	Narrow but firm	Ribs just visible	Rump flat either side of backbone
		Backbone well covered	Croup well defined, some fat
		Spinous processes felt	Slight cavity under tail
3 Good	No crest (except stallions)	Ribs just covered	Covered by fat and rounded
	Firm neck	No gutter along back	No gutter
		Spinous processes covered but can be felt	Pelvis easily felt
4 Fat	Slight crest	Ribs well covered – need firm pressure to feel	Gutter to root of tail
		Gutter along backbone	Pelvis covered by soft fat – felt only with firm pressure
5 Very fat	Marked crest	Ribs buried – cannot be felt	Deep gutter to root of tail
	Very wide and firm	Deep gutter	Skin is distended
	Folds of fat	Back broad and flat	Pelvis buried – cannot be felt

Table 5. Prediction of weight from height and condition score

Condition score	Condition score Height (hands)				
	12h	13h	14h	15h	16h
1	190	240	310	390	420
2	210	285	330	420	470
3	250	345	395	460	505
4	300	370	460	535	570
5	360	460	540	610	670

with diarrhoea, they cannot be absorbed. Changes in the balance of the bacteria in the hindgut will cause diarrhoea, because if the microbial population is not functioning correctly, particles of food are not broken down for absorption, leading to a diarrhoea. Probiotic supplements may assist in the treatment of such diarrhoea by repopulating the hindgut with species of 'good' bacteria. Some people use yoghurt in the horse's feed as a natural probiotic and there are many commercially available Lactobacillus and Streptococcus faecalis preparations. Sometimes when there is no apparent reason for a horse to be losing weight, a probiotic supplement can help. They are also useful when horses are stressed after travelling or competition or after an illness requiring antibiotics.

Disease
Disease can interfere with the horse's ability to eat (e.g. strangles) or cause a disturbance in protein utilisation (e.g. chronic liver disease). Muscle wasting may be the first noticeable indicator of a chronic disease interfering with protein digestion because if the horse is not getting enough protein, it begins to metabolise muscle tissue, which is the biggest reserve of protein in the body.

(a) 0 = very poor.
*Very sunken rump; *Deep cavity under tail; *Skin tight over bones; *Very prominent backbone and pelvis; *Marked ewe neck

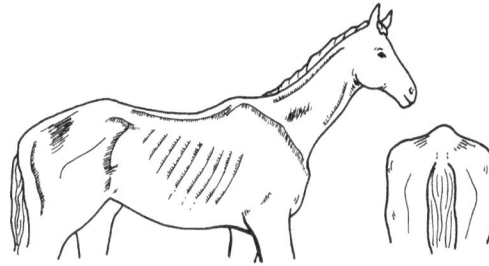

(b) 1 = poor.
*Sunken rump; *Cavity under tail; *Ribs easily visible; *Prominent backbone and croup; *Ewe neck, narrow and slack

(c) 2 = moderate.
*Flat rump either side of backbone; *Ribs just visible; *Narrow but firm neck; *Backbone well covered

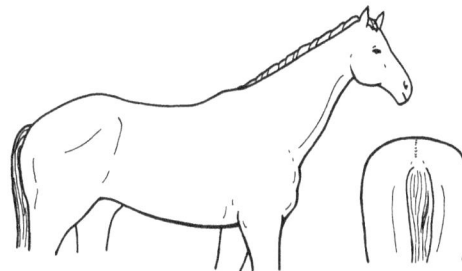

(d) 3 = good.
*Rounded rump; *Ribs just covered but easily felt; *No crest, firm neck

(e) 4 = fat.
*Well-rounded rump; *Gutter along back; *Ribs and pelvis hard to feel; *Slight crest

(f) 5 = very fat.
*Very bulging rump; *Deep gutter along back; *Ribs buried; *Marked crest; *Folds and lumps of fat

Figure 4. Condition scores.

After absorption into the blood, dietary protein and fat go to the liver first and then to where they are needed in the body. If the liver is not functioning correctly, it affects many other body systems and results in weight loss. Liver function can be assessed by a simple blood test.

If the kidneys are malfunctioning, a large amount of dietary protein can be lost in the urine. Horses with kidney problems will usually drink large volumes of water and urinate frequently. Kidney function can also be assessed by a simple blood test.

Internal abscesses or tumours will cause an abnormal increase in the amount of energy required for normal body processes, resulting in chronic weight loss. Horses with chronic obstructive pulmonary disease use more calories because of the increased physical effort required to breathe. A pituitary tumour may speed up the metabolism, causing the horse to use energy even at rest. A heart problem can also cause weight loss through disruption of the blood flow carrying vital nutrients through the body.

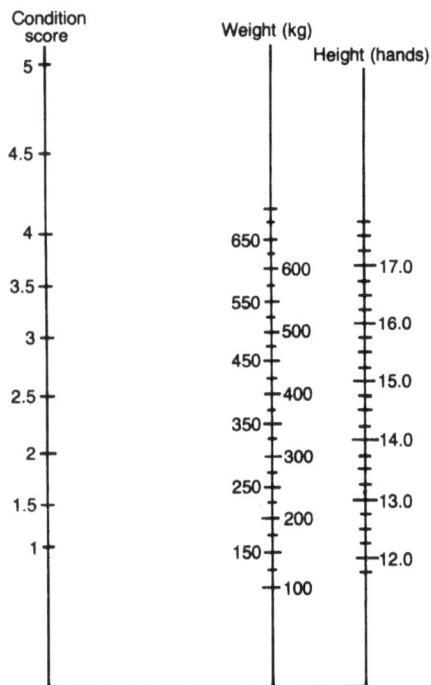

Figure 5. Nomogram for estimation of live weight from condition score and height measurement. A ruler is used to connect the appropriate values of the condition score and height scales, and the weight is read where the ruler intersects the weight scale.

Figure 6. Measurements of girth and length.

Chronic pain is often overlooked as a cause of weight loss in horses. Discomfort not only reduces appetite, but also causes the body to release adrenaline (epinephrine), which puts the body in a state of catabolism (i.e. the break down of energy stores). The presence and cause of low-grade chronic pain is often hard to diagnose.

Environment

If a horse is bored and unhappy in its physical environment, it may lose its appetite and develop inappropriate behaviours such as cribbing, weaving or stall walking, thereby wasting energy and losing condition (see Chapter 2 for management of stereotypies).

Climate can also affect appetite. Horses may be reluctant to eat in very hot weather and will not eat feed that has gone bad after being stored in hot conditions. Working horses use more energy in the hot weather and their energy intake should be adjusted accordingly to prevent loss of condition

If your horse is low down in the social hierarchy of a herd, it may be chased away from the food by the herd leaders. Always space the feed bins for a group of horses and place some extra servings of hay or grain in the paddock to ensure that all horses get a share.

Feeding to gain weight

Horses that have lost weight slowly over a period of time by breaking down fat and muscle stores will look emaciated with poor muscle definition and protruding bones. A thin horse requires enough energy in its diet not only to maintain proper body functions, but also to increase its muscle and fat stores until it has regained the desired condition score. Feeding 2.5% of the current body weight of a thin horse (i.e. maintenance) will only succeed in keeping its weight static; to increase weight and condition, you must set a goal and feed at a rate corresponding to the desired body weight (condition score).

Increase the caloric content of the diet and ensure adequate protein content by increasing the daily intake of the three major ingredients of the horse's diet: fibre, starch and fat.

Fibre

Fibre is the safest constituent of the diet to increase when a horse needs to gain weight. Fibre alone will not be sufficient, but maximising the quality of forage should always be the first adjustment when trying to achieve weight gain. Adding clover hay to the roughage increases the energy and protein content and because horses generally prefer it to other types of hay, it is particularly good for poor doers or picky eaters.

If good-quality forage in the form of chaff, hay or pasture is not available, or the horse does not readily eat hay, use alternative sources of fibre such as lupins, soybean hulls, wheat bran and lucerne chaff or pellets, and in New Zealand you can use beet pulp. Wheat bran is commonly considered as high in fibre, but is actually equivalent to oats. It does have a relatively high proportion of digestible fibre and starch, but because of its high phosphorus content, bran should be used as a supplement to a diet high in lucerne hay or chaff.

Lucerne pellets can be used to supplement average quality chaff or hay. Lucerne cut for pellets is harvested when the digestible fibre of the plant is at its peak. It is always advisable to feed some hay or chaff with the pellets to maintain gut fill and reduce the risk of colic.

Yeast improves fibre digestibility. Some commercial feeds have added yeast, otherwise yeast supplements can be purchased. Some probiotics are also thought to improve digestibility and combinations of yeast and probiotics are also available.

Starch

When a horse cannot maintain or gain weight on roughage alone, grains are added to increase the energy density of the diet. It is tempting to try and increase a horse's weight by simply throwing more grain in the bucket, but there are inherent dangers of feeding excessive amounts of starch (see earlier section).

Stomach ulcers may also develop with a high grain diet, further reducing the absorptive capacity of the gut and causing more discomfort to the horse, which reduces its appetite even further. If the horse loses its appetite for forage, it will lose weight no matter how much grain you feed.

There are supplements that may aid starch digestion. Supplemental chromium may improve the metabolism of starch. Adding enzymes to the diet has not been extensively researched but is based on sound theory. If amylase is the limiting factor of starch digestion in the small intestine, then adding it to the diet may reduce the amount of feed passing through to the hindgut. Although there are a few commercial feeds and supplements that contain enzymes, their efficacy is questionable because enzymes are proteins that are very sensitive to temperature and pH fluctuations, and are denatured by heat or the highly acidic environment of the stomach.

Cool conditioning

One problem of increasing the grain content of the diet is that the horse's behaviour may change. This effect can be reduced by feeding the grain in smaller amounts or by adding fat to the diet. There are commercial feeds that are specifically designed for cool conditioning.

What makes a healthy horse a poor doer?

Metabolism is the speed at which the body uses energy in order to perform normal body functions. A slow metabolism uses very little energy to perform these functions whereas a fast metabolism uses substantially more energy and a higher caloric intake is required to maintain the same weight as the horse with the slow metabolism. In general, metabolism is breed specific, the terms 'hot blooded' and 'cold blooded' are generally indicative of a fast and slow metabolism respectively; for example, Thoroughbreds need to consume more calories per kilogram of body weight than draught horses. There are also variations within breeds; some Thoroughbreds maintain body condition quite easily, whereas others are more challenging when it comes to maintaining weight. Temperament can play a big part in a horse's ability to maintain weight and often goes hand in hand with metabolic rate: nervous or tense horses tend to use more energy than calm ones.

Weight gain

Overfeeding is common in the horse industry because fat horses are more attractive than skinny ones. The obesity of overfeeding can be remedied with one goal: energy expenditure must be greater than energy intake. In black and white, that's diet and exercise.

Feeding to lose weight

- Limit grazing time, particularly during rapid grass growth (such as the spring and early summer).
- Eliminate high-calorie concentrates from the diet. If the quality of the forage is low, supplement with low-intake, low-calorie vitamin and mineral pellets to ensure all critical nutrient requirements are met.

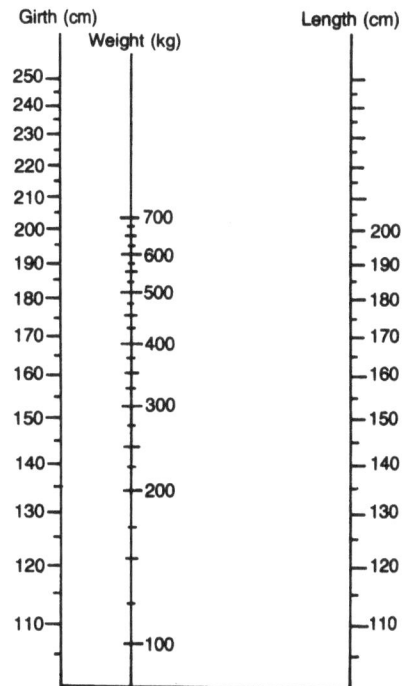

Figure 7. Nomogram for estimation of live weight using girth and length measurements. A ruler is used to connect the appropriate values on the girth and length scales, and the weight is read where the ruler intersects the weight scale.

- Replace legume hay with a grass or cereal grain hay. Legume hays, such as lucerne or clover, contain more calories than grass hays. Feed a high-fibre, low-calorie, long-stem grass hay free of dust, mould and weeds.
- Exclude all high-fat feed additives or supplements for coat conditioning. Instead, use a biotin supplement and a daily grooming regimen.
- Implement an exercise program to increase energy expenditure. To maximize energy output while minimising skeletal stress, exercise should be low intensity and long duration. There are other benefits of daily exercise, including a possible reduction in appetite, and prevention of bone and mineral losses that may occur during calorie restriction.

Consequences of dieting

If a horse is eating less, it may develop gastric ulcers because of increased exposure of the stomach wall to acid. Boredom because of decreased time spent eating may lead to bad behaviours, particularly if the horse is stabled for prolonged periods (see Chapter 2).

Managing a good doer

A common problem of good doers is their tendency to quickly become obese. Increased sweating and the inability to cool down quickly because of excessive body fat acting as insulation decreases physical performance. Obese horses may also experience respiratory difficulties because the increased fat tissue increases the oxygen requirement, beyond what the oxygen intake can supply. Extra weight also predisposes the horse to joint and locomotion problems. Obese horses are also more prone to lipomas, fatty tumours that may develop in the abdominal cavity and strangulate the intestine, precipitating an episode of colic.

The reduction in feed intake required to slim a 'good doer' may seem drastic. You may need to lock it up in a yard or in a paddock with minimal grass, but the metabolism of these types of horses is such that you can keep them locked up for ages and they will not lose much weight, so exercise is very important. The horse needs to be lunged or ridden at a trot or a canter so that it raises a sweat. Be careful with ponies that you do not induce hyperlipaemia in a slimming program (see later section 'Nutritional diseases').

An abundance of pasture in addition to water and a salt block (particularly in the summer) is a sufficient and natural diet for most mature horses. However, horses with increased nutrient demands, such as broodmares, performance horses and young, growing horses, are usually given supplementary feeds, but this may not seem necessary if a horse is a good doer. Does this mean a mare that maintains weight easily does not require additional nutrients during pregnancy? No; the key to managing the nutritional needs of a good doer is the same as managing any other horse: ensure its nutritional requirements are being met by analysing the nutritional value of the forage and balancing the nutrients that are absent with low-intake, low-calorie supplements.

Good doers on an all-forage diet often do not consume sufficient quantities of vitamins and minerals, partly because of the low-quality hay that is fed to overweight horses. The vitamin and mineral content of hay is best determined by chemical analysis, but a non-laboratory method is visual inspection. Simply put, the greener the hay, the higher the nutrient level.

FEED REQUIREMENTS OF DIFFERENT CLASSES OF HORSES

Adult horses

An average adult horse needs to eat 1.5% of its body weight (in dry matter) per day for maintenance. Light, intermittent work increases this requirement up to 2%, most of which will be provided by pasture or roughage. Supplementary feeding of concentrates should be no more than 0.5% of the horse's body weight unless it is performing moderate or hard work.

Pasture

Pasture is the cheapest and most convenient feed for horses. The more good-quality pasture you have available, the less supplementary feed you will need to give your horse (see Chapter 8 for information on pasture maintenance).

The quantity and quality of pasture varies throughout the year and between years. In southern Australia there is a peak of high-quality pasture in autumn and spring, with a shortage of supply in winter, although the quality is good. In the summer, pasture deteriorates in quality and the amount available declines rapidly. In northern Australia pasture quality peaks during summer and declines during autumn as the plants mature. As a result, both the quantity and quality of northern pastures can be lacking during winter and spring. There may be variations because of seasonal rainfall, soil type and pasture species, which means there are certain times in the year when some horses may need supplementary feeding.

Supplementing pasture

Horses kept in a large paddock and ridden only occasionally rarely need supplementary feeding, whereas horses kept on small areas of land may need it regularly. A daily ration of 2 kg of mixed feed, and one-fifth to one-quarter of a bale of pasture hay is a suitable ration for a 500 kg horse kept on a small suburban block and ridden occasionally on weekends.

Horses kept in a large paddock and ridden frequently may need supplementary feeding during winter and summer when less pasture is available. Use the body condition of your horse as your guide. Providing extra hay is the best way to feed a lightly worked horse. Lucerne or clover-rich hay, which is high in protein, is an adequate supplement to dry summer pastures. Good-quality oaten or grass hay, which is high in energy, is an alternative supplement to the short, green winter pastures. Do not overfeed your horse or you may have problems with behaviour and obesity, as well as extra expense.

The most important supplement for horses at rest is salt, which can be provided as a block or sprinkled on the feed as rock salt or trace mineralised salt.

Performance horses

If a working horse has been bred for the job, is well managed, in good health and lives in a suitable environment, nutrition will be the main factor influencing its performance. Energy is the most important part of the diet of a working horse, with the amount required depending on the level of work. A horse in light work requires 1.5–2% of its body weight (dry matter basis) each day, moderate work increases the demand to 2–2.5%, and heavy work to 3% or more. Working horses fed rations that are high in grains should be fed 50–70% less grain on the days when they are not being worked to avoid disorders such as tying up or laminitis, both of which can be fatal (see later). Pasture can provide some of the roughage intake.

Table 6 lists the approximate amounts of roughage and concentrate required for fully handfed horses doing varying amounts of work and Table 1 on page 122 shows the major nutrient requirements, from which you can devise a ration. There are several ration evaluation computer programs, such as KER Microsteed®, which simplify the job of devising a suitable diet. Regular feed intake (i.e. 3–4 meals per day) is essential as loss of appetite can substantially decrease energy or mineral intake.

Nutrient requirements

A mature horse weighing 500 kg needs 68 megajoules (MJ) of energy per day just for maintenance. Each hour of work requires additional energy.

- Walking   1.05 MJ/hour
- Slow trotting, some cantering   10.65 MJ/hour
- Fast trotting, galloping, jumping   26.15 MJ/hour
- Cantering, galloping, jumping   50.15 MJ/hour
- Polo practice and games (strenuous effort)   81.5 MJ/hour

Table 6. Roughage and concentrate requirements (on a dry matter basis) of fully handfed horses

Type of work	Feed per 100 kg live weight	
	Roughage (kg)	Concentrate (kg)
Idle	1.5	Nil
Light (2 hours/day)	1.25–1.5	0.5
Medium (2 hours/day)	1.25–1.75	1.0
Heavy (4 hours/day or racing)	1.0–1.5	1.0–2.0

Digestible energy (DE) is provided by starch, fat, protein and fibre (Table 2). Extra energy demand is usually met by increasing the amount of grain in the ration, but may not always be appropriate. The ration of a performance horse should include a mixture of energy sources, but moderation is the key. Too much starch may lead to colic, founder, or tying up. Too much fat may compromise glycogen storage and too much protein may lead to problems associated with ammonia production. Fibre must be included in the diet to maintain proper hindgut function. The correct mixture of these energy sources in the ration will avoid any problems associated with feeding.

Starch
The primary source of starch is grains, with oats being the first choice because of their digestibility, palatability, high fibre and ease of storage. Other grains require processing for the horse to get the full nutritional value. In hotter environments either higher energy grains or fat are fed to reduce overheating.

Fat
Fat is a less versatile energy source than starch, but is extremely useful for performance horses that may not be able to eat enough grain to supply the high energy requirements of heavy training. Adding fat will increase the energy density of the diet so that less feed is required.

The ration for a performance horse should contain between 5% and 8% fat. Higher levels may affect muscle and liver glycogen storage. Table 7 shows when adding a fat supplement to the diet of a performance horse may be warranted. To meet the energy requirement of a racehorse in training, 6.7 kg of grain must be fed, as well as 6.0 kg of hay, to provide 146.3 MJ DE/day, which may be more grain than the horse's digestive system can safely handle. By substituting 0.5 kg of fat, the amount of grain fed can be reduced by 1.3 kg/day without reducing energy intake.

Protein
Although protein can be used as a source of energy, excessive intake should be avoided in the performance horse for the following reasons.

1. Water requirements increase with increased protein intake.
2. Urea levels in the blood increase leading to greater urea excretion into the gut, which may increase the risk of intestinal disturbances such as enterotoxemia.

Table 7. Fat supplementation in a racehorse ration

Dietary energy source	Digestible energy density	Intake required without fat supplement (kg)	Energy supplied without fat supplement (MJ)	Intake required with fat added (kg)	Energy supplied with fat added (MJ)	Digestible energy required (MJ/day)
Hay	9.6	6.0	57.6	6.0	57.6	
Grain	13.2	6.7	88.7	5.4	70.7	
Fat	35.9			0.50	18.0	
Total		12.7	146.3	11.9	146.3	146.3

3. Ammonia levels in the blood increase, causing problems such as nerve irritability and disturbances in carbohydrate metabolism. Increased ammonia in the urine may also lead to respiratory problems because of ammonia build-up in the stable.

Adult performance horses do not produce any protein-containing products, such as milk, and therefore do not require additional protein supplements. Simply increasing the feed intake to meet the energy requirements of the performance horse results in an increase in daily protein intake and that extra protein will more than adequately meet any requirements over and above maintenance levels. For example, a horse at maintenance fed 7.5 kg/day of a 10% protein ration would receive 750 g/day of protein; increasing the intake to 12 kg/day would increase the protein intake to 1200 g/day even though the protein concentration of the diet is unchanged.

Fibre

Because proper gut function is essential to the health and well-being of the horse, fibre must be considered an essential nutrient. Roughage should be fed at a rate of intake equal to at least 1% of body weight per day. Alternative sources of fibre energy for performance horses are beet pulp and soy hulls.

Older horses

When horses were used for transportation, work and farming, very few lived to their 'golden years' and a horse was considered to be in its prime between 5 and 10 years of age. If a horse had not started working too young, had not been overworked, and had good health care and proper nutrition throughout its life, it might have lived to be a 'very old' horse of 25–30 years of age. Although the natural life span of the horse has not changed, more horses are now living longer and many reach 25 years of age or older. What may have been an old worn-out horse 100 years ago may today be just coming into its prime. Many performance horses have only settled into their work by the time they are in their teens.

What has changed in the past 100 years? First, the workload has significantly decreased as horses are now primarily used for pleasure or competition. Lighter workloads do not put as much wear and tear on the skeletal and muscular systems, and the body has more time to recoup. Second, knowledge of nutrition has improved and many of the nutrients are now commonly provided in fortified commercial feeds. Dental care has become the norm and lastly, parasite control has improved greatly.

At what age is a horse now considered to be geriatric? A general rule is that a horse aged 18–20 years is entering its golden years. Some horses remain in excellent body condition and health until the moment they die, while others deteriorate either quickly or slowly. Because of the physiological changes normally associated with aging, geriatrics may require special adaptations in their health care, environment and diet.

Changes occurring with aging

Four factors negatively affect the ability of senior horses to maintain health and good body condition: decreased nutrient absorption, dental problems, environmental stress and disease.

Decreased nutrient absorption

Intestinal worms decrease the availability of nutrients to the horse because they compete for the nutrients and also damage the intestines, which further affects nutrient absorption. With a consistent, effective parasite control program, horses are more likely to survive to an older age.

There are other factors responsible for the decreased nutrient absorption of the aged horse. The effectiveness of the intestinal mucosa to absorb nutrients, particularly phosphorus, vitamins and protein, decreases with age. Production of amylase, the enzyme necessary for starch digestion, may decrease, allowing too much undigested starch to enter the hindgut where microbial fermentation of the starch will make the hindgut more acidic, which can make the old horse more susceptible to laminitis or colic. Another factor affecting availability of nutrients is the particle size of the food when it reaches the intestinal tract. If the horse is unable to chew its food sufficiently, the food particles will be too large for the digestive enzymes and microbes to effectively break them down and the net result is more undigested food passing through the gut. The decreased efficiency of the digestive tract because of aging cannot be altered, but dietary adjustments can be made; for example, feeding the nutrients in highly digestible forms and as smaller particles.

Protein digestion seems to be a particular problem in the geriatric horse, especially for those with parasitic damage of the digestive tract as well, and if there is insufficient protein absorbed, the body uses its own muscle tissue, resulting in muscle wasting. Production of stomach acid, which aids in protein digestion, decreases with age, and so there is a concurrent decrease in protein availability. Because of these factors, the geriatric horse's ration should be higher in protein than is usual for maintenance, with the grain concentrate having at least 14% protein. Soybean meal and lucerne are excellent protein sources for aged horses.

Dental problems

During normal tooth growth in the horse, the surface wears down and the tooth continually erupts from the jaw. Therefore, the aged horse has shorter tooth roots and the teeth can be easily dislodged. Regular dental care is important, at least twice a year, but the teeth of an older horse should not be floated too aggressively (see Chapter 2). Older horses tend to be less tolerant of pain so teeth problems may bother them more.

Horses that lose their incisors will have trouble tearing grass away from the root, so pasture may be too difficult to eat. On the other hand, pasture is suitable for older horses with less than effective chewing because grass can be digested even if it is not well chewed. Problems may occur if hay is the sole roughage and the horse cannot chew it sufficiently.

Tooth problems may be the reason why older horses are more susceptible to choke, which is more likely to occur if feed is not chewed properly before swallowing. To aggravate the problem, the older horse that chews produces less saliva and therefore there is less lubricant to aid the passage of feed to the stomach. The result is choke, which can become a serious problem. Rations that consist of partially cooked grains or hard extruded components may be difficult for a horse with poor teeth and may result in quidding. A cooked ration, or better still a slightly damp mash, is ideal for older horses with little or no grinding surface on their teeth.

Environmental and herd stress

Older horses do not adapt well to changes in their environment. Relocating an older horse from one paddock or stable to another can be very stressful, especially if it means a change of paddock mates, and there may be detrimental weight loss during the adjustment period. Older horses tend to be at the bottom of the pecking order and are less likely to fight for food. Therefore, when hay or grain is fed to the group, careful observation of the older horse is necessary.

The older horse is less able to tolerate changes in temperature, particularly cold weather. Some of the sensitivity to cold may relate to the reduction of fat stores in the body, which act as insulation, another factor may be changes in the production of the hormones that regulate the body's ability to adjust to external heat and cold. The digestion of fibre in the hindgut produces heat, which will help horses stay warm in the winter, but older horses can find it hard to bite or chew more fibrous feeds. Therefore, adequate shelter is vital for the geriatric.

During cold weather, an older horse may limit its water intake because the cold water lowers the internal temperature, resulting in cold stress. It is not uncommon for old horses to develop colic from self-induced dehydration and subsequent impaction of food. Careful observation of water intake is essential for the welfare of the older horse. Feeding the meal soaked in warm water and/or adding salt to the meal may help increase the water intake. Older horses are particularly suited to expanded forms of feed, which absorb water to make a palatable and digestible 'gruel'.

Pain can make a horse lose its appetite and the principal cause of pain in the older horse is arthritis. The best solution is to allow the geriatric to exercise. Joints become stiff when a horse is kept in a stable for any length of time and then it is more painful to start moving again when turned out into a paddock. It is advisable to keep older horses paddocked all the time, provided there is adequate shelter. Joint supplements and/or some mild anti-inflammatory drugs under veterinary supervision can make the older horse more comfortable. Regular hoof care is essential to avoid unnecessary stresses on joints and to keep the horse mobile (see Chapter 2).

Disease

Age-related disorders and diseases are not uncommon in the geriatric horse. Chronic weight loss may be the result of a chronic infection, atrophy (shrinkage) of the adrenal glands, liver failure or kidney disorders. Other problems are anaemia, lowered disease resistance and allergic respiratory problems (known as chronic obstructive pulmonary disease (COPD) or 'heaves'; see Chapter 6). Tumours, such as melanomas, are frequently observed on the skin, particularly on gray horses, and tumours in the thyroid or pituitary glands can cause the coat to become long and rough (those symptoms are also indicative of Cushing's syndrome, a condition related to the function of the adrenal glands). Older horses with adrenal atrophy or even adrenal exhaustion after a harsh winter will drink excessively and not maintain weight easily. Besides weight loss, signs of kidney or liver failure are poor appetite, lethargy and frequent urination.

If you are worried about the health of your older horse, your veterinarian can take a blood sample to determine if the horse has anaemia, chronic infection, or kidney or liver problems. Anaemia can be treated with B vitamin supplementation; chronic infection and lowered disease resistance may respond to vitamin E and C supplementation. The diet of a horse with liver failure should be low in protein overall, but high in the branched-chain amino acids and vitamin B, and should not have added fat. Kidney problems require decreased calcium in the diet, so lucerne should be avoided. An abnormally high incidence of renal calculi (stones) occurs in aged horses fed completely on lucerne. The management of COPD involves minimizing exposure to the allergens. Keep the diet free of dust and mould by careful selection of the type of feeds, soaking or wetting all feeds before feeding, and use bedding with minimal dust (see Chapter 8).

When any horse becomes ill it can be quite difficult to get it to eat, even more so the senior horse. The food must be highly appetising and nutritious, although this is not the time to make too many changes to the diet. Molasses added to the feed usually encourages a sick horse to eat.

Geriatric diet

Roughage is still a vital part of the diet for the older horse, but may not always be sufficient on its own. Older horses do better on fresh, green grass, even if they have lost some molars, because it is easily chewed and digested. Many older horses pick up weight when the grass is growing and lose it when the pasture is dormant. Signs that a horse is having difficulty eating hay are a low intake or quidding. Chaff requires less chewing or alternatively, other sources of highly digestible fibre should used.

A well-formulated ration for a healthy senior horse should contain grain that has been fully processed to increase digestibility. The concentrates should have a fibre percentage higher than 10% and a protein percentage between 12% and 16%. Use a high-quality protein source such as soybean meal. Fat can be added to increase the energy value of the feed, but liver function should be assessed first (see above). Fat can be fed in the form of oil or sunflower seeds. Rice bran has 20% fat and highly digestible fibre, making it an excellent addition to the diet of the senior horse. The ration must be very palatable to please even the most finicky of seniors. Added live yeast culture may improve fibre digestion and help the horse maintain condition. The vitamin and mineral proportion should be higher than in a maintenance horse feed because of the decrease in digestive efficiency. Feed smaller, more frequent meals.

Older horses may have lower levels of vitamin C in the blood, but the effect of this is still unknown. The vitamin C supply for the horse is normally produced in the liver from glucose and the lower blood levels of vitamin C in geriatrics may be the result of decreased production of the particular liver enzyme that aids in the conversion of glucose. Regardless of the cause, geriatrics with chronic infections or with decreased immunity may benefit from supplemental vitamin C and E. A research study found that supplementing vitamin C (20 g/day) improved the antibody reaction to vaccines in aged horses. Because supplemental vitamin C is not absorbed well by the intestines of the aged horse, more than 5 g/day of vitamin C has to be fed in order to increase the blood vitamin C levels.

Many older horses suffer from glucose intolerance as a result of tumours in the pituitary gland. They can develop very high levels of glucose and insulin after eating, which can lead to increased urination, thirst and other hormone changes. Chromium has been found to reduce the glucose and insulin response to a feed, so supplementation will benefit the older horse.

Brood mares

Good nutrition is the key to successful breeding. By understanding the mare's nutrient needs during the three stages of the reproductive cycle, an appropriate and cost-effective feeding program can be designed.

Stage one is early pregnancy, from conception through the first seven months of gestation. Both dry mares and pregnant mares without foals at foot are in this nutritional category. Stage two is the last trimester of pregnancy, which is from seven months of pregnancy to foaling. Stage three is lactation, which generally lasts for 5–6 months after foaling. The most common mistakes are overfeeding during early pregnancy and underfeeding during lactation. The breeder should aim to maintain the mare in optimum condition throughout the year by monitoring body condition and weight. This is called 'straight line nutrition' and is the best way of ensuring correct development and growth of the foal and the nutritional health of the mare.

Mares should be kept in good to fat condition (condition score 3–4), so the amount of supplementary feed will depend on the available pasture. A suitable pasture for brood mares contains a mixture of grasses and clovers; however, certain varieties of clover, such as Yarloop, should not be included because they contain oestrogen, which can reduce fertility.

When supplementary feeding, each mare should have her own feed bin, well separated to prevent 'bossy' mares from injuring or frightening the others. The bossy or timid mare may have to be removed from the paddock. Feed the mares according to their individual requirements.

Be aware of differences in nutritional requirements between breeds, understand that the season affects pasture quality and quantity, and make condition scoring of the mares a routine management task, acting on the slightest noticeable changes with the aim of always keeping the mare in good body condition.

Pregnant mares
Stage 1: Early pregnancy
To feed the mare correctly during pregnancy, it is important to understand that the foetus does not grow at a constant rate throughout the entire 11 months of pregnancy. It is very small during the first 5 months of pregnancy; even at 7 months gestation, the foetus weighs only approximately 20% of its birth weight and equals less than 2% of the mare's weight. Therefore, its nutrient requirements are miniscule compared with the mare's own maintenance requirements and so the mare can be fed as though she were not pregnant. An ideal diet would be pasture with a vitamin/mineral supplement. Obviously if pasture is poor or scarce, additional hay or chaff should be fed (at least 1–1.5% body weight per day or ad libitum if hay quality is average to poor).

All too often, the pregnant mare is fed as though she were 'eating for two', which can lead to obesity and problems with foaling. Obesity of broodmares is a serious problem, not only affecting fertility and conception, but seriously risking the health of the foetus (i.e. predisposing the foal to developmental orthopaedic diseases after birth). If pasture quality is high while the mare is pregnant, it may even be necessary to restrict grazing by confinement, but for no longer than eight hours to prevent digestive upsets.

On the other hand, the mare in early pregnancy should not go into a negative energy balance and lose condition. Picky eaters, poor doers and lactating mares with foals at foot need to carefully monitored. Mares in poor condition are more likely to lose the pregnancy, and if the mare is producing large quantities of milk, as well as being pregnant, she will lose condition on dry, scarce pasture without supplementary feeding.

Late pregnancy

After 7 months' gestation, the foetus begins to develop rapidly and its nutrient requirements become significantly greater than the mare's maintenance requirements. Energy requirements only increase approximately 15% during late pregnancy, but protein and mineral requirements increase to a greater extent because the foetal tissues being formed during this time are high in protein, calcium and phosphorus (i.e. muscles and bones). During the last 4 months of pregnancy, the foetus and placenta need approximately 77 g of protein, 7.5 g of calcium and 4 g of phosphorus per day. Trace mineral supplementation is very important during late pregnancy. Because mare's milk is quite low in iron, zinc, copper, and manganese, the foetus stores these elements in its liver for use during the first few months of its life. The amount of copper consumed per day should total almost 175 mg/day for a 500 kg mare.

Vitamin A is critically important for the late pregnant mare, especially if the mare is maintained on hay alone without access to green pasture or vitamin A supplementation. The growth rate of the foals will be reduced significantly because the vitamin A content of hay that has been stored for weeks or months is reduced.

Selenium and vitamin E supplementation in late pregnancy will enhance immunity in both the mare and foal. Antibody levels are higher in the foals of mares receiving 3 mg of selenium and 1600 IU of vitamin E each day compared with those receiving the currently recommended 1 mg/day of selenium and 800 IU of vitamin E/day. Selenium may also reduce the risk of retained afterbirth.

Use either complete pre-mixed feeds or supplements containing appropriate levels of the vitamins and minerals. Remember that many commercial feeds designed for broodmares are formulated for Thoroughbreds and feeding rates for other breeds can be significantly less than the recommended rate on the label. Rates of feeding also depend on the size and age of the mare. If you prefer to feed straights, oats are always the first choice for energy, together with adequate vitamin and mineral intake.

Warmblood and draught breeds mares in late pregnancy, particularly, often receive excess energy when they are fed a diet formulated to supply adequate protein and minerals to the developing foal. If the pregnant mare becomes fat during late pregnancy, the feed should be gradually changed to comprise more concentrated protein and minerals so that less can be fed each day.

Some mares may be able to continue on pasture alone with a vitamin and mineral supplement even in late pregnancy and not lose condition. Knowledge of the individual horse and its body condition are the guide to designing the feeding program.

Lactation

The mare must be kept in good body condition during lactation. After foaling, the mare's nutrient requirements increase significantly during the next three months because milk is produced at a rate equal to 1–4% of body weight per day. Mare's milk is rich in energy, protein, calcium, phosphorus and vitamins. Some mares produce so much milk that the foal grows quickly, but the mare begins to lose condition, using her own muscle reserves to supply the protein needed for the milk.

Mares in early lactation may require up to 8 kg/day of grain, depending on the type and quality of the forage. The aim is to use the highest quality forage available and keep the grain proportion of the diet as low as possible to avoid digestive upsets. If straight grains are used, they should be fortified with additional protein, minerals and vitamins, such as in commercial 'balancer' supplements, or the grain portion can be in the form of a complete mixed feed. Grain intake should be increased gradually during the last weeks of pregnancy so that the mare is consuming nearly the amount

Figure 8. Early
lactation
(courtesy of Denise Burrell).

required for milk production by the time she foals. Increasing the grain portion at the time of foaling should be avoided because it may lead to colic or founder.

After nearly three months of lactation, milk production begins to decline and the grain intake can be reduced. It is often beneficial to reduce the mare's grain ration just prior to weaning to help dry up the milk supply.

Milk contains low levels of trace minerals and supplementation is not as important for lactating mares as for late pregnant mares. Research has shown that adding more trace minerals to the lactating mare's diet does not change the composition of the milk. Calcium and phosphorus are the minerals of primary concern during lactation so feed lucerne hay, which has a high calcium content, and possibly supplement as well (see earlier section on the correct feeding of calcium and phosphorus).

Maiden mares

Mares that have been in training are not in breeding condition and may need special treatment. Maiden mares should always be let down slowly from their previous program. They should not be immediately turned out to pasture without a rug and on a reduced ration or they may rapidly lose weight. Begin gradually changing from the performance ration while the mares are still in a yard or in a small group. Rug them before they are turned out in the paddock with mares that are not bossy.

Barren/dry mares

Non-pregnant mares that have been bred before should be maintained in good–fat (3.5) body condition.

Foals

Although many foals are not weaned until they are over 6 months old, mare's milk will not meet the foal's total nutrient requirements after 3 months of age, so the foal must eat pasture or be hand fed. If the foal starts eating from the mare's feed bucket, a fortified feed is important to ensure an adequate intake of vitamins and minerals. If the foal is being creep fed, only the foal's diet needs to be supplemented.

Creep feeding

Creep feeding, although not an essential part of foal rearing, will make weaning easier if begun either 1 month before weaning or after 3 months of age, because the foal is used to eating hard feed. Creep feeding is not force-feeding. The creep feed area is fenced off (e.g. using a 13.1 hh bar) to exclude the mares.

Foal pellets or a mixture of maize, lucerne chaff, soybean meal or milk powder, ground limestone and salt can be fed ad libitum to a group of foals, but the intake of each foal must be carefully monitored. Excessive intake can lead to digestive disturbances or leg deformities (i.e. DOD). Creep-fed foals should be given no more than 1.5–2 kg of grain or concentrates per day starting at three months of age.

Weanlings and yearlings
Principles of feeding the young horse

The weanling has similar nutrient requirements to the yearling, but eats less, so to achieve maximum growth rates the nutrients must be more concentrated, which means weanlings have a greater risk of developmental orthopaedic diseases (DOD) if fed excessive energy (see later).

The feeding program for weanlings presents a challenge. The weanling has the stress of being taken from its dam, mixing with a new group of young horses and sorting out the social order, often at a time of year when weather conditions are bad, but it must continue to grow rapidly to be ready for the annual weanling or yearling sales.

The first factor to consider is the age of weaning. The mare's milk production peaks at two months after foaling, after which there is an increasing gap between the foal's needs and what the milk can supply. A foal can be weaned from four months of age and a recent study showed that weaning at that age was less stressful than at six months old, when stress was judged by a growth setback at weaning. In general, five months of age is thought to be the ideal time to wean, but if the foal is doing too well or not well enough, or has signs of DOD or confirmation defects (e.g. getting erect in the pasterns), it may need to be weaned as early as three months so that you can carefully control nutrient intake. Another factor is how well the mare is doing and early weaning may be needed if the mare is losing condition by continuing to feed an older foal.

The factor that has a big influence on the growth of weanlings is familiarity with hard feed prior to weaning. If the mares are not being fed any supplements or are being fed differently to the weanling ration, it is a good idea to creep-feed your weanlings.

To achieve the optimum growth rate (Table 8), the average 6-month-old, 250 kg weanling growing at 0.9 kg/day will need 2.5–3 kg/day of a weanling feed (Table 9), which equates to 1% of body weight. Because the weanling has a restricted appetite it is important that the feed is palatable. Sweet-feeds are more palatable than pellets or extruded feeds, which encourages weanlings to eat more. Processing of the barley and corn in sweet-feeds will increase the energy content.

Weanlings will need to be fed twice daily, but you need to allow for variation in intake; for example, if you have eight weanlings in a paddock, provide them with nine feeds, either in bins or a trough. Allow two weanling feeds for the 'granny' or 'minder' that you have in the group. Match up the weanlings by age and then you can vary the intake appropriately.

The quality and quantity of pasture available will determine the roughage the weanling needs. The weanling's protein requirements can be met by a combination of young, green pasture and an appropriate grain mix. If pasture is plentiful you can mix some chaff with the grain, but hay is unnecessary, apart from days of rain and cold weather that make weanlings reluctant to graze. It is not essential to feed forage as chaff, but it does increase the safety of feeding grain to a group of weanlings, in which case the amount of chaff should be in proportion to the amount of grain. If pasture is scarce or dry, the weanlings will need supplementary high-quality lucerne or clover hay/chaff, which have higher levels of energy, protein and calcium than grass hay or oaten chaff. A grass or hay diet alone will be deficient in the necessary minerals and vitamins.

The advent of pre-sale X-rays in 2003 has focused attention on DOD and other bone problems in young horses. Although underfeeding leads to reduced growth and/or bone problems, overfeeding causes rapid growth, which can lead to metabolic bone disease and poor conformation. Putting extra weight on a young skeleton is undesirable because the bones are not yet sufficiently developed to carry the heavy weight. If you feed high energy (i.e. high grain) diets to achieve maximal growth you will increase the risk of DOD, but if you do not, you are unlikely to meet the market expectations for growth and condition. However, it is important to remember that nutrition is only one of the risk factors involved in DOD.

The aim of weanling feeding is optimum growth with minimal DOD. With the correct feed and appropriate intake there is no need for added supplements, which may in fact create nutritional imbalances. Key nutrients such as amino acids, calcium, phosphorus, copper, zinc, manganese and vitamin E should be provided by the hard feed. Weanlings that are too heavy or have early signs of DOD should not receive grain, but have their amino acid and mineral needs supplied by a low-calorie balancer pellet as a supplement to forage.

Weanlings that are not going to the sales can be fed more conservatively because they do not have to grow at a maximum rate or look their best at a young age. Again, the best assessment of the

Table 8. Expected growth rates of horses (mature weight 500 kg)

Age (months)	Height (% adult)	Weight (% adult)	Weight (kg)	Daily amount of feed as % of body weight
6	85	50	250	3
12	90	70	350	2.5
18	95	85	425	2
24	100	95	475	1.75
36	100	100	500	1.5

feeding program is the body condition score of the weanling. Weanlings should maintain a thrifty appearance in which the ribs can either be just seen or can be easily felt (i.e. score of 2.5–3.5). Do not let a growing horse get fatter than condition score 3.5 (good–fat).

The amount of grain necessary to maintain a thrifty appearance will vary according to the needs of the individual weanling, and the quality and quantity of the available forage. Feed weanlings as individuals and make necessary feeding adjustments rather than feeding them as a group. If the weanling needs extra energy, supply it as fat or high-quality forage.

Nutrient requirements

Young horses need the best quality feeds to meet requirements and you should seek advice from an equine nutritionist if you need to formulate a specific diet. In general terms you will need:

- an energy source (e.g. oats)
- protein supplement (e.g. soybean meal)
- calcium/phosphorus supplement (e.g. dicalcium phosphate or ground limestone)
- trace mineral and vitamin supplement
- good-quality roughage (e.g. lucerne chaff or clover hay/pasture).

If you are using a mixed feed, make sure it is formulated for the growing horse.

Energy

Oats are usually the cheapest source of energy for young horses and are best fed crushed to weanlings because their teeth are not fully developed and they will have trouble breaking open the grain. Crushed maize, rice and barley are other grains that are good sources of energy for the growing horse.

A general rule of thumb is 1 kg of fortified grain per 100 kg of body weight, up to a maximum of 3 kg/day per weanling.

Protein

Weanlings need 15% crude protein and yearlings need 13% crude protein in the diet. Young, green, grass–clover pasture contains 15–20% crude protein, but this amount falls rapidly as the plants begin flowering and go to seed. Dry summer pasture contains very little protein so yearlings will need supplementation.

Young horses need high-quality protein that contains all the essential amino acids, especially lysine. Animal-derived proteins, such as milk powder, are very high quality and soybean meal is the best quality vegetable protein. Lucerne also has a high level of good-quality protein. Linseed meal is relatively low in lysine and is not a good source of protein for growing horses, although its high oil content will produce a bloom on the coat.

Vitamins and minerals

Calcium and phosphorus are the most important minerals for growing horses. Grains are low in calcium, whereas lucerne and clover hays and high-quality pasture contain more calcium. Dicalcium phosphate and ground limestone are the cheapest supplements and you generally need 30–50 g/day. Unless you are feeding a mixed balanced feed you will need to add limestone to grains, or dical-

Table 9. Nutrient composition of a feed suitable for weanlings

Digestible energy	12–13 MJ/kg
Minimum crude protein	16%
Maximum crude Fat	5%
Minimum crude Fibre	9–10%
Maximum added Salt	1%
Minimum calcium	1%
Minimum phosphorus	0.6–0.8%
Copper	45%
Zinc	130–140%
Vitamin A	12000 IU/kg
Vitamin E	90 mg/kg
Selenium	0.7 mg/kg
Lysine	0.80%
Methionine	0.18%
Cobalt	0.3 mg/kg
Iodine	0.5 mg/kg
Iron	230–250 mg/kg
Manganese	110 mg/kg
Folic acid	3–4 mg/kg
Vitamin D3	1050 IU/kg
Vitamin B1	10 mg/kg
Vitamin B2	12 mg/kg
Vitamin B6	5 mg/kg
Vitamin B12	35 µg/kg
Vitamin K3	5 mg/kg
Biotin	0.3 mg/kg
Niacin	55 mg/kg

cium phosphate if the ration contains lucerne. Bran contains a lot of phosphorus and should not be fed to growing horses. Urinalysis can be used to asses the calcium status of young horses or the calcium/phosphorus balance of the ration can be analysed by an equine nutritionist.

Young horses running in a paddock or regularly being lunged can lose 30–50 g of salt per day in their sweat and urine, especially during hot weather, so provide a trace mineralised salt block or supply a salt supplement.

Trace minerals such as copper, zinc, manganese and selenium are important components in the diet for most growing horses. You should seek expert advice on the best supplement for your situation.

Roughage
To meet the commercial growth requirements of young horses you may have to feed them less roughage than is desirable for good digestive function. Therefore, you must use good-quality chaff or hay to maximize the utilisation of the fibrous feed in meeting the energy requirements and so decrease the amount of starch the weanling has to be fed. High quality, early cut hay will also minimise 'gut-fill' (a pot-belly caused by the fibre and water in the gut), which is often associated with mature hay of high lignin content. Lucerne or clover hay will also supply protein and calcium, as well as fibre, so they are preferred for growing horses and should be fed on demand.

The halter and sales horse
There are two groups of horses that are assessed, and therefore valued, on their looks, conformation and the way they are 'turned out': the halter or 'in hand' horse and the sales weanling or

yearling. Enormous sums of money can be made or lost depending on their preparation (feeding and fitting). There is a common saying that 'fat is a pretty colour' and some sales and halter horses are fed to excess. However, to be really successful in preparing sales and halter horses the 'fitter' must be able to differentiate between fit and fat.

It may seem unusual that the young show horse and the Thoroughbred weanling and yearling are being considered in the same context, but fitting these horses is a combination of superior nutrition, health management, exercise and genetics, tempered with hard work and attention to detail.

The weanling

The preparation of the weanling for a show or sale must take into account the nutrient requirements of the young horse and the critical balance between feed intake and exercise as it affects body condition and soundness.

As described earlier, the weanling feeding program should be based on a balanced ration using palatable, easily assimilated nutrient sources. Too often, all-grain feeds are fed to weanlings to encourage fattening, with little regard to optimum growth and bone development. A high protein diet does not cause bone problems, whereas overfeeding the energy portion and an incorrect balance of minerals will predispose the growing horse to DOD.

Weanlings should have individual feeding programs based on individual performance (growth rate and degree of fatness). The minimum needed is often 500 g of feed/month of age until the time of the sale/show. Feed a 15–16% protein concentrate in addition to high-quality hay, which may need to be restricted just prior to the show/sale to avoid a pot bellied appearance.

The total nutrient profile of the concentrate feed must be considered, not just the protein concentration. If the feed manufacturer has made an error in the formulation or the feed is not used correctly (e.g. 'cutting' a prepared feed), the nutrient/calorie ratio of the mix will be incorrect for weanlings. The nutrient profile of a feed designed for a specific class of horse is not approximate or interchangeable and 'tinkering' with a feed will upset the balance of nutrients. Therefore, feeds formulated for older horses do not have the correct balance of protein, energy, vitamins and minerals for weanlings, even when fed at appropriate amounts. Processing of the grains will increase the energy availability and any feed used to prepare weanlings for the sales or show should contain at least 60 g of added fat from either oil or sunflower seeds; as much as 300 g/day of fat can be fed to achieve a greater energy intake without increasing the starch intake. If the weanling has physitis or another DOD it is preferable to feed a high-fat rather than a high-grain diet. Feeding at least 125 mL of oil and a coat conditioner for 1 month before the sale will condition the coat.

The other important aspects of fitting weanlings are exercise and grooming (see Chapter 2). Routine lungeing for the sales or show weanling can be detrimental, but judicious use of free lungeing ('round-penning'), ponying and hand walking can be very useful for increasing fitness and muscle condition. Daily grooming, rinsing with warm water, braiding or banding of manes and conditioning of tails are all necessary for optimum appearance.

There are several factors to be considered with respect to turning out the weanling to pasture; what works for the sales weanling differs from what works for the show weanling. In general, turn-out for the show weanling works to a limited extent. If hard feed intake is limited and weanlings are turned out on good pasture, they can develop a grass belly; on the other hand, if they are turned out for a short time on good pasture, or in barren paddocks, it can encourage them to exercise. The individual preference will be your guide. Some halter weanlings will tolerate pasture turnout, others will not. Turn-out is a little less critical for the sales weanlings because they can have both a slightly hairy coat and a little belly without detriment.

Weanlings should be dewormed every 30–60 days, alternating between anthelmintics (see Chapter 6).

The last thought is the time of weaning. A rule of thumb is to wean a foal at least 45 days before a sale or futurity or, if that is too early for late foals, wean five days before a futurity. In the first case, 45 days is adequate time for the weanling to recover from the post-weaning slump and get into good shape, and a 5-day pre-futurity weaning avoids the slump.

The yearling

Yearlings are easier to prepare than weanlings because they are at least 12 months of age and therefore less susceptible to DOD.

Yearlings do best on a ration with 13–14% protein and balanced for minerals and vitamins. Feeding rates are extremely variable and depend on the yearlings' growth history, skeletal size, individual metabolism, actual age in months and the availability and quality of forage. Generally a colt requires more feed than a filly does to achieve the same level of condition.

The type and quality of hay fed to weanlings should also be fed to yearlings, unless the yearling needs to lose weight, in which case feed a lower energy grass hay or oaten chaff rather than rich lucerne or clover hay.

Yearlings are more susceptible than weanlings to laminitis and colic caused by starch overload, so always add fat to the yearling ration, as vegetable oil, sunflower seeds or a commercial fat supplement. The high level of fat intake in the yearling is to reduce the amount of grain that must be fed to achieve a specified energy intake. If the yearling needs more than 5 kg/day of hard feed, then the concentrated energy of supplementary fat is required. As much as 750 mL/day of vegetable oil can be fed without a detrimental effect, provided you make the dietary change gradually. For coat conditioning, feed 60 g/day of fat.

In addition to fat, many prep and show rations for yearlings will contain lupins to reduce the amount of starch that the horse consumes while maintaining a relatively high energy intake. Processed feeds alone or as components of the ration are also useful, because although they do not reduce the starch intake, they do ensure most of the digestion takes place in the small intestine, again reducing the risk of starch overload and possible laminitis or colic.

The emphasis of the yearling preparation program is always on the individual in terms of feed intake and exercise (i.e. an individually tailored feeding and fitting program).

Stallions

Domestication of the horse has had the most dramatic effect on the environment and natural instincts of the stallion. The wild stallion runs with a group of mares, ranging far and wide in search of feed, and protects them from capture by another stallion and from predators, so he is usually the fittest horse in the herd.

The domesticated stallion may never go hungry, but he is usually confined in a very small area, fed two meals a day, often with only poor quality pasture to graze, and is locked away from contact with other horses. Little wonder that some stallions have significant behavioural problems. Safe feeding practices that are compatible with good fertility, behaviour and success in the show ring are based on pasture and good-quality hay, and use concentrate feeds and supplements only when necessary.

Body condition scoring

Use body condition scoring (or weight) to monitor the suitability of the stallion's diet. Keep the stallion in the range between moderate and fat (score 2–4). Whereas mares and geldings do not have a crest with a score of 3 (good), stallions do and a marked crest indicates a fat stallion (score of 4).

Underfeeding

In the non-breeding season the underfed stallion does not look his best for inspections by owners of mares or for photography sessions or in the show ring nor does he have enough energy to work well. During the breeding season he may tire easily and not maintain his interest in the mares.

Vitamin A and E deficiencies have been linked anecdotally with reproductive disorders in stallions, but are only likely to occur if the stallion is fed a diet deficient in green forage and without supplementary vitamin A or E for an extended period.

Selenium and manganese deficiencies have also been linked with reduced fertility.

Overfeeding

It is a far more likely that the stallion is overfed energy or protein. A certain level of fat cover is important to provide energy reserves for work and breeding, but obesity is detrimental to fitness and fertility. Stallions that are too fat have a greater risk of dying 'in service', particularly early in the season, and are more likely to lose enthusiasm later in the breeding season.

Overfeeding grain can lead to an increased risk of laminitis, colic and diarrhoea. Many stallions are fed more supplements than they need and are receiving high levels of vitamins and minerals. The imbalance of minerals can create relative deficiencies and vitamin A toxicity from excessive supplementation can occur.

Nutrient requirements

Non-breeding season

During the non-breeding season the stallion has the same nutrient requirements as the average mare or gelding, but stallions often hold their condition better.

Normally, stallions can be maintained on pasture and hay. If pasture quantity and quality is adequate, the stallion may not need any supplementary feed, but a salt block should be provided. In cold weather, hay is the first choice of supplement to pasture. During winter in southern Australia when there is only poor-quality pasture, the stallion needs more hay to produce body heat during digestion and you can feed ad libitum.

If a stallion is losing condition or needs to gain condition, first increase the hay intake up to 8–10 kg/day (i.e. nearly half a bale of grass hay or one-third of a bale of lucerne or clover hay). If this is not sufficient to produce the required change in condition then add a small amount of grain to the diet; 2 kg (4 litres) of oats will supply more than one-third of the energy requirements of the stallion. Add a mineral and vitamin supplement to balance the diet when you are feeding grain.

If you are showing the stallion, both the daily intake and the concentration of energy, protein, minerals and vitamins in the feed will need to be increased.

If you need to condition the stallion's coat you can stable him overnight and during cold weather to reduce hair growth, or you can rug him (remember the stallion with a fine hair coat will need to be rugged anyway when in the paddock during cold weather). Feeding 125 mL/day of polyunsaturated oil, sunflower seeds or full-fat soybean meal will condition the coat. Biotin is a B vitamin that can help increase hoof strength and improve hair condition if fed in high doses. Zinc deficiency, which is not uncommon because many feeds only have low levels, can lead to a rough, dull coat.

Breeding season

During the breeding season the nutrient requirements of the stallion increase, but not as dramatically as many owners assume. They are similar to those of the late pregnant mare or a horse in light work, yet many people feed the stallion as though it were in heavy work.

The breeding stallion will eat approximately 10% more feed, but has a 20% increase in energy, lysine, calcium, phosphorus and magnesium requirements, so the feed may have to be more concentrated. Vitamin A and E requirements increase by over 50%, but the requirement for other trace minerals is relatively unchanged. There is no evidence to suggest that supplementation of the stallion's ration above these requirements increases fertility.

Feed no more than 3 kg/day of any premixed feed to a stallion. Oats can be mixed with a small amount of chaff, but it is more economical to feed most of the horse's roughage intake as hay. Add a mineral and vitamin supplement that has high levels of vitamin E. Most supplements have adequate vitamin A content, but you may need to check the selenium content, if the soils are selenium-deficient. Depending on the salt content of the supplement and the stallion's consumption of a salt block, extra salt may be needed during warmer weather to counteract losses in sweat.

The following is an example of a balanced, simple diet that will supply the needs of a 450 kg stallion during the breeding season if pasture is scarce or poor quality.

Oats	2.5 kg (5 litres)
Oaten chaff	1 kg (7 litres)
Supplement pellet	150 g (1 cup)
Salt	30 g
Lucerne hay	2.5 kg (1 biscuit)
Clover/grass hay pasture	2.3 kg (1 biscuit)

Some stallions are always difficult to handle and others become hard to manage when mares are around, or at shows. There are several feeding strategies to minimise behavioural abnormalities. First, cut down the grain intake and feed a relatively low grain/high roughage diet. Rice-based feeds can be fed instead of oats. Fat is very high in energy, but is not associated with behavioural problems, so you can substitute oil for oats on a 1:3 basis.

NUTRITIONAL DISEASES

There are some health problems in horses that are related to feeding and this chapter deals with their dietary management and Chapter 6 describes the symptoms and treatments required.

Tying-up

Tying-up is the term used for a muscle disease in performance horses that is also called 'azoturia' or 'Monday morning disease'. Recently, two specific causes of tying-up have been identified: recurrent exertional rhabdomyolysis (RER) and a disorder in carbohydrate storage and utilization called equine polysaccharide storage myopathy (PSSM or EPSM).

The classification of tying-up is now based on the frequency of the disease following exercise. Horses that tie-up only a few times in their lifetime are classified as 'sporadic', whereas those that tie-up repeatedly are termed 'chronic'. Sporadic tying-up usually involves a specific muscle group and is a veterinary emergency.

There are preventative measures involving dietary management for both sporadic and chronic tying-up.

Sporadic tying-up

Sporadic tying-up usually requires veterinary attention, followed by stable rest, then hand walking and turn-out after the initial muscle stiffness is relieved. The horse's grain intake must be drastically reduced or even eliminated while the horse is on a reduced exercise program, and may continue to be reduced even when the horse resumes full work.

Prevention involves not overexerting unfit horses, fortifying the diet with salt on a daily basis, and electrolyte supplementation prior to episodes of heavy work and consequent heavy sweat loss. Only feed enough grain to satisfy the horse's energy requirement. Do not overfeed grains because excess energy may cause tying-up. The grain concentrates should be fortified with fat and the levels of antioxidant vitamins and minerals (vitamin E and selenium) in the diet should be checked, particularly in areas where the soil is deficient in selenium.

Chronic tying-up

The proposed causes of chronic tying-up include electrolyte imbalance, hormonal imbalance, lactic acidosis, vitamin E and/or selenium deficiency, hypothyroidism, and muscle ischemia, plus the two recently identified causes, RER and PSSM (or EPSM).

Recurrent exertional rhabdomyolysis (RER)

Treatment of horses experiencing RER requires veterinary assistance. The exact cause of RER is still unknown, but recent research indicates that RER is the result of abnormal muscle contractions caused by abnormal intracellular calcium regulation.

Prevention of further episodes of RER is difficult. Controlling the environment and a well-established daily routine are essential to reduce apprehension and nervousness. Reducing the amount of carbohydrate and increasing the amount of fat and fibre in the diet will help reduce excitability in horses prone to RER; reducing the carbohydrate portion of the diet is recommended generally to prevent tying-up. A high-fat diet will help provide the extra calories required by nervous horses to maintain their body weight. It is recommended that less than 20% of the calories are supplied from carbohydrates and more than 20% from fat. Extensive research in the USA of feedstuffs such as rice bran and soy hulls has shown that using these products in the diet modifies the energy in the diet and can lead to a substantial reduction in the risk and severity of tying up.

Oats are associated with a greater risk of tying-up than other grains, so they should be replaced with barley, lupins, sunflower seeds/oil or rice bran. Because large volumes of oil in the feed can be messy, rice bran and sunflower seeds are more practical options. There are also commercial feeds designed for tying-up and these can be used as part of the preventative feeding program. Check the levels of vitamin E and selenium in prepared feeds. It is thought that horses that tie-up produce more potent free radicals during exercise than 'normal' horses and that consequently they require higher levels of antioxidants in their diets. Adding chromium may assist some horses and is included in some feeds and supplements. If anxiety or nervousness is a major issue, the use of calming agents may help.

Research has suggested that some horses with a history of chronic tying-up have both reduced severity and occurrence of tying-up when the electrolyte balance is controlled through dietary supplementation. Electrolyte status affects the nervous system and consequently muscle contraction. Salt should be fed daily or a salt lick should be available at all times, and commercial electrolyte supplements given after exercise when the horse has lost a lot of sweat.

Polysaccharide storage myopathy (PSSM)

Treatment of PSSM involves many of the same veterinary procedures needed for other causes of tying-up. Prevention of future episodes involves dietary manipulation. Horses with PSSM need to get less than 10% of their calories from starch and sugar, and 20% from fat. The combination of good-quality grass hay, a vitamin/mineral supplement and a fat supplement (e.g. rice bran) will enable many horses with PSSM to work successfully in pleasure activities. Research has shown that up to 2.5 kg/day of rice bran as the only concentrate feed will reduce the risk of tying-up in susceptible horses doing light to moderate work.

In summary, tying-up is a generic term for the symptoms of several different muscle diseases. Although there are many potential causes of the episodes of tying-up, frequently there has been exercise that exceeds the horse's level of fitness; electrolyte depletion, hyperthermia, concurrent lameness or respiratory disease are key triggering factors. Most horses can be successfully returned to competition with management strategies that focus on correct training and diet.

Colic

Colic is a general term for abdominal pain in the horse. Many diseases and disturbances cause abdominal pain, which may be only mild and temporary, but in other cases may be very serious and progress rapidly to death if not treated. It is therefore important that you can recognise the early signs of colic so that veterinary attention is given as soon as possible (see Chapter 6).

Prevention

Prevention of colic involves dietary management and internal parasite control. With regard to dietary management, it is the fibre and starch components of the ration that are important, both the quality and the quantity.

There is a significant decrease in the risk of colic if the horse has access to pasture, but lush, high-moisture spring pasture, which has a low fibre content, can be a risk. Feeding dry hay while horses are grazing lush pastures will reduce or eliminate the risk of colic.

Horses confined to stables have an increased likelihood of colic. First, stabled horses still have the natural desire to continuously consume forage, but they are routinely fed twice daily. Second, stabled horses may not receive adequate amounts of forage for proper gastrointestinal function, especially if their intake has been restricted for the maintenance or reduction of weight.

It is recommended that a minimum of 1 kg of dry forage (hay/pasture) per 100 kg of body weight is provided each day. For horses confined to stables, the selection of lower calorie hays or chaff (grass-type) will mean more hay can be fed and the horse can mimic the continuous feeding behaviour of grazing. Lucerne hay contains more calories than grass hay or oaten/wheaten chaff, so horses are usually fed less of it, which reduces the amount of time spent eating, and it also contains less fermentable fibre than grass hay.

When horses are fed chaff rather than hay they may unintentionally receive less fibre, which is especially important for horses that are always stabled. A 20 litre bucket of chaff actually only weighs slightly more than a biscuit of hay. Another consideration is that chaff is mainly short fibres compared with the long fibres in hay and there is a higher risk of colic in horses fed exclusively on chaff. Feed the rougher cut chaff that has a longer fibre length.

Another factor in the quality of the fibre is its digestibility. Wheat or oat straw have high levels of indigestible fibre (lignin and silica), which give the plants their stiffness and shiny appearance and may cause impaction colic. Rice straw and hulls are even less digestible because they have very high levels of lignin and silica. Horses cannot digest or utilise these poor-quality forages as effectively as cattle and impaction may result from a lack of fermentation of the fibre in the hindgut. The precise amount of poorly digested fibre that can be included in the diets of horses is not known, but the poor physical appearance and performance of horses fed this type of diet should set practical limitations. Be careful feeding forages made from tropical pastures because they can also be relatively indigestible.

The addition of live yeast culture to the diet of mature horses may enable them to efficiently utilise forage of marginal quality. Live yeast culture stimulates the activity of beneficial microbes in the large intestine, particularly the bacteria that digest cellulose, which results in an increase in the fermentation of fibre.

Because horses evolved to digest fibre, the addition of grain to the diet is a potential risk factor for colic, for unknown reasons, but the site of grain digestion suggests a partial explanation. Grains contain large amounts of starch that are normally digested by enzymes in the small intestines and the end products are also absorbed in the small intestine. Factors that influence the rate and extent of grain digestion are the source of the starch (i.e. the type of grain) and the size of the starch granules. Undigested starch passes into the large intestine where it is fermented by the bacteria and one of the end products is lactic acid. The increase in acidity causes bacteria to die and release potentially fatal endotoxins, which potentially leads to colic.

Prevention of colic is based on understanding starch digestion. First, feeding processed grain will increase the digestibility of the starch. Second, limit the amount of grain to no more than 2.5 kg in a single meal. Several small meals are preferable to one or two large meals with a high intake of grain. The use of dietary fat as an energy source can reduce the amount of grain. Forage can contribute much of the energy and protein needed by a horse that is spelling or in light/intermittent work. Finally, recent scientific data on the digestion of oat starch upholds the long-held belief that oats are the best grain to feed to horses.

Never use mouldy feeds and always store feed to minimise the likelihood of mould developing. Mycotoxins, which are the toxic compounds produced by moulds, can cause colic. Other toxic substances occasionally ingested with forage can also cause colic symptoms.

A heavy burden of intestinal parasites is a potential cause of colic, both through damage to the intestinal blood supply during larval migration and decreased motility of the digestive system. Therefore, all horses should be maintained on a regular parasite control program (see Chapter 6).

Australian stringhalt

Australian stringhalt is a very distressing disease for both the horse and its owner, although the horse is not in pain. Horses do recover from Australian stringhalt and can resume their usual activities after a variable period of time. See Chapter 6 for details on symptoms, causes and treatment.

The cause of stringhalt is still unclear, but one of the risk factors is relatively poor-quality paddocks. Flatweed (*Hypochacris radicata*) is commonly found in paddocks in which stringhalt occurs. The disease most commonly occurs in horses that are not receiving enough hand feeding.

Treatment and recovery

Many horses recover without any treatment at all. Surgery can be done, but only to improve the welfare of the horse because the results are very variable. Dietary change is the main form of treatment, together with time.

It is preferable to remove the horse from the paddock and give it some hard feed, but horses have recovered while being hand fed in the same paddock.

High doses of B vitamins, particularly thiamine, are a common treatment, either orally or by injection. Vitamin E is used as a therapy for other neurological disorders and high doses may help boost immunity. Magnesium has also been suggested as a therapy, but there is no evidence to support it.

Hyperlipaemia

Hyperlipaemia occurs in ponies and is caused by a diet that is deficient in energy, resulting in the pony's body fat stores being used as the source of energy.

Treatment

An animal showing signs of hyperlipaemia requires immediate veterinary attention. High-energy feeds and glucose are used to increase the animal's energy intake and 'switch' the pony's metabolism back to normal.

Prevention

Careful supervision is the key to preventing hyperlipaemia. Ponies should be fed to maintain body condition and supplementary feeding must commence before ponies lose too much condition. Record the quantities of feed so that you can monitor and adjust the feeding program. Note that there are variations in the energy requirements of different breeds of ponies, as well as variations in the needs of individuals within any group.

Where no grazing is available, the energy requirements could be met by the daily ration shown in Table 10, based on oats and lucerne chaff/hay.

Table 10. Approximate daily food requirements of ponies not on pasture

	Body weight (kg)	Lucerne hay/chaff (kg)	Oats (kg)
Barren/adult	200	3	–
	300	4.2	–
Late pregnancy	200	3	0.3
	300	4.2	0.4
Early lactation	200	3	3
	300	4.2	4.4

All ponies should have access to clean fresh water. Regular dental care for normal chewing, and regular foot care to ensure freedom of movement, as well as regular parasite control (see Chapter 6) are important.

Monitoring of all ponies is necessary to prevent hyperlipaemia; any noticeable changes in feed intake must be investigated immediately. A decreased appetite, lack of interest or depression may be the first signs of hyperlipaemia.

Anhydrosis

A horse that cannot produce sweat in normal quantities has anhydrosis and may be called a 'non-sweater' or a 'drycoated' horse or a 'puffer' because it puffs and pants a lot after work.

Dietary management

The cause of anhydrosis is unknown.

There is evidence that a hormonal or metabolic imbalance may affect the production of adrenaline, which alters the normal response of the sweat glands. One study found that a supplement containing L-tyrosine, cobalt, vitamin C and niacin was helpful in 80% of anhydrotic horses tested. The effect may be the result of providing amino acids that help to restore the normal response of the sweat glands to adrenaline. Tyrosine may be a dopamine precursor and stimulate dopamine production. Other studies have linked anhydrosis to low thyroid function, which may be related to iodine deficiencies in the diet, and hypothyroid horses may resume sweating when they are treated for this condition.

There is no evidence that a particular feeding program can either cause or cure anhydrosis. Replacing part of the grain ration with fat may help the horse stay more comfortable in hot weather, because less heat is released by the metabolism of fat. These animals should have free access to fresh water, salt and be given electrolyte supplements

Developmental orthopaedic disease

Developmental orthopaedic disease (DOD) is a very significant economic problem for modern horse breeders because the crippling lameness can make affected foals or yearlings essentially worthless.

A single factor that can prevent DOD has not been identified, but there are several that are known to be important (Table 11). Nutrition is an important factor, but is not the only area of management that needs attention. Awareness of the problem has led to the design of feeding and management programs to decrease the incidence and severity of the disease. See Chapter 6 for more details on this disease.

Mineral deficiencies

The currently recommended copper and zinc requirements of the horse are 10 mg/kg and 50 mg/kg, respectively, but research in various countries has found that these levels are inadequate

Table 11. Factors that may contribute to developmental orthopaedic disease in young horses

Factor	Diagnosis	Management
Mineral deficiency	Analyse the ration	Balance ration e.g. add minerals
Mineral excess	Analyse the ration	Balance ration e.g. dilute legume hay with grass hay
Overfeeding		
1. Overall feed intake	Measure growth rate	Monitor condition score or weight Reduce feed to slow growth if horse has access to excellent pasture Increase exercise
2. Energy intake per meal	Measure energy content of the ration	Reduce grain intake Feed 2–3 times per day
Environment	Hardness of ground	Soften ground (by choice of paddock)
Prolonged confinement	Length of time stabled	Minimise time in stable Some daily exercise if permanently boxed Graded return to paddock
Genetics	Soundness appraisal Performance of offspring	Introduce new bloodlines or cross-breeding
Trauma	Observe the horse	Small groups Segregate some horses Treat injuries promptly

for prevention of DOD in foals. Most natural feeds do not contain sufficient copper and zinc, which are both cheap and non-toxic, so high levels of supplementation are suggested; however, as described later, excessively high levels will also predispose to DOD.

Rations for late pregnant mares and young horses should provide copper and zinc intakes of 150 mg/day and 450 mg/day, respectively, so the mineral supplement needs to contain at least 150 mg of copper and 400 mg of zinc, and a fortified concentrate feed should contain 40 mg/kg of copper and 120 mg/kg of zinc (which many do not). In a research project, copper supplementation, together with zinc and selenium, of pasture-fed mares led to a significant reduction in the incidence and severity of physitis and articular cartilage lesions in the weanlings. Interestingly, supplementation of the mares and foals after birth did not reduce the incidence of DOD. If the mare has been supplemented in late pregnancy, there is no need to supplement the foal, which will have built up adequate stores of trace minerals in the liver prior to birth. Milk is a relatively poor source of these minerals.

Other minerals, such as manganese, sulphur and magnesium, are involved in the formation of bone and in the maturation of cartilage, and deficiencies of calcium and phosphorus will drastically affect skeletal development. Be aware of the need for calcium, particularly when high grain diets are fed. The phosphorus content of grains and hay is not readily available, but the addition of live yeast culture to the diet of growing horses and late-pregnant mares can improve both the availability of phosphorus and other minerals for rapid growth and overall feed digestibility.

Chromium can enhance glucose metabolism in young horses, which reduces the insulin and cortisol levels after feeding grain and may be useful in reducing the risk of OCD in young horses.

Mineral excess
Foals consuming high levels of zinc in conjunction with marginally deficient levels of copper have severe risk of DOD because the zinc interferes with the absorption and utilisation of the copper, thus creating a deficiency.

Excess phosphorus increases the risk of DOD, such as when foals are fed very large amounts of grain, bran and protein. If foals are fed phosphorus at 5-fold the recommended rate, then DOD occurs consistently.

High calcium diets have also been implicated in the development of DOD, perhaps by reducing the availability of other minerals and thus creating deficiencies. Excessive calcium supplementation can occur in young horses eating large amounts of lucerne and a calcium-only supplement (e.g. limestone). The diet for the growing horse should have a calcium to phosphorus ratio of 1.1:1–2:1. Have your horse's diet analysed by a veterinarian or nutritionist to check that these minerals are being supplied in the correct amounts and proportions.

Excessive manganese may lead to problems if the diet is marginal for copper. The copper to manganese ratio also needs to be considered if there is a higher than normal incidence of DOD. Some pastures accumulate manganese and will need to be tested.

Overfeeding

Horses that are forced to grow at a rapid rate are more prone to DOD. Fat, young horses will not become good athletes; a recent study of 10 000 Thoroughbred yearlings in the USA and England showed that overweight and obese yearlings performed substantially worse on the racetrack.

All breeders can monitor the body condition of growing horses and should not let them get too fat (condition score 3.5). Remember that exercise is important for young horses and can limit the side-effects of a high-energy diet. When weanlings and yearlings are confined or cannot exercise because of lameness, reduce the amount of grain and slowly reintroduce them to exercise.

One of the new theories for the development of OCD is sensitivity of cartilage to insulin, so management procedures that reduce the post feeding levels of insulin are recommended, such as small amounts of grain in each meal, dilution of the grain with chaff, added fat for energy and added chromium.

Gastric ulcers

Gastric ulcers frequently occur in performance horses particularly, leading to a decline in athletic output. Ulcers can be alleviated or even eliminated with proper management, including dietary management.

The severity of ulcers can range from mild to severe and you should consult your veterinarian if your horse shows symptoms or you suspect it has ulcers (see Chapter 6 for more details).

Prevention and treatment

Sick foals can be given treatment to prevent the development of ulcers, in addition to treatment for the other illness. Providing horses with free-range grazing or continual access to hay is the best way to prevent acid build-up in the stomach. If that is not possible, particularly for horses in heavy training or good doers, the amount of time spent eating forage such as grass, chaff and hay seems to be a critical factor. The feeding schedule of the stabled horse should mimic grazing by having free access to hay. A diet of lucerne hay and grain gives greater buffering of gastric acidity than a grass hay diet.

Turn the horse out to pasture as often as possible. Even a large yard is better than stable confinement. Instead of two meals per day, feed smaller meals at more frequent intervals and include chaff to slow down grain intake and increase saliva production. Reduce grain intake by adding fat and maximise the intake of forage. If a horse with ulcers is spelled on a high forage diet, many ulcers heal spontaneously in one month. Medical treatment of ulcers can be very costly. Prevention is considerably less expensive.

Tropical pastures and 'big head'

Big head is a calcium deficiency of horses and donkeys grazing introduced tropical pasture grasses. Cattle and sheep are not affected as the rumen bacteria break down the oxalates and release the calcium for absorption.

Cases of big head have occurred on pastures of buffel grass, green panic, setaria, kikuyu, guinea grass, para grass, pangola grass and signal grass. Purple pigeon grass is also hazardous. The hazard is greatest when these grasses provide all, or almost all, the feed.

Prevention

Native grasses have not caused the disease, nor have introduced temperate grasses such as rye grass and the sorghums. Some introduced tropical grasses are non-hazardous, such as Rhodes grass, the paspalums, the couches and creeping blue grass. Avoid grazing horses on the introduced species of grasses for longer than 1 month. You can graze horses on these pastures if they have access to other safe pastures.

If only hazardous grasses are available, encourage the growth of a legume component in the pasture to provide a source of feed that is free of oxalates, and feed one of the following calcium/phosphorus supplements:

- 1 kg rock phosphate mixed with 1.5 kg molasses or
- 1 kg of a mixture of one-third ground limestone and two-thirds dicalcium phosphate mixed with 1.5 kg molasses.

Other mineral mixes that provide calcium and phosphorus in the ratio of 1:1 can be used, but are likely to be more expensive than these two mixes. A supplement must be fed weekly to each horse while grazing hazardous pastures. The molasses is used as a carrier and to make the supplement palatable. It can be omitted if the horses will eat all the minerals by other means. Do not be concerned if your horses eat their week's supplement in 2 or 3 days; it contains enough mineral to last them the full week, but you may divide up the weekly amount and feed it each day if you wish.

To provide approximately the same amount of phosphorus and calcium as in the supplement, 20 kg of good-quality lucerne would need to be fed to each horse weekly.

Treatment

The lameness and ill-thrift can be cured, but the swellings of the jaws may not completely disappear if the animal has been severely affected. Double the amount of the mineral supplements for affected animals for at least 6 months to replace the mineral lost from their bones.

Hoof problems

A horse cannot perform to its potential if it has hoof problems that prevent it from working or reduce its performance. Even though most cases of bad feet are the result of genetics and bad mechanics, good nutrition has a role. Supplementary feeds should supply the necessary minerals and vitamins, and the horse should be provided with high-quality hay or pasture. Maintaining good feet is a combination of good farriery, good nutrition, good health care and selecting for horses that genetically have a great foot.

Energy and hoof growth

Meeting the energy requirements for hoof growth and integrity is an important aspect of the diet. If your horse is losing weight it is in a negative energy balance that will affect hoof growth. Research has shown that hoof wall growth was 50% greater in ponies in positive energy balance than in ponies on restricted diets. Many racehorses and lactating mares are in a negative energy balance for prolonged periods so it is not surprising that they have hoof problems.

Protein and hoof growth

The hoof wall is approximately 93% protein on a dry matter basis. Most of the commercial available hoof supplements contain methionine, but it is only one of the amino acids (building blocks) contained in the protein of the hoof, and deficiencies of any essential amino acid is as detrimental as a deficiency of methionine. Horses in training that are on an oaten chaff and oats diet, which supplies poor-quality protein, will benefit from a protein supplement.

Minerals and hoof growth

The health of the foot reflects the health of the horse with regard to minerals. Zinc is involved in the health and integrity of hair, skin and hooves, but adding zinc to a diet that is already adequate in zinc will not result in any dramatic increase in the quality or growth of the hoof. European research has shown that horses with 'bad feet' have lower levels of zinc in the blood than horses with normal feet, which may be related to individual zinc absorption or retention abnormalities. The form of zinc does have some relevance for the horse with bad feet because chelated zinc in the diet may produce results when inorganic zinc does not work. Chelated zinc contains zinc bound to an animo acid, which potentially enhances absorption.

Biotin and vitamins for hoof growth

The vitamin that is most relevant to hoof wall integrity is biotin. It is thought that the normal horse has a biotin requirement of 1–2 mg/day. Studies have shown that supplemental biotin of 15–20 mg/day can have positive effects on the histological characteristics of the hoof wall tissue. However, you need to be patient because it takes 6–9 months to grow a new hoof. Biotin treatment may reduce the incidence and severity of hoof horn defects, increase the tensile strength and improve the condition of the white line.

Supplements incorporating biotin, zinc and methionine have a better effect than biotin alone, which is estimated to work in only 5% of cases. At higher dose rates the combination products are excellent coat conditioners and can be used in the recovery from laminitis. Biotin products are expensive, but if you can keep a horse in work longer, reduce the incidence of hoof-related lameness and extend the life of a set of shoes, it is worthwhile.

Vitamin A in the diet may be important for normal hoof wall growth because it is involved in maintaining the integrity of epithelial cells. Vitamin A is present in green grass, freshly cut hay and supplements, so deficiencies are unlikely except in dry conditions or if no supplements are fed for a prolonged time.

DROUGHT FEEDING

Forage

One of the biggest problems in drought is the lack of forage. If you have been well prepared, you may have a store of hay, but in extended periods of drought, you will most likely run out. Furthermore, hay prices will be at a peak and availability next to non-existent. A good relationship with a farmer or produce store can help, but failing that there are ways of rationing the hay while still maintaining healthy gut function.

Horses need at least 1% of their body weight as dry forage each day (i.e. 5 kg for a 500 kg horse). Any less and you increase the risk of colic, laminitis, gastric ulcers and loss of condition. During a drought, the nutritive value of the roughage is less important than the physical bulk of it, because poor-quality, low-energy roughage can be supplemented with hard feed.

Figure 11. Drought
(courtesy of Jane Myers).

It may be more economical to buy the large, square or round bales of hay rather than the small bales and bulk buy whenever possible. Do not feed mouldy or uncured hay, but other than that, even sub-standard long-stem grass hay will suffice when nothing else is available.

When hay is scarce, you need alternatives to supply the fibre and roughage in the diet. Unconventional forages such as sorghum stubble, pea straw and good-quality silage can be used. Silage is a nutritious feed for horses because it is high in energy, so it can be fed in small amounts diluted with other hay or forage types. Only buy the quantity that can be eaten in a few days, to prevent it going mouldy, and introduce the horses to it slowly to allow the digestive system to adapt. There is a risk of botulism with silage that is not made specifically for horses, so check that the bale has not been pierced and that the silage has been cured correctly.

When using unusual types of forage, carefully check their nutritive value and adjust the rest of the diet to ensure the ration is balanced. If necessary, consult a horse nutritionist for assistance or advice on using an unfamiliar type of forage. In terms of hard feed, there are some high-fibre ingredients that can increase the bulk of the diet and supplement the limited hay intake.

Grain and seed hulls

The polished hulls of lupins, soybeans and sunflowers can be used to increase the roughage content of the diet. Although these by-products have little digestible fibre or energy, they do add bulk and can replace some of the chaff. Hulls can be dusty, so dampen them prior to feeding and you may need to add some molasses to increase palatability and intake. The hulls should be mixed well with the grain and chaff. Preferably, do not use oat or rice hulls; oat hulls are sharp and can pierce the horse's gums and cause mouth ulcers and rice hulls are indigestible and can cause impaction colic.

Bran and pollard

Although they are traditional feeds for horses, bran and pollard have high phosphorus and low calcium levels. Bran has very little digestible energy and is only really useful for adding bulk to a ration. Pollard has slightly more energy than bran, but has even lower levels of fibre and must be carefully balanced for calcium. If either bran or pollard is used as the bulking agent, then a calcium supplement must be added and the feeds should not be used for an extended period of time.

Commercial feeds

Commercial fibre mixes are available in the form of sweet-feeds with a high proportion of added chaff. The sweet-feeds have a vitamin and mineral premix added that balances the deficiencies of hay and grains, and because the feed manufacturers can usually maintain their supplies of chaff, these feeds are a good alternative where local supplies of roughage are low or poor quality.

Chose a feed with a fibre content of 10–15% or more to simply increase fibre and bulk. Do not alter the manufacturer's recommended feeding rate and if you have any questions about the maximum feeding rates for your horse, contact the manufacturer for guidance.

Bread

During drought some people use the old bread discarded by bakeries as a cheap way of feeding their horses. Although bread may provide extra dietary energy, it should only be used in moderation and with an awareness of the dangers associated with feeding it. Bread is high in energy (carbohydrates) and very low in fibre, so it should only be fed in small amounts as part of each meal. Feeding too much bread may lead to the behavioural and clinical problems commonly associated with high grain diets and the risk of laminitis and colic is increased. Bread also contains a large amount of gluten, so when it is wet it forms a sticky ball that can cause choke, colic and possibly impactions. Therefore, if you have to feed bread, feed it stale rather than fresh, and mix it well with other fibre sources.

Waste paper

In desperation, some people have fed their horses paper or cardboard during drought as a fibre source. The fibre in paper and cardboard varies in digestibility for horses, with stiffer material being less digestible. The ink on used paper is not be an ideal feed! Cardboard is very difficult to chew, so large portions may be swallowed whole, risking choke and impactions. Paper should only be used as a last resort and always with caution.

Energy

Most horses in drought conditions will need supplementary feeding to supply their energy requirements for maintaining body condition. Pasture will no longer be available as an energy source and the available forages will vary in the amount and quality of their energy content.

Most horse owners use grains to provide any shortfall in energy from forage. The dangers of too much starch (i.e. too much grain) in the diet have been discussed earlier, but suffice to say that when providing hard feed, minimise the amount of grain and use other energy sources such as fats and digestible fibre. Whether you decide to make up your own grain mix or buy a commercial pre-mixed feed, always introduce the new feed gradually.

The safest grain is oats, followed by processed barley and corn. Less common grains such as wheat and triticale (wheat/rye mix) are often fed to horses during times of drought to increase both the bulk and the digestible energy content of the feed. Wheat should be processed prior to feeding and fed in moderation because its high gluten content may result in a glutinous mass in the mouth or stomach. Its high starch/low fibre content may increase the risks of starch overload in the hindgut. Most horses can digest triticale if it is fed as part of the total diet.

Low-starch grains, such as lupins, can supply fibre and protein and minimise the carbohydrate content of the diet. Up to 2 kg/day of lupins is suitable for a 500 kg horse, but use only as required to avoid excessive intake of protein and of the alkaloids found in some types of lupins. Whole or cracked lupins should be soaked overnight prior to being fed so that they absorb moisture and expand.

Commercial feeds can be a convenient, safe and economical balanced diet during a period of drought. Prices may increase slightly, but not by the same amount as forages. The feed you choose should have an energy content suitable for the level and type of work that your horse is doing; for example, if you are resting the horse, a feed designed for racehorses or broodmares is not suitable whereas a feed with a high chaff content will provide sufficient energy and bulk. If in doubt about the type or quantity of feed for your situation, contact the manufacturer or an equine nutritionist.

Fat supplements

Fat supplements, such as oil or rice bran, are considered as sources of 'cool' and 'safe' energy that can replace some of the grain during drought feeding. Add 1–2 cups (250–500 mL) of vegetable oil or 0.5–1 kg rice bran to the hard feed ration. Another advantage of using fat as the energy source is that the price of fats and oils remains quite stable during a drought, whereas grains become expensive because of their scarcity.

Molasses and honey

These simple sugars do provide energy, but should be used cautiously. Overwhelming the digestive system with sugar is the same as overloading with grain or rich pasture. Sugars do not contain any fibre and the readily absorbed energy can cause a rapid increase in the blood sugar levels with possible undesirable effects on behaviour. If sugars are to be used as an energy supplement, they should be mixed with a feed containing sufficient bulk and fibre. Molasses and honey will increase the palatability of feeds, which can be very useful if unfamiliar feeds have to be used; however, sugar will ferment in hot weather, so it needs to be stored correctly.

Protein

Most classes of horse, except the adult, spelling horse, will need a protein supplement when poor quality, low-protein forage has to be used during a drought. The most important are pregnant and lactating mares, young growing horses and old horses. Good-quality protein is vital for healthy development and growth of the foetus and young horse. Signs of a protein deficiency are muscle wasting and a pot belly, especially in young horses.

Soybeans and full-fat soy

Full-fat soy contains nearly 38% protein and soybeans contain 40–48% protein. Soybeans contain almost all of the essential amino acids, including lysine, in sufficient amounts. Extruded full-fat soybean meal is popular for feeding young growing horses, providing 'cool' energy in the form of fat as well as good-quality protein. Feeding rates of up to 600 g/day may be required during drought.

Canola meal

Canola meal can be used to replace soybean meal as a protein and fat supplement. Up to 600 g/day may be required in a drought management program.

Lupins, beans and peas

Lupins, bean and peas all have similar protein levels and a moderate lysine content, but beans and peas are less palatable than lupins and may be 'sorted' out of the feed by finicky eaters. They all must be cracked before feeding to enable maximum digestion. Lupins can be fed at up to 2 kg/day, and more when other protein sources are limited. Some types of peas are toxic for horses.

Sunflower seeds

Although sunflower seeds are deficient in lysine, an essential amino acid, they contain some fat, which adds energy to the diet. If the drought feed is low in protein, up to 500 g/day may be fed.

Cottonseed and linseed meals

Cottonseed meal can be fed during a drought, but not to young or breeding horses because of its low lysine content. Linseed meal is another protein supplement that has a poor lysine content and is often very expensive as well.

Vitamin and mineral supplements

Vitamin and mineral supplements will be required if the hay and grains are deficient, especially vitamins A and E. Use a broad-spectrum multivitamin product when drought-feeding horses that are spelling or in light to moderate work. Horses in heavy training will require a specific supplement.

Common health risks associated with drought feeding

You may have to make frequent changes to your horse's diet during a drought as different ingredients become unavailable and a new source of a nutrient is found. Frequent changes to the diet, unfamiliar ingredients and more hard feed in the ration increases the risk of colic, gastric ulcers, laminitis and worms.

Colic

Provide as much roughage as possible and provide as much energy in the form of fibre and fat to minimise the quantity of grains. Introduce all new feeds and ingredients, including different types of hay and chaff, gradually. If you have to feed a large amount of grain for energy, feed little and often; feed no more than 2.5 kg of hard feed (excluding chaff) each meal and provide three or even four meals per day. Provide plenty of fresh water, whatever the cost, even if you have to move your horses to a new area.

Gastric ulcers

Gastric ulcers are precipitated by feeding a diet high in grains and from the stress of work or travel. Again, minimise the amount of grains in the diet during a drought and feed small meals with plenty of forage. Use chaff or one of the chaff substitutes mentioned earlier to slow down consumption and feed as much hay as you can ration.

Laminitis

Horses are at high risk of laminitis with drought feeding practices. Always change the diet very slowly, feed small meals and use digestible grains such as oats or heat-processed grains to reduce the risk of starch overflow to the hindgut.

Parasites

Dry conditions do not favour survival of the larval stage of intestinal worms, so their concentration in the paddock is less during drought, but if there is a lack of forage horses may start to eat droppings and graze the rough areas of the paddock. Continue to maintain a regular worming program during drought to ensure that the horses get the maximum benefit from their feed and regularly remove droppings from the paddock or yard.

Wood chewing

Horses may begin eating tree bark or the fences if there is insufficient forage in the diet. If possible, make more forage available, either as hay or chaff, or use one of the alternatives.

Taking stock

At the beginning of a drought, it is usually impossible to know how long it will last, how severe it will become and how widespread the effects will be. It is possible that you will experience a drought of some description at some stage, so becoming familiar with the weather patterns and average rainfall of your areas will help you to prepare to some extent for dry periods.

Throughout a drought, particularly a long one, continually take stock of the situation and reassess your ability to keep your horses healthy and well fed. If it is becoming too expensive, you may have to sell some of your stock in order to sustain the rest. It is unlawful to allow horses to starve and become malnourished (see Appendix 1). If you have competition or breeding horses, you may decide to spell them or agist them on a property that has some pasture in order to reduce their feeding requirements and thus reduce the associated costs.

In very severe droughts where horses cannot be fed or watered sufficiently, and the animals cannot be sold, the question of euthanasia may arise. Although this is a drastic measure of last resort, if no other option is available, humane destruction will prevent unnecessary suffering of the animals and legal action against you.

Once the drought has broken

When the rain comes, everyone breathes a collective sigh of relief, but the problems do not end with the first patter of the life-giving drops. Paddocks that have become dry and dusty will take time to fully recover and until then, the dangers to your horse of fresh, new grass will also have to be carefully managed. The temptation is to put the horses back into a paddock that is lush with new growth and feel the glow of satisfaction from watching them tuck into nature's gift. But as with all dietary changes, the introduction to new pasture must be done slowly to avoid the risk of colic and laminitis. If possible, start by allowing just a couple of hours each day on the new pasture to assist both the changeover in diet for the horse and the recovery of the pasture.

During the first months after the drought has broken, build up the period of time spent on pasture, removing the horses to a yard or a different paddock, so that the new pastures will not become depleted immediately. If you can, selectively graze your pasture by using temporary fencing.

Beware of the 'green drought' in which your pasture may appear green, but there is insufficient grass. You will need to continue with supplementary forage until there is sufficient pasture for the horse to graze consistently.

You can gradually wean the horses off any extra hard feed and resume your usual management systems. However, for those horses that have lost body condition during the drought despite your best efforts, continue to provide enough hard feed to recover body condition before changing the ration.

PROHIBITED SUBSTANCES

Drugs in sport are a hot issue and not just for human athletes. Performance horses are subjected to a many regulations, depending on the organisation that oversees the competitions, and so it is not possible to detail all the individual rules and regulations regarding the use of drugs and medications for each horse organisation. In addition, the list of banned substances and the methods of detection change on a daily basis.

The accuracy of testing has increased dramatically since testing procedures began; for example, the sedative acepromazine, which in 1960 could not be detected in the urine of horses, can now be detected for up to 5 days after its administration. Modern testing equipment and techniques can measure levels as low as several parts per billion (ppb), which is a strong deterrent against the deliberate use of performance-enhancing drugs.

But can the honest horse person be caught with a positive drug test? In some cases the answer is yes. It is possible to inadvertently administer a forbidden substance via environmental contaminants or natural products. A therapeutic drug may have an unusually prolonged excretion time in a particular horse. It is a common misconception that herbal products do not contain any drugs and therefore do not contravene the rules of racing or Federation Equestrian Internationale (FEI). Certain herbs have been classified as 'doping' by a number of horse sport regulatory authorities; for example, rosemary and caffeine are considered stimulants and are banned under FEI rules. Herbal products that contain ephreda or valerian are classed as illegal substances. Racing authorities warn trainers not to use herbal products (prior to racing) because prohibited substances may be detected.

Regulatory authorities recognise that many substances, or their metabolites, that produce a 'positive' test result are present at trace levels and are most likely derived from legitimate therapeutic medications or are of a dietary, environmental or endogenous (from within the horse) origin. Previously, a positive result of any magnitude was deemed an offence, but today there are approved thresholds for certain medications within the racing and FEI regulations. Some of the more common contaminants are listed here.

Hordenine

Hordenine is a naturally occurring plant alkaloid produced during the sprouting phase of barley. It is closely related to epinephrine, know commonly as 'elephant juice'. Brewer's grains, malt combings, distillers dried solubles and sprouted barley may all contain hordenine. There is no threshold allowance for hordenine, so do not include any of these products in the diet of a performance horse.

Dimethyl sulfoxide (DMSO)

DMSO is found in the urine of all horses and is believed to be almost entirely of dietary origin, most likely from lucerne hay. As a drug, DMSO is used as a topical anti-inflammatory and analgesic (pain relief) agent. A threshold of 15.0 micrograms per millilitre of urine or 1.0 micrograms per millilitre of plasma is permitted.

Salicylate

Salicylic acid (salicylate) is a normal metabolite in horses' urine and occurs naturally in several plants, notably lucerne and in the bark of some trees such as willows. The drug is used as an anti-inflammatory agent (aspirin), in topical creams and powders to treat ringworm and skin problems and as an active ingredient in liniments and medicated washes. Authorities allow for natural sources of salicylic acid with a maximum tolerance level of 750 micrograms per millilitre of urine or 6.5 micrograms per millilitre of plasma.

Morphine

Morphine occurs naturally in a number of pasture species and in poppy seed, which is used in certain bakery products. Poppies grow wild in Australia and New Zealand and their prevalence depends on seasonal and geographic considerations, so you should check your oats for the presence of poppy seeds and seek veterinary advice if you see them. There is no threshold for morphine.

Arsenic

Arsenic occurs throughout nature and is found in the urine of all horses; however, it has been popular as a stimulant in small amounts and to slow horses down in larger amounts. The international threshold for arsenic is 0.3 micrograms per mililitre in urine.

Caffeine

Caffeine is one of the xanthine alkaloids, which are widely distributed in nature and include theophylline (from tea) and theobromine (from chocolate). In addition to these sources, many so-called 'natural' products also contain xanthine alkaloids. Guarana is 60% caffeine and is commonly marketed as an 'energy uplifting' product. Feeding or administering products containing guarana extract will likely result in a positive swab. Because of the widespread distribution of caffeine, some regulatory authorities allow a low threshold in plasma and urine samples, but this is more the exception, with most sport horse organisations having a zero tolerance to caffeine in swab samples. An oral dose of caffeine can be detected for up to 10 days after administration.

Theobromine

Cocoa bean meal is a source of the alkaloid theobromine, as are the products of cocoa bean such as chocolate. Feeding 20 'M & M' chocolates with peanuts would produce a detectable concentration of theobromine in the horse's urine! Shipping of feed products in containers commonly used to transport cocoa bean meal or cocoa husks is another potential source of contamination. Given the high risk of inadvertent administration of theobromine, a threshold of 2.0 micrograms per millilitre of urine has been set by the major regulatory authorities.

Bicarbonate

Sodium bicarbonate has long been popular as a buffer to delay the onset of fatigue brought about by lactic acid accumulation. A maximum threshold for sodium bicarbonate in plasma has been set by several regulatory authorities. It should be remembered that horses vary in their reaction and excretion rates for additives and drugs and so it is not possible to set a definitive safe dose that is applicable to all horses. Bicarbonate is a common additive in ruminant diets and further illustrates the dangers of using feeds other than those specifically designed for horses. Other alkalising agents such as citrate can also produce an effect that may exceed the allowable threshold.

Vitamins

It should also be noted that under the rules of racing, vitamins administered by injection are also classified as a prohibited substance; however, no positive swabs have been 'called' in Australia.

This list is by no means exhaustive and simply serves to highlight some of the more common sources of inadvertent contamination of the diet of a performance horse with substances that may produce a positive drug test.

So, what measures can the concerned rider take in order to avoid, or minimise the risk of contamination?

1. Buy prepared horse feeds only from reputable manufacturers who operate under stringent quality control standards. The feed should be accompanied by documentation that lists the composition and comprehensive details of the nutrient content. A bag tag that states only the protein, fat and fibre percentage is not sufficient. Full disclosure shows that the technical and production teams at the mills are operating under best practice and are aware of the problems of prohibited substances and take these into account in the manufacturing, transport and storage of raw materials and finished products.

2. Whenever possible, you should also purchase your grains and hay from reputable merchants who understand the quality issues relating to performance horses. It is unwise to purchase hay or grain from farmers who may have stored these products next to feed or supplements intended for other animal species. Check oats for poppy seed contamination.

3. If you are competing at the elite level of FEI events where drug testing is being conducted, keep small samples of feed (250 g) and supplements (30 g) with their bag tags and/or a record of the batch number, from about one month prior to the commencement of the competition period. Ensure that samples are taken near the bottom of the feed bag where fine material may accumulate. These samples should be stored in cool, dry environment away from sunlight for 6 months. If a positive test is discovered, you will have a reference sample of feed.

4. Do not allow chocolate or coffee to contaminate your feed room or be fed to your horse.

5. Do not store feed for other species in any area used for preparing or storing horse feed.

6. Do not feed anything to your horse unless you are confident about the contents and composition of that feed/supplement. All products should be completely labelled with the ingredients and analysis.

7. Many substances can be absorbed through the skin and are then detected in tests. Be careful of liniments that contain oil of wintergreen, which is methyl salicylate.

8. Seek veterinary advice if in doubt regarding the cessation of treatment prior to competition.

It is the responsibility of riders, trainers and veterinarians to be aware of the regulations relating to a particular area of equine sport. It is the responsibility of the riders and trainers to ensure that they do not feed, inject or apply anything to their horses that may contravene these rules. Ignorance is not a legitimate plea against a positive drug test.

WHERE TO GO FOR MORE INFORMATION

Kentucky Equine Research has a website www.ker.com. The website of the Rural Industries Research and Development corporation (www.rirdc.gov.au) summarises local research on nutrition and general health relevant to Australian conditions.

AgResearch has a collection of information sheets, called AdFacts, on a range of pasture and weed-related topics for New Zealand conditions: see www.agresearch.co.nz/agr/publications.asp

CHAPTER 6

HORSE HEALTH

HEALTH ASSESSMENT

Prevention of ill health is better than cure, so it is important to:

- check your horse as often as possible
- ensure that stables and paddocks are safe
- keep a health record of your horse (e.g. past injuries etc.)
- have a good routine of feeding, watering, exercise, worming, vaccination (tetanus and strangles), hoof care and dental care
- know your horse's normal vital signs (i.e. the six health indicators: body temperature, heart rate, respiration rate, hydration status, mucous membrane colour and gut sounds)
- constantly evaluate the horse's appetite, body condition, coat, nasal discharge, faeces, urine and behaviour.

You can make a better assessment of the severity or cause of your horse's illness if you can make a basic clinical examination and you know the horse's normal vital signs. You can then discuss your findings with your veterinarian over the phone, which will help determine the urgency of a visit.

Health indicators

Body temperature

To take the temperature of the horse you need a thermometer, which you can purchase from your vet, a pharmacy or a saddlery shop. Thermometers are either digital or mercury (you must shake a mercury thermometer to force the mercury column down before using it or you may be mislead into thinking that the horse has a fever). If you are not used to taking a temperature it is easier to use the digital type.

Grease the bulb end of the thermometer with petroleum jelly or cooking oil. Stand close to the horse and to the side. Lift the horse's tail and gently insert the thermometer into the rectum. Push the thermometer against the rectal wall. Do not let go of the thermometer. If it is a mercury thermometer remove it after approximately one minute. Digital thermometers will beep when they are ready to be removed.

The normal temperature range for a horse is between 37°C and 38°C. Foals and small ponies are at the higher end of the normal scale than older and larger horses. Abnormally low or high temperatures need the immediate attention of a vet.

Figure 1. Taking a horse's temperature (courtesy of Jane Myers).

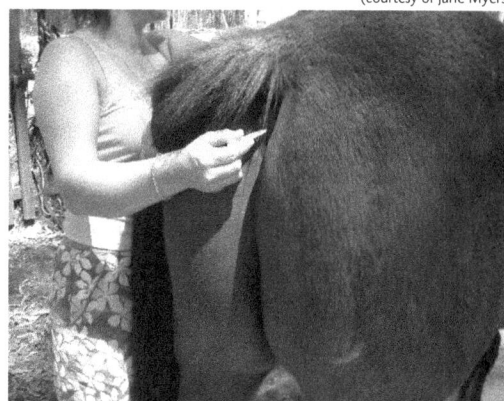

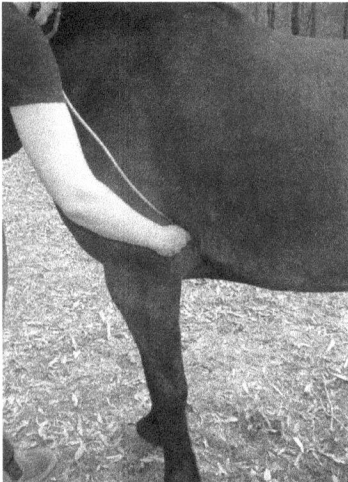

Figure 2. Measuring the heart rate with a stethoscope (courtesy of Jane Myers).

Heart rate

Heart rate can be measured with a stethoscope (purchase from vet, pharmacy or saddlery) or by hand.

To listen to the heart with a stethoscope, stand on the near side of the horse near the girth. Approximately one handspan up from the base of the chest (sternum), push the stethoscope in behind the muscle above the elbow (the brachial triceps; see Chapter 4, Figure 2) and press it against the ribs. Often the act of putting the stethoscope on the skin excites the horse, which increases the heart rate, so you may have to wait until the horse (and its heart) settle. Count the beats for one minute. If the horse will not stand still for one minute, count the beats for 15 seconds and multiply the result by four.

You will hear two heart sounds, which correspond to the heart muscle contracting (systole) and then relaxing (diastole). An approximation of the sounds that you hear are 'lubb' (systole), 'dupp' (diastole), although they may be indistinguishable at high heart rates.

The normal heart rate for a horse ranges from 30 to 40 beats per minute (bpm), but this increases when the horse is eating, excited, fevered, anaemic, hot, exercising etc.

Young horses have a higher heart rate than older horses:

- foals at birth are 80–120 bpm
- older foals are 60–80 bpm
- yearlings are 40–60 bpm.

You need to know your horse's normal, resting heart rate, which will become lower as the horse becomes fitter; for example, a very fit endurance horse can have a resting heart rate of 27 bpm. A horse that is fit will return to its resting heart rate sooner after exercise than an unfit horse.

To measure the heart rate by hand, which is more correctly termed 'taking a pulse rate', you use an arterial pressure point. These are found on the inside of the lower jaw, 10 cm (4 inches) below the eyes, the inside of the forearm just in front of the elbow, on the pastern above the heels (if a bounding pulse can be felt it is a possible sign of laminitis or a hoof abscess) and under the tail near to the body.

Respiration rate

Respiration is the alternating inhalation and exhalation of air. The normal respiration rate for a horse is 8–20 breaths per min. The rate of respiration is measured by counting flank movements either by sight or by placing a hand on the ribcage. Normally you can see a horse breath in, but not out. If the horse is making an effort to breath out it is a sign of a respiratory problem. The respiration rate should never be higher than the heart rate.

Younger horses and smaller horses breathe faster than older and larger horses. Pregnant females also have a higher rate. The respiration rate is increased by exertion, fever, pneumonia, changes in the acidity of the blood (acidosis or alkalosis), anaemia and infectious diseases.

Synchronous diaphragmatic flutter ('thumps') occurs when the diaphragm contracts in unison with the heart rate and is caused by electrolyte disturbances when under stress from exercise.

Hydration status

If the horse loses more water from its body than is absorbed, it is dehydrated, which leads to a reduction in the circulating blood and relatively dry tissues. Dehydration is caused by decreased water intake and/or too much water lost through sweating, diarrhoea or urination. The pulse becomes weaker and other signs of dehydration are sunken eyes, dry gums, darker gums and the 'skin-tenting' test.

The skin-tenting test involves pinching a fold of skin on the neck between the fingers. The skin is let go and it should return to normal by the count of two. It should not stay up like a tent; if it does, this is a sign of significant dehydration. If the horse is also not drinking, then veterinary attention should be sought quickly.

Mucous membrane colour and capillary refill time

If the horse is healthy the mucous membranes will be pink and the capillaries will fill quickly with blood. To determine the mucous membrane colour, look at the horse's gums, which are naturally a shade paler than a human's. Too pale indicates anaemia or shock, yellow shows jaundice, blue is a lack of oxygen, and darker red indicates dehydration and serious metabolic disease.

To check the capillary refill time, press the gums with a fingertip and release. The colour should return by the count of two; any longer may indicate shock.

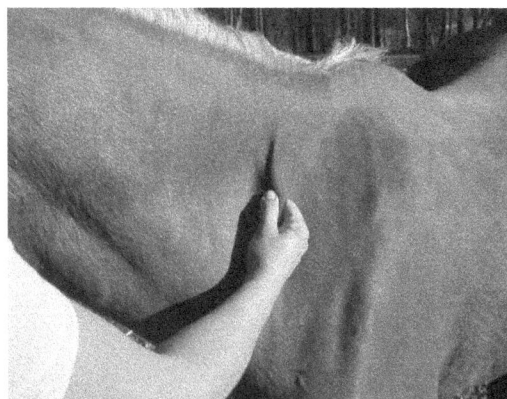

Figure 3. Skin-tenting test to determine hydration status (courtesy of Stuart Myers).

Gut sounds

Gut sounds should always be audible and if not present may indicate colic. Place your ear or a stethoscope against the barrel behind the last rib. Listen on both sides. If there are no sounds call a vet.

Other indicators

Appetite

Horses vary in their normal appetite, some are naturally very greedy and others are very finicky. It is important that you know what is normal for your horse so that you have with a comparison for when the horse is ill.

Body condition

You should be constantly assessing your horse's body condition. There is a scoring system for body condition and it is a good idea to practice using it. Ideally your horse should have good body condition, which is a score of 3 (see Chapter 5).

Coat

The coat of a healthy horse lies close to the body and shines (some colours shine more than others e.g. bays and blacks shine more than greys and roans). If the coat is standing up, it usually means that the horse is cold, but it can indicate illness. The coat hair stands up because the horse is attempting to trap air and warm itself. In the short-term, the condition of a horse's coat is not affected by illness, but a chronically unwell horse has a 'staring' coat and/or very little body condition. Diet can also affect the condition of the coat.

Figure 4. Checking the capillary refill time (courtesy of Jane Myers).

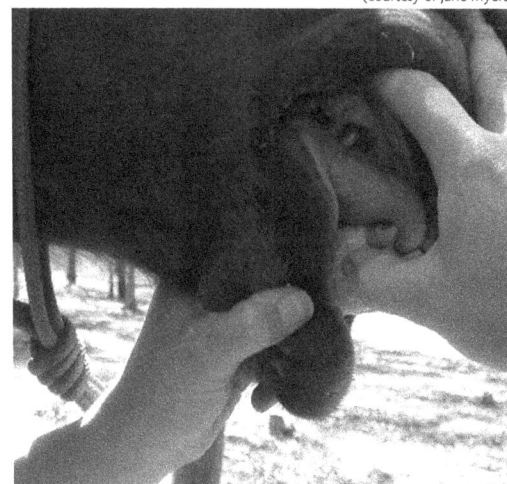

Nasal discharge

It is quite normal for a horse to have a trickle of liquid in the nostrils, but it should be clear and not sticky. Yellow or green sticky mucus indicates illness.

Behaviour

The behaviour of individual horses varies enormously, so again it is important that you know what is normal for your horse. A healthy horse should be alert, interested in what is going on around it, sociable with other horses and be forward moving when ridden. A healthy horse will, when turned out with other horses, occasionally have a run around, lie down to sleep in the sun, play with other horses etc. (see Chapter 2 for more information about horse behaviour).

When a horse is ill it will not display its normal behaviour. It may stand with its head down or lie down more than normal or not at all. The sick horse looks 'tucked up' (the flanks are sucked in). The horse will either not be alert or may be excessively anxious, it may keep looking at its flanks and/or paw the ground (signs of abdominal pain). Other behavioural signs of ill health include excessive chewing on objects or sudden aggressive behaviour.

Basically, any changes in normal behaviour should be investigated. Checking the temperature, pulse and respiration is the minimum examination if you suspect that your horse is ill. If you are concerned about anything you should call your vet.

FIRST AID AND WHEN TO CALL THE VETERINARIAN

First aid is the initial treatment applied if a horse has been injured or is developing a condition or disease. These are the steps that you can take to improve the situation, or at least prevent it from worsening, before a vet examines the horse. A veterinarian should be called for serious conditions, but if the owner is sensible and follows the basic techniques of first aid, there are only a few true emergencies that require immediate veterinary attention. These will be dealt with later in the chapter.

First aid techniques

- Assess the situation and then take steps to prevent it worsening.
- Attend to any bleeding.
- Assess the horse and call a vet if necessary.
- Attend to any swellings and pain.
- Make the horse comfortable.

First aid kit

The following items should be kept together as a kit:

- 2 rolls of 75-mm adhesive dressing (e.g. Elastoplast®)
- 2 rolls of 75-mm elastic bandage
- a twitch
- 1 roll of cotton wool
- 1 packet of gauze swabs
- iodine disinfectant solution (e.g. Betadine®)
- antibiotic powder
- blunt-ended scissors
- tetanus antitoxin.

The telephone numbers of your usual veterinary clinic (plus those of other clinics/vets if the usual vet cannot be contacted in an emergency) should always be displayed in a prominent place near the phone.

Emergencies requiring immediate veterinary attention

- Shock
- Foaling difficulties
- Arterial bleeding
- Evisceration
- Fractured limbs
- Snakebite

Shock

Shock may be caused by severe internal or external bleeding, extensive wounds, or traumatic accidents such as being hit by a car. Horses that are in shock are usually subdued, have pale mucous membranes and a fast, weak pulse or a heart rate of more than 70 bpm. Any bleeding should be controlled by bandaging and the horse should be kept warm, calm and stationary.

Foaling difficulties

Any mare that has been unsuccessfully trying to foal for 30 minutes after her water bag has broken should be considered an emergency. Keep the mare on her feet and walking around while waiting for the vet to arrive.

Arterial bleeding

The bleeding from most wounds is venous: it either oozes or runs out in a continuous stream and can be easily controlled with a pressure bandage. Arterial bleeding spurts out. Bleeding from small arteries in the lower leg can usually be controlled with a pressure bandage. Other arteries, however, have to be surgically closed to stop the bleeding. In any case of arterial bleeding, apply a firm pressure bandage, keep the horse quiet and call a vet immediately. (See later section on bandaging.)

Evisceration

Evisceration is when the intestines of the horse fall out of its abdomen. It is very uncommon, but may occur after a stake wound to the abdomen or, very rarely, after a horse has been gelded.

If the eviscerated horse is down, keep it down and quiet. Use a twitch if necessary. If it is standing, keep it still and improvise a sling with a clean sheet around the abdomen to hold the intestines off the ground. Try to stop any more intestines falling out, and keep those that are out as clean as possible until the vet arrives.

Fractured limbs

It is very important to secure the horse and keep it still to prevent aggravation of the fracture because it may be possible to salvage the horse if no further damage is done. Put a heavy, support bandage around the area of the suspected fracture. (See later section on bandaging.)

Snakebite

It is very rare that someone actually sees a horse being bitten by a snake. If you suspect that the horse has been bitten, do the following immediately.

- Keep the horse quiet and in the shade.
- Apply ice to the bite and hose off with cold water.
- Apply a pressure bandage if you do not have ice or after hosing.
- Seek veterinary advice.

Common conditions requiring prompt veterinary attention

- Choke
- Allergic reactions
- Tying-up
- Colic
- Laminitis

Tying-up, colic and laminitis are discussed later under Nutritional Diseases.

Choke

If something blocks the horse's oesophagus, making it impossible for the horse to swallow, food and water will come out the nostrils and the horse will stop eating.

Allergic reactions

Allergic reactions can cause swellings or lumps under the skin that may require veterinary treatment (see Plate2, Figure 11). If the horse is having trouble breathing, it is an emergency.

NUTRITIONAL DISEASES

There are several health problems in horses that are related to feeding, such as tying-up, colic, laminitis, stringhalt, hyperlipaemia and anhydrosis. The developmental orthopaedic diseases are the result of incorrect nutrition and genetics, gastric ulcers are related to stress and restricted access to grazing and a particular problem in northern Australia (big head) is related to tropical pasture species and requires nutritional and management solutions. If the diseases have not already been described, this section details the symptoms and treatments required and Chapter 5 deals with the dietary management.

Tying-up

Tying-up, azoturia or Monday morning disease are terms used to describe the symptoms of several diseases involving muscles. Tying-up can threaten the career of performance horses and is an extremely frustrating condition for the horse owner. For years, all horses that tied-up following exercise were thought to suffer from the same disease, but the treatment and prevention protocols that seemed to work for some horses did not help others. As a result, confusion and controversy developed regarding the cause and treatment of tying-up. Recently the many different disease conditions that result in the common symptoms of tying-up have been studied and tying-up is now classified according to the frequency of episodes. Horses that have one or two episodes of tying-up in a lifetime are considered 'sporadic' cases. There are many potential causes of these episodes, but most frequently it is exercise that exceeds the horse's level of fitness; other key triggering factors are electrolyte depletion, hyperthermia, concurrent lameness or respiratory disease. Horses that have repeated episodes of tying-up are classified as 'chronic' cases and two specific causes have been recently identified: a muscle contraction disorder (recurrent exertional rhabdomyolysis, RER) and a disorder in carbohydrate storage and utilisation (equine polysaccharide storage myopathy, PSSM or EPSM).

Sporadic tying-up

Sporadic tying-up usually involves a specific muscle group and should be considered a veterinary emergency if the horse is sweating profusely, reluctant to move or has dark urine. The veterinarian may administer treatments to relieve anxiety and muscle pain, as well as intravenous fluids and electrolytes to replace fluid loss and lessen the myoglobinuria (muscle proteins in the urine) that may damage the kidneys. Further treatment strategies include stable rest, followed by hand-walking and turn-out after the muscle stiffness is relieved, and adjustments to the diet, particularly the grain portion (see Chapter 5 for more details). The duration of the resting period has not been firmly established by research, but monitoring the serum levels of muscle enzymes may be a useful guide. Following an episode of tying-up any training regime should be resumed gradually and consistently to prevent further muscle damage.

Prevention of further episodes involves not overexerting unfit horses and dietary management (see Chapter 5 for details). Horses that are lame or have an upper respiratory tract infection have an increased risk of tying up.

Chronic tying-up

When horses have repeated episodes of tying-up, the disease is considered chronic. Many different breeds of horses have been reported as susceptible, including Quarterhorses, Thoroughbreds, Standardbreds, Paints, Morgans, Arabians, draught breeds and Warmblood horses. Fillies and mares are more susceptible and nervous horses are thought to have a higher risk of chronic tying-up. The proposed causes of chronic tying-up include electrolyte imbalance, hormonal imbalance, lactic acidosis, vitamin E and/or selenium deficiency, hypothyroidism, and muscle ischemia, as well as RER and PSSM.

Recurrent exertional rhabdomyolysis (RER)

Preliminary genetic research and breeding trials indicate that RER as an inherited trait in Thoroughbred horses and is common in nervous fillies and mares of Arabian, Standardbred and Thoroughbred breeding, although it also occurs in colts and geldings and 'calm' horses. Suscptible horses often develop the condition when they are excited, stressed and/or after a period of stable rest.

The exact cause of RER in horses is still unclear. It was believed to be similar to lactic acidosis in racing horses, but recent research has shown that the muscle lactate concentrations are low, not high, in these horses when tying up occurs and that the condition usually occurs during aerobic (slow) work. The most recent theory is that RER is an abnormality in the intracellular calcium regulation of muscle contraction because muscle tissue from affected horses has been shown experimentally to have an increased sensitivity to contraction.

Treatment of RER requires veterinary assistance to make the horse comfortable and prevent further stress and muscle damage. Administering sedatives to a nervous horse prior to exercise is common practice, but cannot be done for competition events. Medications that regulate the intracellular sodium and calcium levels can be prescribed. Daily exercise is essential. Beginning approximately 24 hours after an RER episode, horses should be hand-walked or turned-out each day. Prolonged stable rest seems to be counterproductive and may predispose the horse to further episodes after training resumes. A gradual return to full training of affected horses can begin when the levels of serum muscle proteins return to normal. Thoroughbreds may need more fast work and less slow work whereas Standardbreds should be jogged a little faster, but for less time.

Prevention of further episodes of RER is difficult. Controlling the environment and a well-established daily routine are essential to reduce the apprehension and nervousness of susceptible horses. Dietary therapy is also warranted (see Chapter 5 for more details). If anxiety or nervousness is a major issue, then a regular routine, changing the diet and using calming agents may help.

Polysaccharide storage myopathy (PSSM)

Horses with PSSM store more than the normal amount of muscle glycogen, not because these horses cannot use muscle glycogen for energy production, but instead there appears to be the creation of an abnormal form of glycogen (filamentous polysaccharide) that is metabolised at a much lower rate by the horse and thus accumulates in the muscle.

PSSM has been identified in Quarterhorses, Paints, Appaloosas, draught breeds, Warmbloods, and a few Thoroughbreds. It is diagnosed from a muscle biopsy showing the distinctive feature of abnormal glycogen storage and has been recently reported in Australia. Horses with PSSM differ from horses with RER by being calm rather than nervous. They typically have a history of tying-up associated with the onset of training when the animal is still relatively unfit. Horses with PSSM have the classic symptoms of tying-up, including long-term elevation of the level of muscle enzymes in the serum.

Treatment of horses with PSSM involves many of the same veterinary procedures needed for other causes of tying-up. Treatment protocols also attempt to minimise the occurrence of future episodes through dietary manipulation (see Chapter 5 for more details). Daily activity, riding or lunging, together with pasture turn-out, is essential for minimising future episodes, but, as with other causes of tying-up, it is essential that the training program is increased slowly, taking at least three weeks to build up to 30 minutes on the lunge with a 15 minute break before riding. Confinement in stables for more than 12 hours/day appears to increase the incidence of tying-up in these horses.

Colic

Colic is a general term for abdominal pain in the horse and is one of the most dangerous and costly equine medical problems, estimated to occur in 1 of every 10 horses each year. The most common potential causes and risk factors associated with colic in horses have been identified, but as any equine veterinarian would attest, there seem to be countless situations that can precipitate colic. Many diseases and disturbances cause abdominal pain, which may be only mild and temporary, but in other cases it may be very serious and progress rapidly to cause death if not treated. It is therefore important that you can recognise the early signs of colic so that veterinary treatment is given as soon as possible.

Risk factors for colic

- Change in diet
- Change in hay type and intake
- Grain intake greater than 2.5 kg/day
- Eating sand
- Spoiled feed
- Restricted water access
- Change in activity
- Transport
- Change in housing
- Parasites
- Administration of wormer
- Previous history of colic
- Breed

Factors that are difficult for horse owners to control include the breed and age of the horse, together with the geographic region of the country where the horse is kept. Other factors such as internal parasite control and the quality and quantity of fibre and grain in the diet can be managed to reduce the risk of colic.

Symptoms of colic

Whatever the cause of colic, the outward signs are similar, especially in the early stages. The horse becomes very restless: it paws at the ground, stamps, kicks at its belly, and frequently lies down, only to get up again almost immediately. In some cases the first signs of colic are failure to eat, lethargy or depression and sweating because of the pain. The heart rate will increase, which can be used as a guide to the severity of colic. If the onset is sudden and severe (an acute case), the mucous membranes become darker and congested. The horse may not pass any droppings, but in some cases it may pass more.

A horse with abdominal pain looks at its flank, rolls, and may lie on its back. It may lie down carefully, but 'slump' onto the ground and then be very slow to get up, or it may assume abnormal postures, often sitting like a dog or standing with the hind legs stretched out in a 'saw-horse' attitude.

In summary, the most common signs shown by horses with colic are:

- rolling   44%
- pawing   43%
- lying down   29%
- getting up and down   21%
- flank watching   14%
- lip curling   13%
- backing into corner   10%
- kicking at belly   7%.

The bouts of pain, and the appearance of associated signs, are often intermittent, especially in the early stages. In the most severe cases, the pain is continuous, and signs of shock may appear in addition to profuse sweating, laboured breathing and violent uncontrolled movements that can cause the horse to injure itself.

Causes of colic

Severe abdominal pain is usually caused by distension of the stomach or intestines (by fluid, gas or food) or a reduction or cessation of the blood supply to the intestinal tract.

It is very important that the particular cause of the colic is accurately determined for correct treatment and prevention. Some of the more common forms of colic are listed here.

Types of colic

Acute colic has severe, continuous pain and rapidly progresses to shock. It is generally fatal. The causes of acute colic include:

- twisting of the bowel, which creates a blockage
- strangulation of the bowel
- intussusception (telescoping of the bowel into itself)
- blockage of the small intestine by a large, dry mass of food or a foreign object
- gastric dilation/grain engorgement
- impaction of the ileo-caecal valve
- major disorder of the intestinal blood supply
- distension of the intestine by gas (flatulent colic).

Sub-acute colic is the most common type of colic. There is intermittent pain and it is seldom fatal. The causes of sub-acute colic include:

- impaction of the large intestine
- hypermotility of the gut (spasmodic colic), which is often precipitated by excitement, show preparation, large drinks of cold water when hot after work, sudden changes of feed, overfeeding or parasites
- urinary tract disorders
- sand
- temporary disorder of the intestinal blood supply
- mineralised aggregation in the large bowel.

Chronic or recurrent colic can be the result of:

- verminous arteritis. Migrating redworm larvae damage the blood vessels supplying the gut, which leads to blockage or constriction of the blood vessels with the result that sections of the intestines do not receive sufficient blood supply.
- mild impaction of the large intestine by indigestible coarse roughage, heavy feeding, poor condition of the molar teeth etc. This is the most common cause of recurrent colic.

Conditions resembling colic

Laminitis, urinary disorders, tying-up, pleuropneumonia, some poisonings and diarrhoea are all conditions that may include symptoms resembling colic.

Treatment of colic and when to call the vet

The treatment of colic depends on its cause. Treatment may be medical, using drugs or mineral oil drenches, or surgical in the case of physical blockages and intestinal emergencies. If a horse with colic needs surgery, early treatment will increase its chance of recovery.

If signs of colic appear and persist for longer than 30 minutes seek veterinary help. If possible, check the horse's heart rate and mucous membrane colour before you phone the vet. Stay with the horse until the vet arrives and try to prevent the horse from injuring itself or complicating a simple colic by rolling (the intestines may twist, resulting in severe colic).

Horses with colic often lie down because it is the most comfortable position. They can be left in this position as long as they do not attempt to roll. Do not administer anything by mouth.

Prevention of colic

Always check that there are not any foreign objects in the horse's feed and avoid giving horses a large feed or drink immediately after strenuous exercise. Always make changes to the feeding program gradually over 7–14 days and do not let a horse gorge itself on grain. Irregular or excessive feeding and exercise are often the cause of spasmodic colic. (See Chapter 5 for dietary management and prevention of colic.)

Poor control of intestinal parasites (worms) is another common cause of colic. A regular program of preventative treatment for the horse and reducing its worm intake (manure control: see Chapter 8) will minimise this problem. Your vet can advise on a worm control program.

Regular dental care to prevent your horse's teeth becoming too sharp or worn will ensure that it can properly chew its feed and thus reduce the likelihood of large masses of roughage building up in the intestinal tract (see Chapter 2 for horse dental care).

Laminitis (founder)

Symptoms of laminitis

Laminitis usually affects both front feet. Sometimes only one foot is affected, but this is usually because one front leg has had to take more weight, for example, after the other front leg has been injured. It is possible, but rare, for laminitis to affect the hind feet without affecting the front.

The horse stands with its forelimbs stretched out in front of the chest and the hind limbs placed under the abdomen, in an attempt to relieve the weight on the front feet. There will be heat in the feet and a bounding pulse can be felt in the main artery to the foot. The horse will be reluctant to move and may lie down.

Laminitis is a breakdown in the bond between the sensitive and insensitive laminae in the foot, which leads to rotation of the pedal bone and even penetration of the sole (see Chapter 4 for a description of the parts of the hoof). The bond between the laminae breaks down because the cells are not receiving sufficient nutrients from the blood supply. The laminae anchor the pedal bone to the hoof wall against the pull of the deep digital flexor tendon and when they break down, the pedal bone tips downward towards the sole. In long-standing (chronic) cases, the rotation of the pedal bone will lead to disturbances of hoof growth, with lines appearing in the hoof, a concave wall with the toe turned up and a sole in which the pedal bone is very close to the ground. These horses are often sore or lame after minimal exercise and they need continual corrective shoeing to be able to function (see Chapter 2 for details about shoeing and hoof care).

Figure 5. Acute laminitis (courtesy of Reg Pascoe).

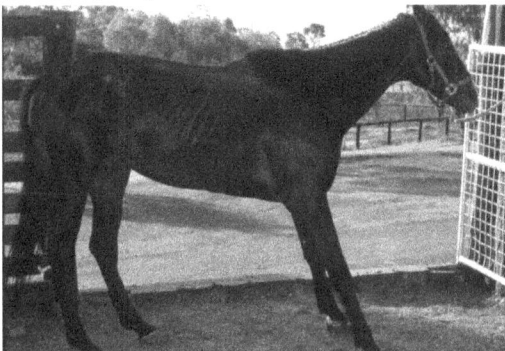

Causes of laminitis
- Grazing unlimited amounts of lush pasture.
- Gorging on starch-rich grains.
- Obesity and lack of exercise.
- Intestinal disorders resulting in acute diarrhoea.
- Retained afterbirth in recently foaled mares.
- Surgery, especially if one leg is taking extra weight.
- Non-weight-bearing lameness in one leg.
- Some drugs.

The recent discovery of fructans in rich, lush grass has shed new light on the involvement of pasture in the occurrence of laminitis. It is now believed that if the horse is unable to digest the fructans it will cause the micro-organisms in the hindgut to react, which in turn affects the circulation to the feet. The concentration of fructans in the grass changes according to such factors as the amount of sunlight, temperature and water availability. In addition, the concentration of fructans changes during the day, being highest between 9.00 am and 3.00 pm.

Therefore, until more research is conducted, the management of horses that are prone to laminitis should include preventing them from grazing grasses that are high in fructans. The horse with acute or chronic laminitis needs a high-fibre, low energy diet without any grain (see Chapter 5 for more details on the correct feeding of horses).

Treatment of laminitis

Mild laminitis: an experienced owner using a combination of anti-inflammatory medication, a weight loss program and exercise can sometimes manage mild cases.

Acute laminitis: prompt veterinary attention is essential. Delayed treatment may lead to the development of chronic laminitis and the irreversible changes in the hoof that may necessitate the destruction of the horse. The vet will use drugs, administer paraffin oil via a stomach tube to flush out the gut, and provide a temporary 'frog support' for each of the affected feet, which is very important and can also be done with corrective shoes such a heart bar shoe. Keeping the horse in a sand yard will provide some support for the sole and frog. Following the medical treatment, some horses will benefit from walking and your vet will advise you on a suitable program of short walks. Although some horses with acute laminitis like to stand in the mud it is preferable to stand the horse's feet in a bucket of warm water to increase the blood flow to the laminae.

Severe cases may require surgical resection of the front of the hoof wall or the creation of a drainage point in the hoof wall. It may even be necessary to surgically cut the check ligament or deep flexor tendon within the foot to reduce the rotation of the pedal bone.

In some severe cases, the pedal bone will continue to sink or rotate until its tip has penetrated the sole or created an abscess under the sole. These horses need prolonged and intensive treatment and in many instances this may be uneconomic and contrary to the horse's welfare. Laminitis is very painful!

Chronic laminitis: it will usually be necessary to X-ray the horse's feet to determine the degree of rotation of the pedal bone. Hoof trimming to shorten the toe and lower the heel will then allow a more normal pedal bone/hoof wall relationship to develop. These horses will often have disturbances of hoof growth and need good hoof care and supplements that promote hoof growth (e.g. bitoin, zinc and methionine). Some horses will always have sore feet.

Prevention of laminitis

Prevention is vital because laminitis is a serious condition. During spring you must record your horse's body condition (see 'Condition scoring' in Chapter 5), so that you will notice the early changes in weight that are a prelude to founder. The crest should not get too large or hard, and you should always be able to feel the ribs. If the horse or pony is getting too fat, take preventative measures. Reduce the feed intake by placing the horse in a paddock with less feed or even in a yard or a ploughed paddock (with low-energy grass hay provided) for a period of the day. Exercise will also help to more quickly reduce the horse's weight; without work it is a slow process to regain normal (good) body condition.

When horses are being fed large amounts of grain, reduce their feed intake when they are not working hard, which will help prevent the occurrence of both laminitis and tying-up. The risk of laminitis will be minimised if the roughage content of the diet is kept high.

After a mare has foaled check that all the afterbirth has been passed within 6 hours of foaling and if not, seek immediate veterinary help.

When a horse has a very severe lameness for another reason that causes it to put all its weight onto one front leg, watch closely for the development of laminitis. Ask your vet for advice in this situation.

Australian stringhalt

Australian stringhalt was first recorded more than 130 years ago. It is a very distressing disease for the horse, and the owner, although it does not involve severe pain. Recovery from Australian stringhalt is possible and horses can return to their usual activities after a variable period of time.

Symptoms of stringhalt

The classic sign of Australian stringhalt is exaggerated lifting of each hind leg as the horse moves forwards or backwards. The severity of the diseases varies and is worsened by cold, excitement and variation in daily routine. In very severe cases the horse starts to walk and one leg becomes stuck up under its belly, as though it was glued there. The horse is unable to move backwards at all and may only be able to move forward using an exaggerated 'bunny hopping' motion in which both hind legs are lifted and the front legs are used as paddles. In mild cases, exaggerated lifting of the leg is only obvious under special circumstances such as when the horse goes backwards, is upset, turns sharply sideways, or in cold weather. Mild cases are often first noticed when a horse is unloaded from a float.

The high-stepping gait is accompanied by muscle wasting in the gaskin area just above the hock, between the legs, and often on the thighs. In an unusual case there may be muscle wasting in and around the shoulder area, accompanied by changes in the forelimb gait.

Another important consequence of stringhalt in that the horse may become a 'roarer' because of laryngeal paralysis. Some degree of laryngeal paralysis is very common with stringhalt and the horse may need surgery to restore the laryngeal function.

Horses that are severely affected with Australian stringhalt are often very frustrated by their inability to move around the paddock, but there are no signs that they are in pain. Some owners report that the horse seems unusually nervous after contracting the disease, but others do not notice any behavioural changes.

Cause of stringhalt

The high-stepping gait and the muscle wasting are the result of degeneration of the nerves in the horse's hind legs. Unlike many other diseases of the nervous system, the nerves and muscles regenerate over time and so horses can recover from Australian stringhalt.

The reason for the nerve degeneration is still unknown, despite much research over recent years. However, some of the risk factors have been identified: the disease principally occurs in larger, adult horses during late summer and autumn, in dry years and when grazing relatively poor-quality pasture. Flatweed (*Hypochacris radicata*) is often found in the paddocks in which horses develop stringhalt. The disease is most common in horses that are not receiving a significant amount of hand feeding. It occurs sporadically and the average incidence is approximately 10–15% of a group of horses. Just because one horse in a paddock contracts stringhalt, it is not inevitable that the others will also develop the disease, although it is a good signal for some management changes to minimise the risk. In some cases the disease has recurred in the same paddock in successive years, but most often it occurs as an isolated incident. If the horse has stringhalt once, it is not likely to get it again. There appear to be some geographic areas of Australia where the disease is more common and Thoroughbreds and draught breeds appear to have a greater risk of developing the disease than other breeds of horses.

In summary the cause(s) is unknown, but could be toxic factors in plants or fungi, or nutritional deficiencies of energy or vitamins that cause the nerves to degenerate. It is likely that the cause is multifactorial.

Treatment and recovery of stringhalt

It is common for the signs of Australian stringhalt to be more severe during the first month after the disease is diagnosed, and then there is a gradual recovery, usually over 6–9 months. Some horses have recovered within 2 weeks whereas others have taken 2 years. Many horses can resume full work some months before they are completely recovered if they have only minimal alterations in gait.

If possible, remove the horse from the paddock and give some hard feed, otherwise hand-feed while the horse remains in the paddock. Horses have recovered while remaining in the same paddock and because the disease is seasonal, it is likely that the causative factors may only persist for a short time. Horses can be safely put back into an affected paddock after an autumn break. You may need to sedate a horse affected by stringhalt in order to reduce the severity of the signs while you are moving it.

There are many treatments that have been used for stringhalt, but some horses recover without any treatment at all. Surgical removal of some of the extensor tendon over the hock has been popular, but with very variable results and is only really justified when surgery is essential to improve the welfare of the horse.

Two new medical treatments have been recently reported by researchers in Australia and may be used in the future to reduce the severity of the clinical signs of stringhalt. Vitamins may also help (see Chapter 5 for details). Remember that the horse with stringhalt is not in any pain or distress and that recovery, given enough time, is the rule rather than the exception.

Hyperlipaemia

Hyperlipaemia is a disorder of ponies in which there are high levels of fat in the blood. It is not an infectious or contagious disease and occurs when ponies are not getting enough energy from their feed, causing body fat stores to be used as the source of energy.

Symptoms of hyperlipaemia

Hyperlipaemia only occurs at certain times of the year, mostly when pasture is scarce or poor quality, which is usually late summer in southern Australia and winter in northern Australia. Most cases of hyperlipaemia involve mares in late pregnancy or early lactation when their energy requirements are highest and they are most susceptible to a decrease in energy intake. Cases of hyperlipaemia have also occurred in dry mares that have been stressed by prolonged transport. Any factor that causes a decrease in food intake may lead to hyperlipaemia; for example, colic, intestinal parasites and laminitis treatment or prevention.

The symptoms vary and early signs may be only minor behavioural changes such as disinterest. This is followed by depression and loss of appetite, progressing to muscular weakness, tremors and incoordination. Some animals may have oedema (fluid swelling under the skin) between the front legs and along the abdomen. Many of the signs are the result of internal blood clotting caused by the high levels of fat in the blood. Extensive fatty infiltration of the liver occurs, and in some cases the liver may rupture. Cloudy, milk-like blood plasma is a characteristic feature of the disease.

Treatment of hyperlipaemia

If the disease is not detected early in its clinical course, 80% of affected animals will die despite intensive treatment. An pony showing signs of hyperlipaemia requires immediate veterinary attention. High-energy feeds and glucose are used to increase the animal's energy intake, to 'switch' the pony's metabolism back to using sugars as the primary energy source, rather than body fat. A 5–7 day period of intensive veterinary treatment may cost more than $200 per day.

The second phase of treatment is 5–6 weeks of intensive care during which the pony must continue to be fed a high-energy ration and any changes in feed intake and general condition must be carefully monitored. Because most affected ponies die, every effort should be made to prevent the disease.

Table 1. Progression of the clinical signs of hyperlipaemia

Time	Clinical signs
Approximately 10 days	Decreased appetite
	Depression, dullness
	Loss of bodyweight
	Muscle weakness
	Jaundice (yellowish discolouration of mouth and eyes)
	Muscle tremors
	Swelling between front legs and under belly
	Mild abdominal pain
	Staggering gait
	Complete loss of appetite
	Diarrhoea
	Death

Prevention of hyperlipaemia

Ponies should be fed to maintain body condition and supplementary feeding must commence before they lose too much condition (see Chapter 5). Segregation of ponies according to age, type and body condition, and careful observation of weight changes within these groups is essential.

Growth, work and lactation are the main factors influencing a pony's nutritional requirements. Owners need to know that there are variations in the energy requirements of different breeds of ponies, as well as variations in the needs of individual animals.

Anhydrosis

A horse that cannot produce sweat in normal quantities has anhydrosis and may be called a 'non-sweater' or a 'drycoated' horse, or a 'puffer' because these horses puff and pant a lot after work. Horses that sweat lightly or only in patches, such as under the mane, in the saddle area, and on the chest, are known as 'shy' sweaters. The condition seems to occur most often in Thoroughbreds, but Quarterhorses and other breeds may also be affected. A veterinarian can make a definite diagnosis by injecting adrenaline under the skin. In a normal horse, the area around the injection site will rapidly produce sweat whereas the anhydrotic horse has a slow, weak, or nonexistent sweating response.

Symptoms of anhydrosis

The conversion of stored energy to fuel for exercising muscles is not an efficient process, with most of the energy being lost as heat. A horse performing treadmill exercise in hot, humid conditions will show a rise in body temperature as rapid as 0.2°C every 10 seconds. It is estimated that during the cross-country phase of a three-day event a horse can produce enough heat to raise its body temperature more than 15°C. In a normal horse approximately 65% of body cooling comes from evaporation of sweat, with another 25% attributed to respiration. A horse that is unable to sweat can rapidly develop heat stress, which affects the brain and can lead to incoordination, unwillingness to continue exercise, collapse, convulsions, and death. Horses that survive severe heat stress may have permanent brain damage.

Up to 25% of horses in hot and humid climates may show some degree of anhydrosis, including some that are not working. There does not appear to be a correlation with sex or coat colour. Often the problem is not noticed until the horse is moved to a region with hot, humid weather. However, anhydrosis can also develop suddenly in horses that have been trained in warm climates. Typically, such a horse might perform on an extremely hot day and sweat excessively. Several days later, sweat production is greatly reduced or absent and then other signs of the syndrome appear:

reduced appetite, sluggish performance, dull coat, and hair loss on the face and croup. Loss of performance and weight occur in chronically affected horses. Core temperature rises quickly with exercise, and pulse and respiration remain elevated after work as the body attempts to lose heat from the lungs.

Treatment of anhydrosis

The cause of anhydrosis is unknown. Intensive electrolyte supplementation seems to trigger a return to normal sweat patterns in some cases, and non-sweaters often resume sweating when they are moved to a cooler climate or when temperatures moderate. Acupuncture has been effective for some horses.

One study has found that a hormonal or metabolic imbalance may lead to high levels of adrenaline being produced and affecting the normal response of the sweat glands. Examination of the skin of some anhydrotic horses has found a high percentage of sweat glands that are abnormal, atrophied or plugged. Other studies have examined the use of supplements (see Chapter 5 for details).

Suggested management practices include:

- moving the horse to a cooler climate
- stopping work for 2–3 weeks at the onset of signs and proving other relief measures
- spelling the horse during the hottest and most humid weather
- hosing with cold water as needed before and after exercise or periodically during hot weather
- training and exercise during the coolest part of the day
- scheduling the horse to perform in competitions in the early morning
- Installing stable fans, misting machines or air-conditioning
- access to shady areas, dams, or creeks for paddocked horses
- avoiding stress and excitement
- avoiding excess energy intake that gives a horse too much fat cover
- vigorous grooming to remove old hair and stimulate the sweat glands.

Ideally, horses from cool regions that have to perform intense exercise in hot, humid conditions should be moved well in advance of the event to allow them to adapt slowly to the climate. Up to three months of increasingly demanding exercise may be necessary before the horse gains maximum thermoregulatory efficiency. Among the physiological changes required are an expansion in plasma volume, increasingly stable cardiovascular function, and alterations in the sweating pattern.

There is no evidence that a particular feeding program can either cause or cure anhydrosis, but some dietary changes may help keep the horse comfortable in hot climates (see Chapter 5 for details).

Developmental orthopaedic disease

Developmental orthopaedic disease (DOD) is a very significant economic problem for modern horse breeders because the crippling lameness can make affected foals or yearlings essentially worthless.

The conditions included in DOD are epiphysitis (physitis), wobbler syndrome, osteochondritis dessicans (OCD), osteochondrosis, contracted flexor tendons, angular limb deformities (ALD) and juvenile arthritis. These are not new problems and it is unclear if there has been an increase in incidence, or if they are being diagnosed more accurately. There is not one factor that can prevent DOD, but several are known to be important. Some of these are listed below, and other nutritional factors are included in Chapter 5.

Genetics

Research in several countries and breeds has identified sires and sire lines in which the occurrence of DOD is significantly greater than average, so the genetic component of the disease must be considered. The exact mechanism of inheritance is unclear, but each case of DOD should be evaluated from the genetic standpoint and either cull the mares or carefully choose the stallion. The genetic influence may relate to rapid growth, body type, conformation, weight, early development or other factors.

Environment

Environmental conditions, such as running on hard surfaces, may contribute to physical or mechanical trauma to the growth plate of the bones. It is important to restrict the exercise of weanlings and yearlings with physitis or OCD by confining them in small paddocks or yards. However, foals and weanlings confined in stables for extended periods have a higher incidence of OCD because there is less stress on the bones, which in turn reduces bone strength. When the foal is let out into the paddock the combination of weaker bones and increased exercise may result in OCD. Only confine foals and weanlings in stables when it is really necessary or allow some free exercise every day while they are confined. A graded return to the paddock is very important.

Biomechanics

Another theory is that DOD results from excessive biomechanical forces exerted on otherwise normal cartilage. These forces disrupt the blood supply to the cartilage and prevent its conversion to bone. There are several possible reasons why normal cartilage cannot withstand these forces.

- Specific joints have periods of vulnerability when they are particularly susceptible to damage because there may be an inadequate amount of underlying bone to support the weight and force exerted on the joint. For example, stifles and hocks seem to be most vulnerable at 6–8 months of age. Excessive force may damage the underlying bone and the vascular supply to the cartilage so that it does not ossify and forms a lesion, which does not become obvious until later when it is severe enough to cause lameness or swelling.
- Foals with genetic potential for rapid skeletal and muscle development may simply develop more muscle mass than the rapidly growing bone can support.
- Foals that do not have the genetic potential for rapid growth become fat because of excessive energy intake and overload the joints.
- A conformation defect creates an uneven distribution of force on the joint surface.
- Foals that have been confined because of illness do not develop enough bone to support their weight. When they are finally returned to the paddock, the bone is too weak to support a normal amount of exercise and the joint cartilage collapses.
- Changes in management that greatly alter the foal's exercise patterns may overload the joint.

Until a cure or the exact cause of DOD is found, use good management and selection procedures, and feed a well-balanced diet in a conservative manner. Remember that the management of the late-pregnant mare, wet mare, foal, weanling and yearling is important for minimising the incidence and impact of DOD.

Gastric ulcers

The incidence of gastric ulcers is extremely high, particularly in performance horses, with 80–90% of active racehorses and up to 60% of those involved in eventing, showing and Western competition being affected. There is an association between ulceration and nervousness, but it is unclear whether this is a cause or effect. Ulcers are also a problem in foals, particularly those that are sick for another reason. Most domesticated horses will have an ulcer during their lifetime and although it may not cause any observable problems in some horses, in others ulcers are both painful and costly in terms of treatment and loss of performance. However, ulcers can be alleviated or eliminated with proper management.

Development of gastric ulcers

The majority of gastric ulcers form in the upper portion of the stomach that is lined with cells similar to those lining the oesophagus. This section of the stomach has little protection from the irritating secretions produced in the lower part (hydrochloric acid and pepsin) that help to break down the ingested food. Although the lower, glandular section of the stomach has buffers and a mucous lining to protect it, the whole system is in a delicate balance that can be altered by a change in eating habits, feeds, and possibly exercise-induced stress. Most horses, allowed the opportunity to graze free-range, will rarely develop ulcers, but many performance and pleasure horses do not routinely graze or even eat hay ad libitum. Therefore the irritating substances secreted by the lower stomach can affect the upper region, leading to ulcers. Equine gastric ulcers are not caused by the bacteria, *Helicobacter pylori*, that commonly causes ulcers in humans.

Foals develop ulcers when the stomach produces more acid than the mare's milk can neutralise. Ulcers can occur in healthy foals, but if a foal's intake of milk is reduced because of disease, such as diarrhoea, or the stress of separation from its dam, the acid build up in the stomach can lead to perforating ulcers and death. As the foal matures and begins to graze, the problem of ulcers is lessened because the grass or hay in the upper portion of the stomach provides the necessary buffer to the acid.

Research on feeding programs and their effects on ulcers in confined horses have shown that animals with continual access to hay have significantly lower levels of acid in their stomachs. Forage consumption stimulates saliva production, which helps to protect the upper region of the stomach. Conversely, horses that have not been fed for 24 hours have a much greater level of acid in their stomachs.

Stress may be another factor in the development of ulcers. Weanlings are almost always able graze pasture, yet many develop ulcers. There is some concern that training stress may be a contributing factor to ulcers in performance horses. Horses in training are confined for much of the day and fed high-energy diets, which increases the production of gastrin, a hormone that stimulates gastric acid. Although horses produce more saliva when eating hay than when eating grain, some racehorses provided with a constant supply of hay are still prone to ulcers. The stress of hard training or travel may be a factor. One study showed that horses that were galloped on a treadmill had a much higher incidence of ulcers than horses on the same feeding program that were doing slow trotting work. This may be related to the stress of training or to acid splashing around in the stomach.

Symptoms of gastric ulcers

Many foals and horses with ulcers show no signs; however, even mild ulceration can be associated with a loss of performance. Severe ulcers in foals causes signs such as salivation and teeth grinding. You should consult your veterinarian if your horse shows symptoms or you suspect it has ulcers. Although 40% of horses with ulcers will show some of the signs associated with ulceration, there is not a direct correlation between the symptoms and the severity of ulceration.

Signs of gastric ulceration in horses and foals include:

- picky eater
- weight loss or poor condition
- dull or sour attitude
- rough coat
- colic
- lack of 'attention'
- drop in performance

- lying on back (foals)
- salivation (foals)
- diarrhoea (foals)
- teeth grinding (foals).

Examination of the stomach with an endoscope will definitively diagnose gastric ulceration, but diagnosis is often based on clinical signs or response to treatment (increased appetite and condition).

Preventing and treating gastric ulcers

Time spent eating forage such as grass, chaff and hay seems to be a critical factor (see Chapter 5 for more details). Sick foals that are not drinking well can be given prophylactic ulcer treatment. If grazing is not available, including an equine antacid in the diet might help buffer gastric acidity. If a horse with ulcers is spelled on a high-forage diet, many ulcers heal spontaneously in one month.

Treatment involves either inhibiting gastric acid secretion or neutralizing the acid produced. Three types of drugs are used to inhibit gastric acid secretion: histamine type-2 antagonists, such as cimetidine and ranitidine (which has a 3-day withholding period prior to racing), proton-pump inhibitors, such as omeprazole (which is not yet available in Australia), and prostaglandin analogues. These drugs will cure gastric ulcers in 3–4 weeks, but they are very expensive. The response to treatment will be better if the horse is spelled from work in a paddock, but that may not be possible. Lower doses may be effective in preventing recurrence, because if the ulcers are treated while the horse stays in work they usually recur quickly. Less expensive versions of these drugs have been marketed as compound ulcer medications, but research has shown them to be less effective.

An alternative to suppressing acid production is to neutralize it and protect the lining of the stomach from the acid. Antacids can give some symptomatic relief from the discomfort associated with ulcers. Human antacids need to be given frequently in large volumes by stomach tube and may have adverse side effects such as incoordination. There are commercial equine antacids available through veterinarians and other suppliers. Although antacids will not cure the ulcers, they prevent gastric acid accumulating in the stomach and buffer excess acidity thus preventing the ulcers from recurring. Prevention is less expensive than treatment.

Tropical pastures and 'big head'

Big head is a calcium deficiency of horses and donkeys grazing introduced tropical pasture grasses that contain crystals of calcium oxalate, which prevent the horse from absorbing the calcium from the grass during digestion.

The signs of big head include lameness, ill-thrift and swollen jaw bones. It can develop within 2 months of being put on hazardous pastures, but commonly takes 6–8 months. Mares and foals are more susceptible than stallions and geldings.

Treatment involves grazing and pasture management and feeding a calcium/phosphorus supplement (see Chapter 5 for details). The lameness and ill-thrift can be cured, but the swellings of the jaws may not completely disappear if the animal has been severely affected.

WOUNDS

Wounds are common in horses and because horses are very susceptible to tetanus, all horses should be vaccinated every four years for this disease. Any horse that is wounded and has not been vaccinated within the previous four years will require a tetanus antitoxin as part of the treatment.

Wounds are a major problem in horses that are kept in paddocks. Despite the best facilities, horses can still manage to injure themselves. If the facilities are not appropriate for the holding of horses (e.g. barbed wire, mesh or loose wire fencing), there is a risk of major injury. Horses can also suffer injuries to the skin and deeper tissues during competition.

Wound management

The major problem of wound management in horses is that if the edges of the wound become separated, the wound will heal very slowly, with the possibility of either scarring or 'proud flesh'. Many wounds of horses are not able to be sutured (stitched), which again means there can be a long recuperation, even for seemingly minor wounds. The costs of bandages, drugs and wound treatments can run into hundreds of dollars.

All wounds should be cleaned as soon as possible with cold running water. If there is cut or torn skin, place it back where it was

Figure 6. (a). Wound immediately after the accident. (b). Wound 16 days later.
(Courtesy of Reg Pascoe.)

before the injury and hold it in place with a dressing, padding and a bandage. It is usually not possible to bandage wounds to the upper legs or body. If the wound is large or may leave the horse lame or with a gait abnormality, seek prompt veterinary attention. If a wound needs to be stitched it should be done as soon as possible and certainly within four hours of the accident. An small, unstitched wound will heal well if kept clean and free of infection.

Lower leg wounds, even seemingly minor ones, should be seen promptly by a vet because these injuries are more difficult to treat for a number of reasons. First, they tend to swell more, which separates the edges of the wound, so it is important to correctly bandage these wounds. Second, the skin of the lower leg has a poor blood supply and consequently if there is a flap of skin it often dies, leaving an open wound that will take a long time to heal, with an increased likelihood of proud flesh.

The outcome of a wound depends on a number of factors:

- the site of the wound; head wounds heal more quickly than leg wounds, but hock wounds are really bad news!
- the type of wound e.g. abrasion, contusion (bruising), laceration or puncture.
- the size of the wound.
- the extent of damage to deeper structures including muscle, tendons, joints, bone, blood supply and nerves.
- the orientation of the wound: vertical wounds heal more quickly than horizontal wounds.
- whether or not there is infection, especially if deeper structures such as joints or bone have been damaged.

Suturing

If a wound can be sutured and the sutures stay in place, then the healing process will be relatively rapid, between 14 and 28 days. However, many wounds cannot be sutured for the following reasons.

- The delay between the injury and suturing is more than 4–6 hours.
- There is disruption to the blood supply to the wound, which may cause the skin to die and the sutures to pull out.
- Excessive movement of the wound as the horse moves
- Excessive swelling around the wound.
- Contamination of the wound, leading to infection.

Wounds with a good blood supply, in an area that experiences less movement and with minimal swelling, such as wounds to the head or body, will heal far more quickly and easily when sutured than wounds on the legs. Most leg wounds cannot be successfully sutured unless they are seen

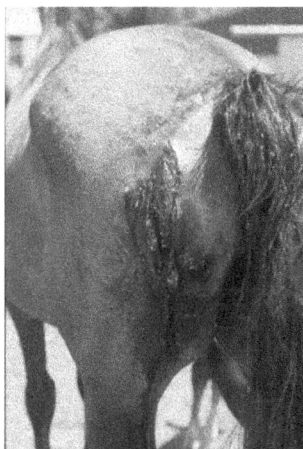

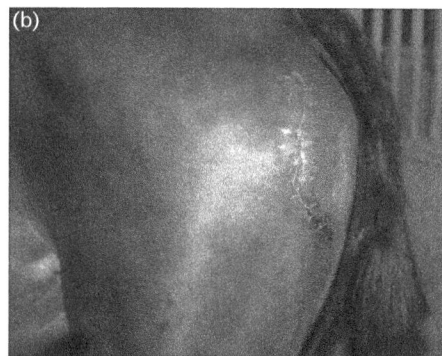

very soon after the injury. Even then, do not insist on the wound being stitched because very often it will break open again a few days later and you have to deal with a larger wound.

Delayed wound healing

Most leg wounds will have delayed healing because they cannot be sutured and will therefore be susceptible to the development of proud flesh, which also slows down healing. Such wounds will usually leave a scar, although the extent of the scarring will depend on the individual horse and the management of the wound.

In the initial stages of healing it is desirable to bandage the wound to reduce contamination and possible infection, and to encourage the growth of the pink, granulation tissue that will fill the gap between the edges of the skin. It is overgrowth of this granulation tissue above the level of the skin that is called proud flesh. Covering the wound will stimulate growth of normal granulation tissue and removal of dead tissue from the wound. In this phase of wound healing, bandages usually need to be changed daily, but if the wound is clean and not weeping the bandage can be left on for 2–3 days.

Unless you have been specifically advised by your vet to use a spray-on, pour-on or powder wound treatment, you should avoid these compounds because they may slow down healing.

Once the granulation tissue has filled the gap between the edges of the wound, it is important that it not protrude above the skin surface. The surest way of removing the excess granulation tissue is by resecting it with a scalpel, which must be done by your vet, or you can use an agent, such as an antibiotic/corticosteroid cream, under the bandage to reduce granulation growth. This is preferable to the use of chemical agents, such as copper sulphate, that are reputed to reduce proud flesh.

There will be a better cosmetic appearance and a smaller scar if the wound is kept bandaged throughout the healing process. Always bandage a wound in which the underlying tissues, such as tendon or bone, are exposed to the air and remove any proud flesh. Alternatively, the wound can be left open and unbandaged and in many instances, the expense and complication of constant bandage changes, and the difficulty of bandaging the site, such as the hock, means that this is the preferred approach and the wound will heal with minimal intervention.

In summary there is not a magic formula for good healing of wounds. Consult your veterinarian about the progress of the case and be prepared for a long process if the wound is on the horse's leg.

SPRAINS AND BRUISES

It is important to control any swelling and inflammation for the first few days after an injury that results in a sprain or bruise. Use cold water or, preferably, ice packs and apply for a minimum of 20 minutes at least three times daily. As well as reducing the swelling and inflammation, ice is very effective in minimising pain. For the rest of the time, the affected area should be kept bandaged if possible.

Some rest from work will usually be required to allow the injury to heal. Your vet is the best person for advice.

RESPIRATORY DISEASES

Symptoms of respiratory diseases

Coughing and a nasal discharge (runny nose) are the signs of a respiratory disease, which may be contagious or an allergy to such things as stable or feed dust.

If your horse has a runny nose or cough, stop exercising or working it. If the horse is having trouble breathing, not eating or the cough/runny nose symptoms persist for more than a few days, seek veterinary attention. Pneumonia can lead to permanent damage to the lungs or even death.

Virus infections

Viruses, including equine herpes virus, equine adenovirus, and equine rhinovirus, can cause respiratory infections and fortunately that the most important and severe of these diseases, equine influenza, does not occur in Australia or New Zealand. Viral respiratory infections are commonly diagnosed in young horses in spring and autumn, although horses can be affected at any time of the year.

The typical signs of a viral respiratory infection are a fever, cough and a runny nose. The horse may be slightly lethargic or depressed and often will not want to eat. The lymph nodes under the jaw are usually enlarged, but not to the same degree as in strangles. The nasal discharge is initially clear, but soon becomes sticky and may be purulent (yellowy-green) in the later stages. In an uncomplicated case, the horse will improve in a few days although it may still cough for a few weeks when exercised.

A viral infection does not have any specific treatment and antibiotics are only useful if a secondary, bacterial infection develops. If this occurs, the nasal discharge, cough and fever will persist for more than three days and you should seek veterinary treatment. In uncomplicated cases, recovery will occur without specific treatment; however, you should stop working the horse, put it in a dust-free, well-ventilated environment, feed damp feed and isolate it to prevent transmission to other horses. It is also advisable to clean and disinfect gear such as twitches, headcollars or bridles that have been used on the infected horse. Drugs that open up the airways and mucolytics, which help dissolve mucus in the lungs, may help recovery in complicated cases. Vitamins C and E may help to stimulate the horse's immune system and enhance recovery from a viral infection. They are not an essential part of the treatment, but can be given to horses recently affected by a 'cold' and may help in the treatment of some persistent cases.

Working the horse too soon after a viral infection may lead to permanent damage to the horse's respiratory tract. Preferably, the horse should be rested for 2–4 weeks after the infection, but this can include light work to maintain the horse's fitness.

Strangles

Strangles is caused by a bacterium, *Streptococcus equi*, and is the most important infectious respiratory disease for Australian and New Zealand horses. Outbreaks usually occur in groups of horses, particularly young horses, when there is frequent movement of horses to and from a property. In most cases the disease is cured with a simple, veterinary treatment program and is prevented by vaccination.

Strangles is transmissible from horse to horse or via contaminated equipment or facilities. The typical symptoms are easily recognised: the horse will have a fever, be lethargic and depressed, have a thick, yellow, purulent nasal discharge and large swellings under or behind the jaw. These swellings are abscesses in the lymph nodes and they will continue to swell until they burst or are drained. A mild form of strangles also exists in which the abscess formation is not as prominent, but the horse still has a thick nasal discharge and a moderate fever. In some severe cases, the horse will have generalised abscesses, including in the lungs, abdomen and other sites ('bastard' strangles). A typical case can be diagnosed from the clinical signs, but there also should be laboratory confirmation of the cause of disease. Your veterinarian will take a swab sample of the either the nasal discharge or material from the abscesses for laboratory testing.

Although only a few infected horses will die, strangles is a significant, performance-affecting disease and bastard strangles is a severe condition. Some horses with strangles will recover uneventfully without treatment, but others need drainage of the abscesses and antibiotic treatment. All cases should be isolated to prevent transmission to other horses. If you think your horse has strangles, then you should seek treatment.

Prevention of strangles involves a combination of vaccination and isolation practices. It is hard to prevent the disease occurring when large numbers of horses congregate together, as they do on studs and agistment properties. The vaccination will not provide complete protection against the disease, but will reduce the spread in an outbreak and should also reduce the severity the disease if your horse develops it. The vaccine must be administered each year (annual booster) and you should give a booster before taking the horse to a high-risk location such as a stud or an agistment property (see later section on vaccination).

Isolation of incoming horses to a property for a period of 7–14 days will substantially reduce the risk of introducing strangles (see later section on quarantine), and isolation of your horse from others at events will also help in this regard. Unfortunately, there is a carrier state in which horse does not show any clinical signs, but can transmit strangles and this makes strangles hard to control.

The strangles bacterium can persist in the environment for several months unless the contaminated area is cleaned and disinfected using iodine or chlorohexidine-based disinfectants.

Pneumonia

Pneumonia can be either a complication of a viral respiratory infection or a discrete disease. It will affect performance and can be life threatening if not treated successfully, so you should seek prompt veterinary attention.

The affected horse is likely to be not eating, lethargic, have a fever, a cough and an increased respiratory rate with heaving of the flanks and/or flaring of the nostrils. It may have a discharge from the nostrils. Because of the pain in its chest, the horse may be reluctant to move and may stand with its legs apart.

Pneumonia needs intensive veterinary treatment. A horse with suspected pneumonia should not be worked and must be spelled for a considerable period during recovery. Pneumonia and pleuritis are common complications after long-distance travel (see the section 'Travel Sickness' in Chapter 9).

Chronic coughing

Chronic coughing can be the result of returning a horse to work too soon after a viral infection, or it can be an allergic reaction to dust. The horse will cough when being worked and may show some exercise intolerance, but will be well in other respects.

Horses will usually recover if given correct treatment, a rest from work and a relatively dust-free environment such as a paddock rather than a stable. If they must remain in a stable, then dusty hay should be avoided or dampened and wood shavings, shredded paper or rice hulls are preferable to straw as bedding. Always remove the horse from the stables when you are mucking out to reduce its exposure to dust.

Prevention of chronic coughing can be achieved by providing adequate rest for a horse that has had a viral infection and seeking veterinary treatment if the signs persist for more than a few days. Minimise dust and improve the ventilation in the horse's environment. If the ventilators or windows of the stable are blocked or clogged they should be opened and cleaned. The top door of the stable should be shut only in exceptional circumstances.

Heaves/chronic bronchitis

This is a relatively rare condition, but can affect some older horses. The horse develops an allergy to stable or feed dust and will show exercise intolerance as well as coughing and breathing difficulty while at rest. Seriously affected horses develop a 'heave line' in the flank area from the constant exertion of the abdominal muscles during breathing. Most horses will improve when placed in a paddock rather than a stable. There is a range of drugs that your veterinarian can use to treat the problem.

DIARRHOEA

Diarrhoea is the passage of loose or liquid faeces at an increased frequency. It is one of the more serious problems that affect horses and is a difficult and frustrating disorder to manage. If the diarrhoea does not respond to conventional therapy, the horse may die or progress to a chronic state. Mild diarrhoea, which may occur in response to a feed change, can resolve without treatment, but if the horse is depressed, lethargic and not eating it should be examined and/or treated by your veterinarian.

Whether or not an animal develops diarrhoea depends upon the balance between absorption and secretion of water by the cells lining the bowel. A healthy 500 kg horse will drink 25–50 litres of water each day, depending on weather conditions, the amount of work, and type of feed. The large bowel will usually absorb this fluid, but any decrease in absorption or increase in secretion by the bowel can result in substantial volumes of fluid lost as diarrhoea.

The movement of water across the intestinal lining is such that water always follows salts. A disturbance of the fluid exchange between the body tissues and bowel contents results in:

- a loss of body salts (sodium and potassium)
- changed bowel movements
- poor fluid absorption by the large bowel.

Cause of diarrhoea

In the absence of detailed and expensive laboratory tests, most veterinarians attribute diarrhoea in horses to enteritis (inflammation of the large intestines), usually caused by:

- poor composition and quality of the diet
- sudden changes in diet composition and quality
- bacteria
- viruses
- changes to the normal gut bacteria by administration of antibiotics
- heavy infestation of small strongyle worms
- peritonitis
- sand

Diet changes

A change of feed onto lush green pasture or lucerne hay may precipitate mild diarrhoea that will firm up with time or another minor feed change. Horses grazing poor pasture, particularly late summer pasture that is dry and short, may ingest more soil and develop a watery post-defecation discharge, but in all other respects be normal. This form of diarrhoea, called 'sanding', occurs when increased soil intake irritates the large intestine. Supplementary feeding and changes in the management of the horses are usually all that is needed.

Bacteria

The *Salmonella* species are highly infectious bacteria and the most common agent causing acute diarrhoea. Affected animals should always be isolated and need intensive veterinary treatment. Salmonellosis is contagious and may affect humans as well.

Colitis X

Colitis X is a fatal haemorrhagic enteritis that occurs in all types of horses. As the name implies the cause of the disease is unknown, but it is likely that stress, exhaustion, low resistance, endotoxaemia and enterotoxaemia are contributing factors.

Intestinal worms

Heavy infestations of large and small redworms (strongyles) may damage the intestine and lead to severe diarrhoea, which if left untreated can become chronic and unresponsive to treatment (see later section on control and treatment of parasites).

Stress

The role of stress in outbreaks of diarrhoea in adult horses cannot be over-emphasised, because many cases can be traced to a previous stressful experience such as transport, major surgery, recent drenching, hard exercise, and movement to a new property or stabling.

Symptoms of diarrhoea

Abdominal discomfort (colic) frequently precedes the onset of diarrhoea and is typified by restlessness, noisy gut sounds and loss of appetite. The general behaviour of the horse will indicate the severity of the disease.

The body temperature will rise early in the course of the disease, but may fall as shock and dehydration develop. Dehydration may be evident in the horse's lethargy, dry gums, darker mucous membrane colour and reduced urination. Dehydrated horses will also show 'skin tenting'. Tail soiling is evidence of projectile diarrhoea.

Treatment of diarrhoea

At the first signs of severe diarrhoea, call your veterinarian. If the horse's droppings are soft or similar to a cow-pat you may wish to observe the horse for a while before calling the vet, which is a reasonable course provided the horse is bright and alert, and eating.

Treatment of acute diarrhoea includes giving large volumes of intravenous fluids and electrolytes, pain relief, intestinal absorbents and broad-spectrum antibiotics, even if the cause is not salmonellosis or colitis X. This intensive care is very expensive and time consuming.

Diarrhoea in foals

Diarrhoea is a frequent problem in foals, with up to 75% affected at least once before weaning. It requires early and thorough investigation because foals with diarrhoea can deteriorate quickly and even die. Fortunately, most cases of diarrhoea in foals do not last long and are not life threatening.

Causes of diarrhoea in foals
- Overeating, ingestion of fibrous material or foreign matter, eating of faeces.
- Milk intolerance.
- Immunodeficiency: foals that have not absorbed sufficient protective antibodies from the mare's colostrum are susceptible to infections.
- Viral and bacterial infections.
- Parasites.
- Foal heat diarrhoea.

Foal heat diarrhoea

Foal heat diarrhoea is the most common form of diarrhoea in foals and is generally self limiting. It usually develops in foals that are 6–14 days old, whose dams are in heat for the first time after foaling. The foals remain active, continue to suckle vigorously and gain weight. The body temperature remains normal. Treatment is not required.

The cause of foal heat diarrhoea is much debated, but there is general agreement that it is related to changes in the intestinal microbial population and increased amounts of roughage reaching the underdeveloped large intestine.

Symptoms of diarrhoea in foals

Early warning signs are lack of suckling causing the mare to 'bag up', a quiet foal, or a foal that spends too much time sleeping.

When to call the vet

Severe diarrhoea in foals can quickly become a serious problem, so do not delay in seeking veterinary help. If the diarrhoea is mild and the foal is bright and alert, you should confine the mare and foal for close observation. Milk may worsen the diarrhoea, so you may wish to stop the foal suckling and give it an electrolyte/glucose replacement fluid.

The main aim of veterinary treatment will be to correct any dehydration using oral and/or intravenous fluids. Severe cases will require time-consuming intensive care.

General management of foal diarrhoea

Diarrhoea will cause scalding and hair loss around the foal's rump, and attract flies, so keep the area clean and use petroleum jelly (e.g. Vaseline) to prevent scalding. Some types of foal diarrhoea are contagious, so it is desirable to isolate the mare and foal and to clean and disinfect the stable/yard before it is occupied by other mares and foals. Disinfect your hands and clothes before handling other foals.

Prevention of foal diarrhoea

This is the most important aspect and each of the following measures should be carried out.

- Ensure all foals have received colostrum so that they can develop adequate immunity.
- Minimise stress to foals.
- Keep mares and foals in the 'cleanest' paddock, one that has been rested to allow decontamination of parasites and infections.
- Do not keep foals in dusty yards for long periods.
- Worm the mare just after foaling and clean up the droppings.
- Observe foals closely after birth and for the next few weeks.
- Isolate foals with diarrhoea to prevent transmission.
- Do not overfeed mares prior to and immediately after foaling.
- Seek prompt veterinary attention for any abnormalities in newborn or young foals.

See the section 'Care of the newborn foal' in Chapter 7 for more information.

EYE PROBLEMS

Good sight is vital for all working horses; the safety of the horse and the rider depends upon it. Signs of an eye problem include partial closure of the eyelid or changes in the appearance of the eyeball. Eye problems require prompt veterinary attention.

Common eye problems

Ulcers

An ulcer is the result of damage and erosion of the surface of the eyeball (cornea). Ulcers are caused by infections, foreign objects, such as chaff or grass seeds, in the eye, or by contact (horses that have been struck in the eye with a whip may develop an ulcer). Ulcers are very painful for the horse, which will cause it to keep its eyelid partly closed, there will be some discharge from the eye and the ulcer may be visible. Often the cornea is cloudy. If left untreated, an ulcer may heal satisfactorily, but it can lead to scarring of the eye that may impair vision.

Place the horse in a darkened stable until a vet has examined the eye. Special stains are used to determine the size of the ulcer, which will determine the

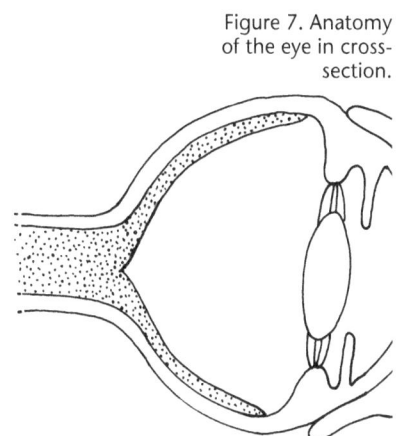

Figure 7. Anatomy of the eye in cross-section.

treatment required for rapid healing. A severe, deep ulcer can lead to a rupture of the cornea and loss of sight. Treatment usually involves putting ointment into the eye every few hours or in severe cases a 'drip' treatment system can be surgically implanted into the upper eyelid. Do not use any old eye ointment that you have left from treating a previous eye problem because eye ointments that contain cortisone/corticosteroids will make the ulcer worse.

Uveitis

Uveitis is inflammation of the whole eye is inflamed and is very painful. The cause of uveitis is often unknown, although in some cases it may relate to infections or trauma.

A horse with uveitis will have a closed eye, some discharge from the corner of the eye, a closed pupil, reddening of the conjunctiva (inner eyelids) and often a cloudy cornea (see Plate 2, Figure 16). Early and aggressive treatment of this problem is essential, although some cases become recurrent and will flare up later. In severe, recurrent cases, the horse's sight may be threatened.

Conjunctivitis

Conjunctivitis most commonly occurs in summer when dust, flies and wind irritate the lining of the horse's eyelids. It is more common in horses with 'open' eyes in which a lot of the white (sclera) is not covered by the eyelids. The horse will show a reddening of the conjunctivae and a discharge from the eye. Usually both eyes are affected.

Conjunctivitis may resolve untreated if the predisposing condition is modified. It is desirable to clear up the discharge from the eye because it will attract more flies and further irritate the horse. You may need to use eye ointment if the discharge becomes purulent.

Eyelid injury

Although uncommon, wounds in the eyelids, even if small, require veterinary attention. It is important that the wound is sutured to restore a continuous eyelid margin. A small wound will heal if it is not stitched, but there may be a gap or a ledge in the eyelid that predisposes the eye to other problems such as conjunctivitis or corneal penetration. Repair of an eyelid wound will not generally leave a scar.

Blocked tear ducts

Normal drainage of the eye occurs via the tear duct located inside the corner of the eye and out through the nostril, which is one reason why horses always have some dampness around their nostrils. If the tear duct is blocked, the liquid will overflow the eyelid and run down the side of the horse's face, which is unsightly and can attract flies. It is a simple problem to diagnose and treat, involving sedation of the horse and flushing of the tear duct.

Eye trauma

If a horse has been knocked or kicked around the eye, there may be swelling of the eyelids and around the eye, a discharge from the eye, or blue or white marks on the cornea, which indicate damage to the surface of the eye.

Because of the risk of uveitis or corneal damage that will compromise sight, eye trauma requires prompt veterinary attention. A seemingly minor knock can set up a process that leads to loss of the eye.

Testing a horse's sight

You cannot ask a horse if it can see and a blind horse will know its way around a familiar environment. To check a horse's vision you need to blindfold one eye and lead the horse into an unfamiliar obstacle course. A horse that has vision problems in the uncovered eye will bump into or trip over objects. Check the other eye in the same way. A 'menace' test can also check vision: an object, such as your finger, is slowly moved towards the eye and a blink indicates the horse can see the finger. Do not touch the eyelids, eyelashes or eyeball and do not move air against the eye when doing the test.

LAMENESS

Lameness is a problem for all horse owners at some stage. It is a sign that a horse has an injury that causes pain when it moves or places weight on the affected leg. If the cause of lameness is apparent, you may be able to administer first aid (see later). If not, confine the horse to prevent further damage to the leg and call the vet if mild lameness persists for more than two days. Horses with severe lameness need prompt veterinary attention as they are usually experiencing significant pain.

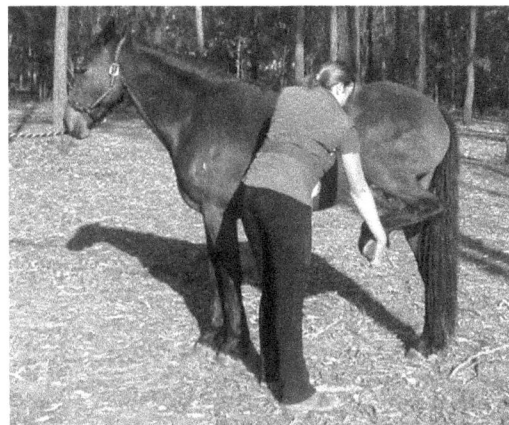
Figure 8. Flexion test for a hind leg (courtesy of Stuart Myers).

Most lameness affects the forelimbs and 90% of problems occur in the knee and below. In the hind limbs 80% of cases of lameness involve the hock or stifle.

Different breeds and disciplines have different predispositions for lameness. Standardbreds are predisposed to hock problems and Quarterhorses are predisposed to ringbone. Racing predisposes horses to knee chips whereas trail riding often leads to stone bruises in the hooves. Horses worked on hard surfaces are more likely to have navicular disease. Some problems that affect young, growing horses, such as shin soreness or locking patellas (knee caps), rarely trouble adult horses.

Any signs of lameness should be investigated immediately and any hoof or foot problem treated correctly. The lameness may need examination and treatment by a vet, a farrier or both. Make a habit of regularly handling your horse's legs and feet because then you will notice any abnormality such as swelling or heat. Never neglect cracks, wounds or punctures to the feet.

A lame horse may have a shortened stride, feel uneven when ridden or nod the head up and down when trotted. If the horse is trotting on a hard surface a difference may be heard in the way the horse puts the hoof to the ground when bearing weight on the lame leg.

Locating the lame leg/s

When examining the lame horse care must be taken to avoid causing further pain and harm. Look for obvious signs first while the horse is at rest. Does the horse appear to be favouring a leg? It is normal for a horse to rest a hind leg; however, it should willingly take weight on the rested leg. It is not normal for a horse to rest a front leg. If the lame leg is not apparent pick up and pick out all the feet (see Chapter 2 for instructions for picking up a foot).

Inspect the legs and feet for:

* any signs of injury such as blood, swelling, heat or a bounding pulse.
* signs that the horse is unwilling to put full weight on a limb when the opposite foot is picked up.
* objects that may have caused injury to the foot, such as stones, nails or wire.

Further investigation can involve having the horse 'trot out' on a hard, level surface. The horse is trotted on lead away, towards and past the observer. It is vital that the horse trots on a loose rein so its head is free to move up and down.

If the horse is lame in a front leg, the head goes up when the lame leg hits the ground (as the horse attempts to put as little weight as possible on that leg), and the head goes down when the good leg takes weight. If the horse is sore in both front legs it will have a stiff, 'proppy' action. Sometimes lunging a horse will demonstrate lameness that is not evident when moving in a straight line and working on a soft rather than a hard surface may reveal different degrees of lameness.

Hind limb lameness is harder to detect than forelimb lameness because the head nodding may not be as obvious. The horse may nod its head down when taking weight on the sore hind limb. From the side view, short stepping in the lame leg may be seen; from behind, the hindquarter on the side of the lame leg may rise higher than the other side (again, as the horse attempts to put less weight on the painful leg).

Once the lame leg has been identified, recheck for:

- heat in the leg and foot
- swelling
- a strong digital pulse
- pain when a joint is palpated or flexed.

Determining exactly where the problem is once the lame leg has been identified can take a lot of skill. Mild cases of lameness can be managed by you, depending on your experience, but otherwise a veterinary examination is required.

The vet will carry out the preliminary checks again, and if the site of the lameness is still undetected further diagnostic tests are required.

- A flexion test. The leg is held in a flexed position for one minute and the horse then immediately trotted out by a handler; the vet will watch for lameness in the first steps taken after release.
- Hoof test using hoof testers.
- Diagnostic imaging using ultrasound, thermography, radiography and/or scintigraphy.
- Nerve blocks.

The history of the horse is important for making a diagnosis. Keeping a record of the following factors will be helpful for the vet.

- Has the horse been lame in the past?
- When was the horse last shod and does the horse have any abnormal shoe wear?
- Have there been any recent changes in the management of the horse?
- Does the lameness increase when being ridden?
- Does the lameness disappear/change with exercise?
- How old is the horse?
- What type of work is the horse doing now?
- Is the horse stumbling?
- Is the horse undergoing any treatments?
- What surface does the horse usually work on?

First aid for lameness

The basic first aid for lameness is rest and confinement to prevent further damage and allow the natural repair process to begin.

If something has penetrated the hoof talk to your veterinarian before pulling it out. If there is a puncture wound, keep the entry hole open until the puncture has been cleaned and drained.

If the lameness has had a sudden onset and there is swelling in the leg then hosing, bandaging and application of ice packs may be useful. If the lameness is severe with swelling indicating a fracture or major damage to tendons or a joint, a heavy support bandage should be applied to immobilise and support the leg (see later section on bandaging and poulticing). Poulticing may help reduce swelling or encourage an abscess to drain.

Treatment of lameness

Treatment of lameness will be determined by your veterinarian in consultation with you and will include combinations of:

- rest
- confinement

- elimination of infection
- controlled exercise
- surgery
- medical therapy
- bandaging
- poulticing
- corrective shoeing.

Prevention of lameness

Lameness can usually be prevented or reduced by:

- choosing safe ground to work on
- not pushing an unfit horse beyond its capabilities
- not riding an unshod or poorly shod horse too hard
- feeding a balanced diet
- prompt veterinary attention when necessary
- not working a horse that shows early signs of lameness
- not using anti-inflammatory drugs that allow the horse to be worked while masking the pain
- taking good care of wounds.

Common hoof problems

Foot abscess

A foot abscess is a very common form of lameness that can result in severe lameness, even holding the hoof off the ground. It results from a penetrating wound of the sole or white line that introduces bacteria. Subsequently, an infection begins between the sole or wall and the sensitive internal tissues of the foot. As the infection progresses up or around the wall, it becomes very painful. A poultice may help drain the abscess through the entry point or the abscess may burst out at the coronary band. The latter will usually lead to a horizontal crack in the hoof wall as the wall grows down. Horses with a foot abscess should be given a tetanus antitoxin if they are not vaccinated for tetanus. Veterinary attention is needed.

Thrush

Thrush is an infection of the frog, characterised by a dark-coloured discharge from the grooves and associated disintegration of the horn.

Thrush may not always cause lameness, but it is invariably associated with poor stable management and lack of exercise. It is a preventable disease. Stalls and looseboxes that are not regularly mucked out, failure to pick out the feet and irregular shoeing encourage the development of thrush.

Local treatment comprises paring away all loose, under run and diseased tissue of the frog, keeping the grooves clean and dry, and applying a drying agent, such as Lotagen.

Seedy toe

Seedy toe is a cavity in the toe, which results from separation of the sensitive and insensitive laminae at the white line and is filled with a crumbly type of horn and compacted soil.

Treatment involves scooping out the degenerate horn and removing the under run wall, then packing the cavity with cotton wool and Stockholm tar, or Lotagen. To protect the bearing surface of the hoof and retain the dressing within the cavity, the horse needs a plain shoe with a wide web at the toe. In extensive cases a pad may need to be fitted to give extra protection. In unshod horses the wall over the cavity is pared away to prevent accumulation of dirt, which will lead to a deficit in the bearing surface of the wall until the new hoof above the seedy toe grows down.

Corns

A corn is a bruise of the sole in the angle between the wall and the bar. They are most frequently seen on the inner heels of the front feet and are rarely seen in hind feet. They are especially common in horses with wide, flat feet. Although they may result from a stone getting under the shoe, they are more often caused by pressure from the heels of shoes that have been left on for too long, or from shoes with short heels that have been fitted too narrow and close.

In the majority of cases all that is required is to remove the cause of the corn, pare away the discoloured horn to below the level of the hoof wall to relieve the pressure and fit an ordinary shoe, making sure that the heel of the shoe rests on the wall and the bar. The horse may need a shoe that will relieve the pressure on the corn and protect the seat of the corn.

Sole bruise

Sole bruises are caused by walking or working on stones or hard, rough, uneven ground. It is common in horses with flat feet or thin rough soles, and in unshod horses with hooves that do not protect the soles from contacting the ground. If the bruising is severe, lameness can result and the bruise may need to be cut out. Having the horse shod with a pad will often help. Mild cases will respond to a few days rest and care in choosing the surfaces over which the horse is ridden.

Hoof wall cracks

Hoof wall cracks commence at the ground surface and extend a variable distance up the hoof. They are usually seen in horses with brittle hooves or hooves that are too long. Severe cracks can allow infection to enter the hoof. To prevent the crack from extending, a horizontal groove should be cut, approximately 10 mm in length, across the top of the crack. The pressure at the extremity should be relieved by cutting a small inverted V. Toe clips will often help prevent expansion of the crack. As hoof cracks are often the result of dry, brittle or overly long hooves, the correct use of hoof dressings and hoof growth supplements, as well as correct trimming, are needed.

PARASITES

Intestinal worms

Types of worms

The intestinal worms that affect horses in Australia and New Zealand are the large and small strongyles (redworms), roundworms and pinworms. Stomach bots are the larvae of a type of fly, but they do not cause a major problem to most horses. Other worms, such as tapeworms, thread-worms and lungworms, rarely cause problems.

The common signs of worm infestation are tail rubbing, pale gums, ill-thrift, colic and poor coats. Severe infestation can also cause diarrhoea or sudden death.

Large strongyles

Large strongyles are grey-coloured, blood-sucking worms, approximately 20 mm long, that are often referred to as redworms or bloodworms. The most important of the group is *Strongylus vulgaris*.

The eggs from adult worms living in the horse's large intestine are passed out with dung onto pasture where if the conditions are right they hatch and develop into infective larvae. When eaten by a horse the larvae follow a complex migration, passing through the intestinal wall to the inner lining of small arteries, then moving up these to the large arteries supplying the digestive tract. The larval migrations last 5–7 months and the larvae eventually return to the intestine where they become adults.

The migration of *Strongylus vulgaris* larvae can cause serious weakening of the walls of arteries and the formation of blood clots. When these clots break away and block smaller arteries supply-

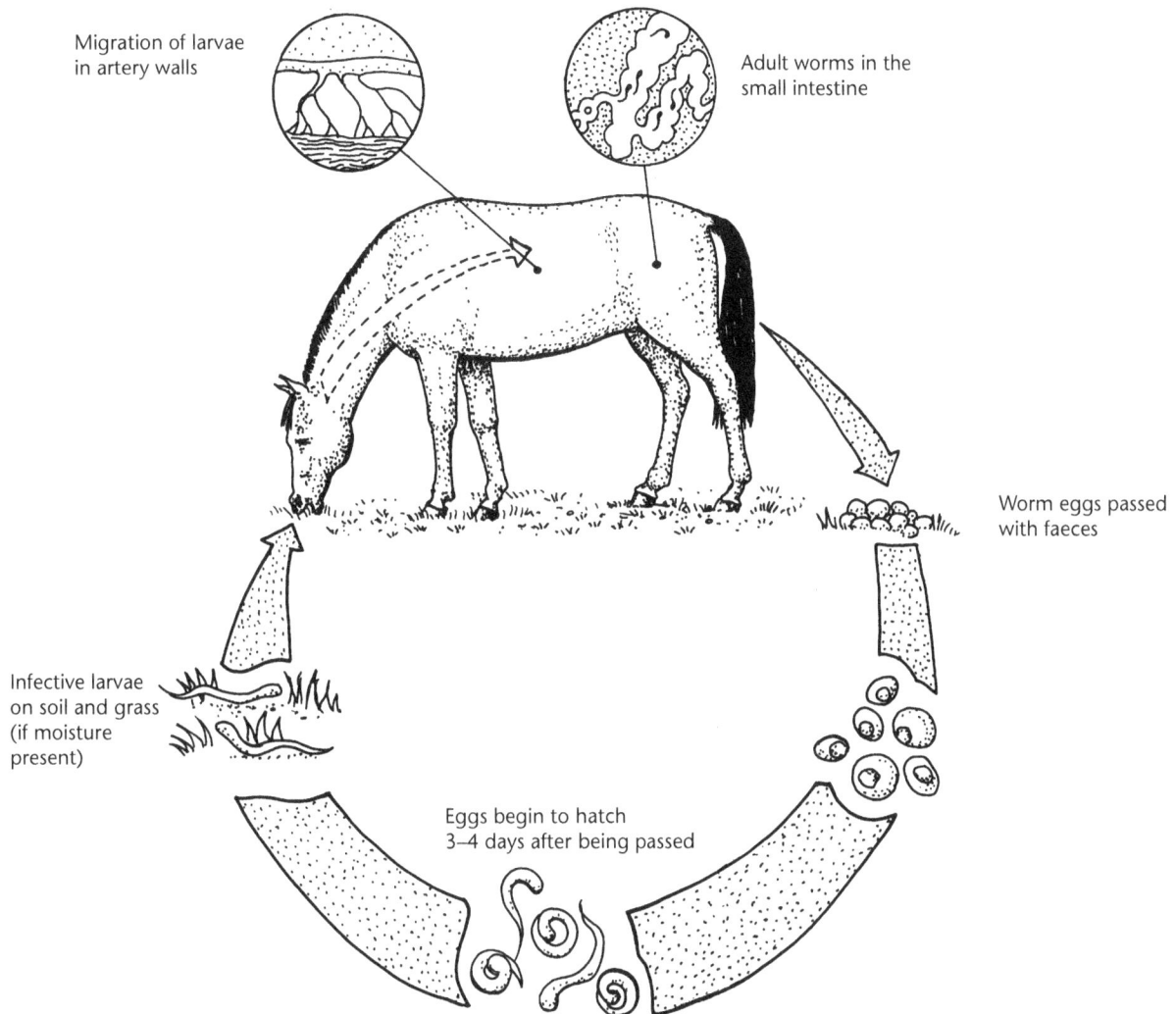

Migration of larvae
in artery walls

Adult worms in the
small intestine

Worm eggs passed
with faeces

Infective larvae
on soil and grass
(if moisture
present)

Eggs begin to hatch
3–4 days after being passed

Figure 9. Life cycle
of the large
strongyle worm,
Strongylus vulgaris.

ing sections of the intestines, colic commonly results. On rare occasions, a weakened section of artery wall may rupture while a horse is exercising vigorously, resulting in sudden collapse and death. Other side effects include weakness.

Although horses usually pick up worm larvae from pasture, any grassy areas around yards and stables that have been contaminated with manure will also be heavily infested.

Small strongyles
Small strongyles are small, redworms usually found in the large intestine. The larvae migrate into the gut wall, causing the formation of small nodules. Heavy infestations of small strongyles can cause ill-thrift and diarrhoea. These worms have developed resistance to many of the common drenches and they are now a great problem to horses. The immature stage hibernates in the gut wall through winter, and then emerges in spring to cause substantial damage to the gut and high worm burdens.

Roundworms
Roundworms are a problem in foals and yearlings; older horses develop resistance to the effects of this parasite.

Roundworms are creamy-white, almost the thickness of a pencil and up to 50 cm long. You may notice the worms in droppings after worming young horses.

The adult roundworm produces eggs that have a considerable resistance to adverse conditions and they can survive for several years on pastures and in yards and stables before hatching into larvae. This means eggs from one year's foals can infect next year's foals.

Once ingested, the larvae migrate through the liver and lungs before becoming adults in the small intestine. These very active worms irritate the intestinal lining and can cause intermittent scouring (diarrhoea), ill-thrift and a pot belly. Pneumonia in foals is a common sequel to the lung damage caused by the migrating larvae. Heavy burdens of adult roundworms can cause intestinal obstruction, colic and occasionally death.

Pinworms
Pinworms are 10 cm long and greyish-white. The eggs are eaten with feed and hatch into larvae in the small intestine. The adult female lives in the large intestine and lays eggs near the horse's anus, causing intense irritation. The commonest indication of pinworm infestation is persistent tail rubbing. Pinworms are most commonly a problem of stabled horses.

Although pinworms are not a particularly serious parasite, horses that continually rub their tails can develop abrasions and unsightly bare areas.

Tapeworms
Tapeworms rarely cause severe clinical symptoms, but may occasionally cause colic, ill-thrift and diarrhoea. Problems usually arise in late summer and autumn following infections in late spring and early summer. Although rare, acute symptoms or sudden death can occur from blockage and then rupture of the bowel. Not all worming products eliminate tapeworms so once a year make sure you use a product that does as part of your worming program.

Threadworms
Threadworms may cause diarrhoea in young foals that are infected through the mare's milk. Treatment of the mare on the day of foaling and rotation of foaling paddocks will assist control.

Lungworm
Lungworm infestation is only a problem when horses are grazed with donkeys. Affected horses may show ill-thrift or coughing, but clinical problems are seen more commonly in donkeys.

Control, treatment and prevention of worms
An effective worm control program will reduce the incidence of disease, particularly in foals and growing horses, and correspondingly lower your veterinary bills. Modern anthelmintics (worming pastes and drenches) are highly effective and easy to use, but unless they are used correctly the parasites will develop resistance to the drugs. The responsible horse owner will develop a worm management programme that incorporates targeted preventative treatment with property and animal management.

You can minimise the worm burden of your horses by:
- collecting the droppings in small paddocks, twice a week if possible
- giving young horses the cleanest paddocks
- using feed bins and hay racks rather than feeding on the ground
- harrowing paddocks
- rotational and/or mixed grazing
- removing bot eggs from hair on the legs and body in autumn
- not overstocking horse paddocks

Worm treatment
- Use an effective product.
- Calculate the correct dose by weighing the horse or using the Coprice Horse Weight Estimator (see Chapter 5).

- Ensure that the entire dose is administered.
- Use a different product every 12 months (seek veterinary advice).
- Effective summer treatment (winter in northern Australia) may have significant benefits.
- Do not overtreat or use several different treatments in a 12-month period (this is a waste of money and will select for drug resistance in the worms).
- Do not undertreat as this may allow paddock contamination and the development of clinical problems in the horse.
- Use 'Wormtest' or faecal egg counts to determine the need for treatment.
- Check for resistance after worming with Wormtest or an egg count 10–14 days after worming.
- Treat foals from 6 to 10 weeks of age.
- Treat mares just after foaling.
- Treat for bots in late autumn/winter.
- Treat for tapeworms in late winter/early spring, if necessary.
- Seek veterinary advice on your treatment and control program.

Wormtest

A 'Wormtest' kit that will help you determine the need for drenching is available from your vet, produce store, saddlery or the NSW Department of Agriculture (see the end of the chapter for contact details). Using 'Wormtest' or having a faecal egg count carried out by your vet will tell you if the horse needs treatment, what the appropriate treatment interval is for the particular worm and whether the drug you have used has worked.

Bots

The bot is the larval stage of the bot fly, not an actual worm. The flies are active in summer, laying eggs on the lower body of the horse, particularly on the legs. The eggs are pin-sized and yellow in colour.

When the horse bites and rubs the affected area, it ingests the larvae. After burrowing in the gums for approximately three weeks, the larvae are swallowed and attach themselves to the stomach wall for several months before being passed out in the dung to develop into adult flies.

The egg-laying activities of the adult flies cause considerable annoyance to horses in summer. Although bots cause mild ulceration at their attachment sites in the stomach, and on very rare occasions these ulcers may perforate, infestations generally present no problem to most horses. The significance of bots is often exaggerated.

Although a tedious task, it is possible to remove eggs from the horse's coat by scraping or clipping, but this should be done in a yard, not where horses graze. Treatment for stomach bots should be given in early winter when fly activity has ceased and the larvae are in the stomach.

Table 2. Suggested worming strategy for grazing horses in temperate Australia

Month	Adults	Pregnant mares	Foals	Yearlings
August/September	'Wormtest'	Day of foaling: place in clean paddock*		Treat
October/November	Treat	Treat	Treat	Treat
December/January	'Wormtest'	'Wormtest'	Treat	Treat
February	Harrow paddocks	Harrow paddocks		Harrow paddocks
March	Treat	Treat	Treat when weaned; clean paddock*	
April/May	'Wormtest'	'Wormtest'	Treat	'Wormtest'
June/July	Worm**/Bot	Worm on arrival at stud**	Worm**/Bot	Worm**/Bot

Plan intervals between treatments using 'Wormtest' or faecal egg counts.
*Place into a clean paddock that has been rested or had droppings removed twice.
**Treat annually with a suitable product for tapeworms if tapeworm infestation occurs on your property.

Eggs laid on hairs hatch after
5 days into infective larvae
that are ingested

Bot larvae attached to
stomach lining

Larvae passed in
faeces (spring)

Flies active in
summer and early
autumn

Adult emerges in
spring

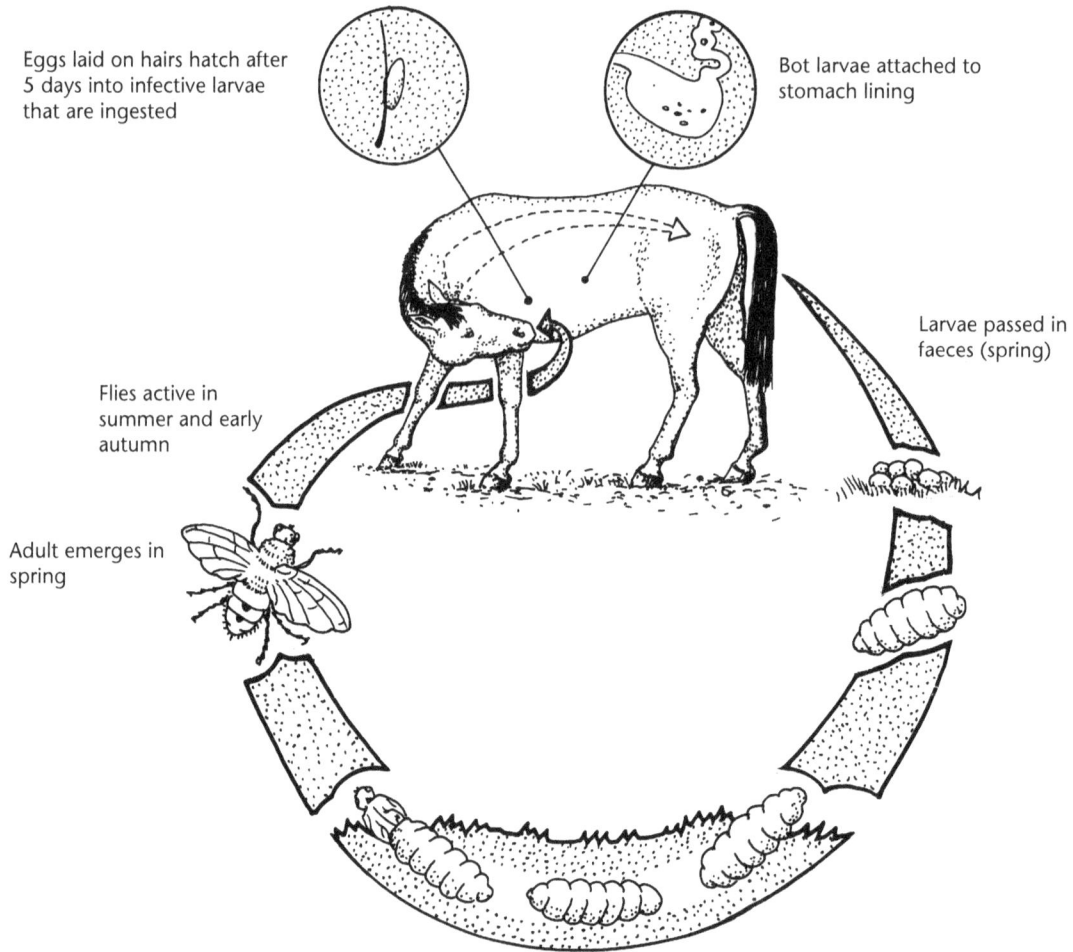

Figure 10. Life cycle
of the bot fly.

SKIN DISEASES

The appearance of the common skin diseases of horses can vary greatly, especially after treatments have been given, so you should consult your veterinarian whenever necessary. Each condition is listed in Table 3 according to its usual appearance, but a condition may appear under two
or more headings. For example, horses with rain scald may show both 'hair loss' and 'weeping
sores with scabs'. An alphabetic list of the skin diseases is also provided.

Bacterial acne (also called saddle boils, folliculitis, contagious acne)

Cause
Bacterial acne is the result of infection of the hair follicles with bacteria, usually *Staphylococcus
aureus* (golden staph).

Description
Hair follicles develop into small, very painful abscesses, mostly under the saddle cloth and
harness. Because acne can be confused with other conditions, such as fly bites, ticks, feed allergies
or rain scald, you should consult your veterinarian.

Treatment
The infection is contagious so you should disinfect all cloths, harness and grooming equipment
that have been used on the affected horse. Wash the affected areas of skin daily for one week with

Table 3. Skin diseases of horses classified by their major symptoms

Symptom/appearance	Possible cause
Itching	Buffalo fly bites
	Ear mites
	Leg mange
	Lice
	Onchocercal dermatitis
	Queensland itch
	Ringworm
	Stable fly bites
	Summer sores
	Tail itch
Lumps in the skin	Bacterial acne
	Fibroma
	Melanoma
	Mosquito bites
	Neurofibroma
	Nodular necrobiosis
	Queensland itch
	Sarcoid
	Stable fly bites
	Ticks
	Urticaria
Growths on the skin	Sarcoid
	Squamous cell carcinoma
	Summer sores
	Swamp cancer
	Warts
Hair loss	Leg mange
	Lice
	Onchocercal dermatitis
	Queensland itch
	Rain scald/greasy heel/mud fever
	Ringworm
	Tail itch
Weeping sores	Big leg
	Fly worry
	Melanoma
	Sarcoid
	Squamous cell carcinoma
	Summer sores
	Swamp cancer
	Girth galls
	Saddle sores
Weeping sores with scabs	Bacterial acne
	Buffalo fly bites
	Leg mange
	Lice
	Onchocercal dermatitis
	Queensland itch
	Rain scald/greasy heel/mud fever
	Sarcoid
	Photosensitisation
Blisters/Generalised swelling	Coital exanthema
	Bee and wasp stings
	Big leg
	Spider bites

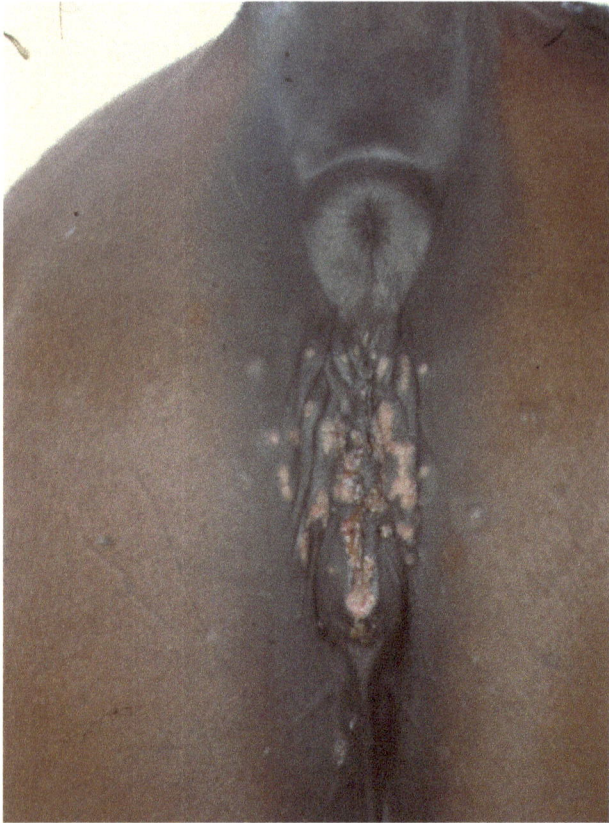

Figure 1. Coital exanthema (courtesy of Reg Pascoe)

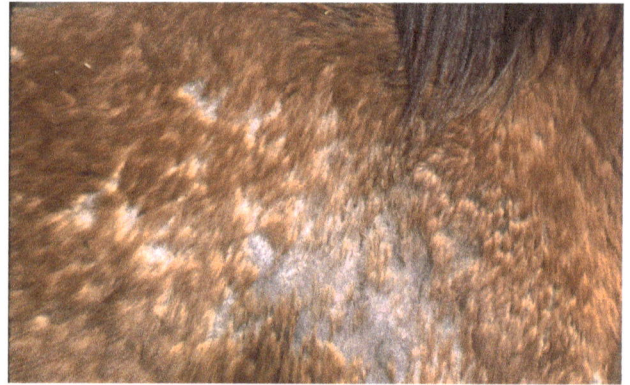

Figure 2. Lice infestation (courtesy of Reg Pascoe).

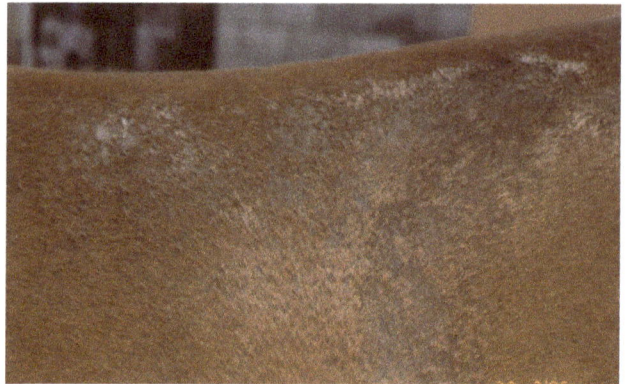

Figure 3. Rain scald (courtesy of Reg Pascoe).

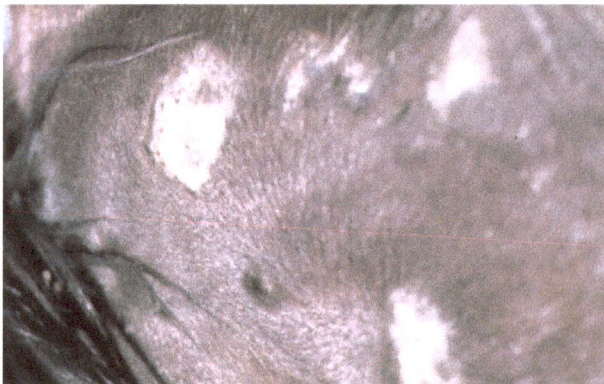

Figure 4. Onchocercal dermatitis (courtesy of Reg Pascoe).

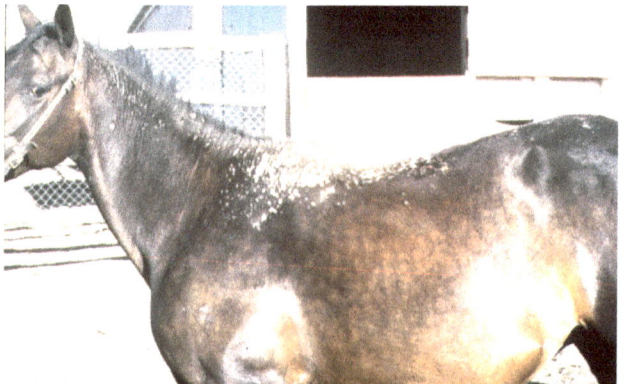

Figure 5. Queensland itch (courtesy of Reg Pascoe).

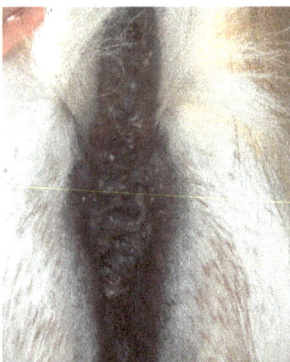

Figure 6. Melanoma
(courtesy of Reg Pascoe).

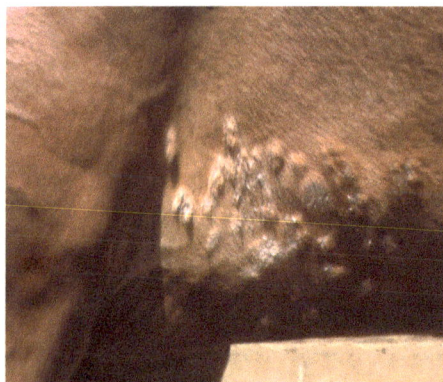

Figure 7. Ringworm (courtesy of Reg Pascoe).

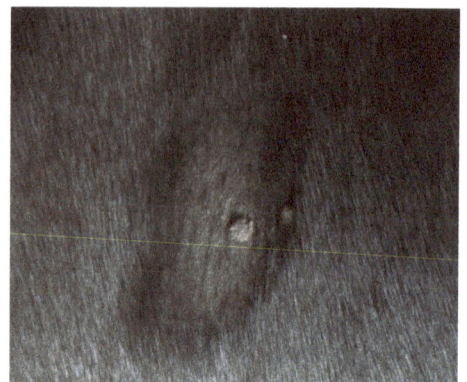

Figure 8. Simple sarcoid (courtesy of Reg Pascoe).

PLATE 2 – HORSE HEALTH   211

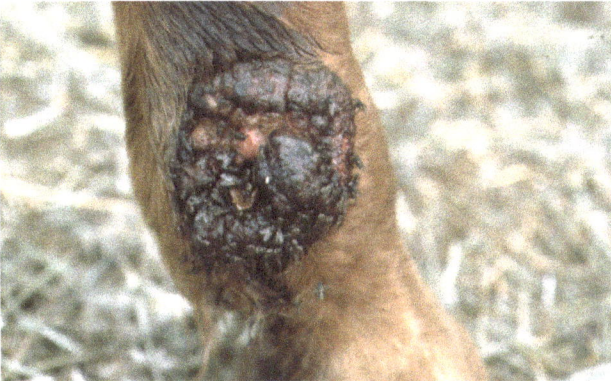

Figure 9. Invasive granulomatous sarcoid (courtesy of Reg Pascoe).

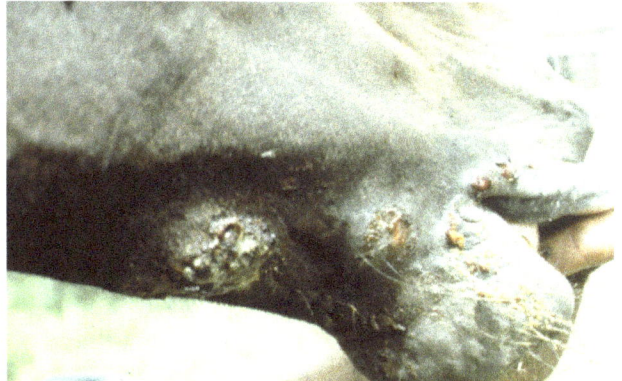

Figure 10. Squamous cell carcinoma (courtesy of Reg Pascoe).

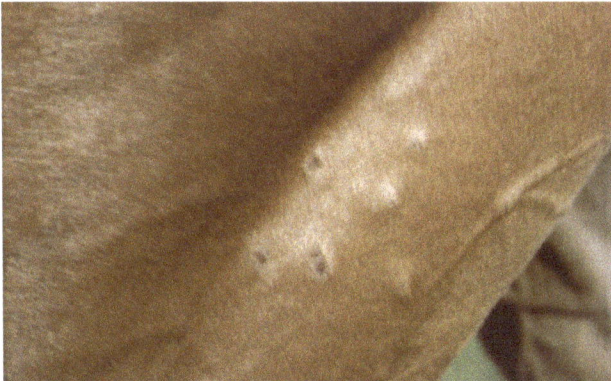

Figure 11. Stable fly bites (courtesy of Reg Pascoe).

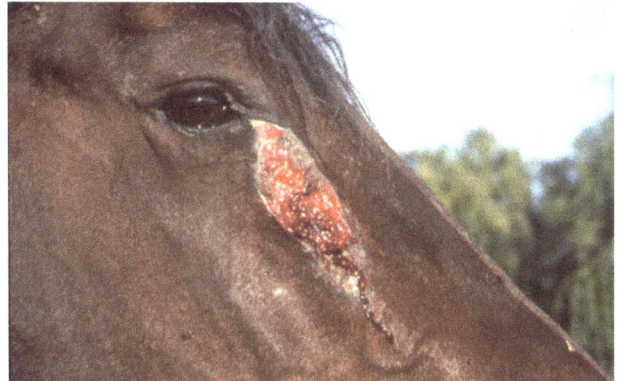

Figure 12. Summer sores or habronemiasis (courtesy of Reg Pascoe).

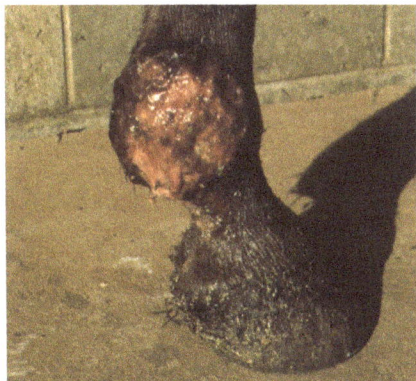

Figure 13. Swamp cancer (courtesy of Reg Pascoe).

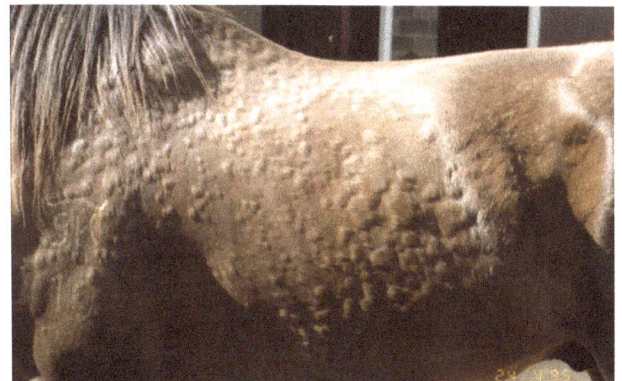

Figure 14. Urticaria (courtesy of Reg Pascoe).

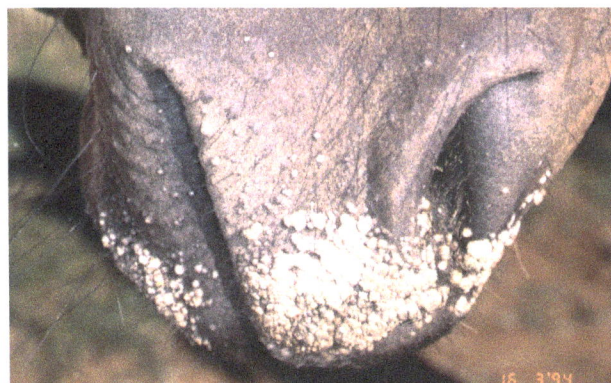

Figure 15. Warts (courtesy of Reg Pascoe).

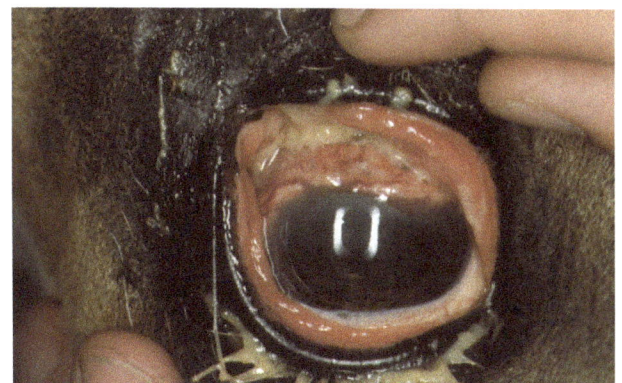

Figure 16. Uveitis (courtesy of Reg Pascoe).

antiseptic, such as an iodine-based wash (e.g. Betadine, Iovone, Vetadine) or chlorhexidine-based wash (e.g. Hibitaine, Chlorhex). Make sure that all harness and tack is stored hygienically.

Bee stings

Cause
Stings from the honey bee, *Apis mellifica*.

Description
Bee stings can be seen in the horse's skin, usually surrounded by a swollen patch. If the horse is stung by a swarm the area can swell enormously, which can cause problems around the eyes, nose, throat or mouth.

Treatment
Consult your veterinarian immediately if the horse is having trouble breathing. Stings must be removed carefully to avoid squeezing the venom sac. The horse may need treatment to reduce the swelling.

Big leg (ulcerative lymphangitis)

Cause
Several different bacteria cause big leg.

Description
Big leg occurs when infection spreads from a skin or hoof wound to the lymph vessels draining the leg. The infection localises along these vessels, causing swelling of the limb with the formation of abscesses and subsequent discharge. The leg is swollen or enlarged from the pastern to the elbow or stifle, with numerous small points of pus discharge. The limb is painful to touch.

Treatment and control
The wounds must be cleaned and specific antibiotic injections for the type of bacteria must be given.

Buffalo fly bites

Cause
The blood-sucking fly, *Haematobia exigua*.

Description
The buffalo fly has spread south from Queensland and northern coastal New South Wales. Its bite can cause irritation and sores on the ears, eyes and abdomen. Heavy infestations can lead to severe mutilation.

Treatment and control
Insecticides can be sprayed onto the head and abdomen. Many of the synthetic pyrethroid agents have a residual effect.

Coital exanthema

Cause
Coital exanthema is the result of a herpes virus (EHV-3), one of three types of herpes virus that affect horses (Plate 2, Figure 1).

Description

This is a venereal disease that involves the penis of the stallion and vulva of the mare. The virus is transmitted during mating. Some horses are carriers and may only show signs of disease if subjected to a stress such as foaling or another illness. Following infection, the sheath of the penis or the lips of the vulva swell and redden. Watery blisters appear and rupture, becoming ulcerated. The ulcerated areas form scabs and heal in 2–3 weeks. Mares that are infected during service may still conceive.

Treatment

Neither mares nor stallions should be used for breeding until the vesicles have healed. Mild antiseptic washes will help healing. The disease may recur in the next breeding season in some horses, but this is rare.

Ear mites (psoroptic mange)

Cause

The ear mite, *Psoroptes canniculi*.

Description

A horse with ear mites will shake its head, be hard to handle around the head and may have a discharge from the ear canal. These mites may also cause itchiness around the tail.

Treatment

If ear mites are confirmed as the cause of the problem, the infection can be treated with insecticidal ear drops obtained from your veterinarian.

Fibroma

Description

Fibroma is a tumour of the fibrous tissue just under the skin that can occur anywhere on the body. It can appear as a simple swelling that is slow growing, or as a cauliflower-like growth. If it grows rapidly it can form ulcerating sores. Positive diagnosis is made from a biopsy.

Fibromas are often difficult to remove surgically, particularly if treatment is delayed and they become large. They often involve other structures such as nerves, blood vessels and tendons.

Treatment

Surgical removal is the usual treatment, but a drug, cimetidine, shows promise as a useful treatment. Early diagnosis by your veterinarian is important to achieve a satisfactory cure.

Fly worry

Cause

Species of flies including the house fly, *Musca domestica*, and the bush fly, *M. vetustissima*.

Description

Flies settle on moist areas of the body such as the corners of the eyes, the vulva and the top of the penis. Large numbers of them can cause irritation, which can become a serious problem, especially around the eye where it can lead to conjunctivitis.

Treatment and control

Insecticidal and fly repellent creams or sprays can be used around the affected area. Use a fly veil to protect the eyes. If the horse has conjunctivitis consult your veterinarian.

Leg mange (chorioptic mange)
Cause
The mite, *Chorioptes bovis*.

Description
Leg mange affects mainly the lower limbs of draught-type horses, but in heavy infestations can extend to the upper limbs, armpits and groin. The mites cause extensive irritation and the skin becomes moist with a bloody exudate. Affected horses repeatedly stamp their feet when being examined. Leg mange is easily confused with 'greasy heel'.

Treatment
Leg mange can be difficult to treat, requiring continued application of insecticides. Seek veterinary advice.

Lice
Cause
Biting lice (*Damalina equi*) or sucking lice (*Haematopinus asini*).

Description
Lice infestation occurs throughout Australia and New Zealand, reaching a peak in late winter and spring. Horses in poor condition and with long, winter coats that are not groomed are more prone to heavy infestation.

Biting lice live on skin scurf and are usually found on the back and head. Sucking lice tend to be found under the mane, tail and around the fetlocks, although they can occur almost anywhere on the body. Horses with heavy infestations of biting lice can become anaemic. Lice cause intense itching and irritation of the skin. Horses bite and rub the skin, causing the hair to fall out and leaving red, weeping areas (Plate 2, Figure 2).

Treatment and control
Horses can be treated with sprays or powder insecticides, such as Malathion. Rugs and harness should be sprayed and left unused for 14 days and a second treatment 17 days later is necessary. Be careful with pour-on treatments because these can cause additional irritation and hair loss, and hair colour change.

Melanoma
Description
Melanomas are tumours formed from the cells that produce pigment in the skin (Plate 2, Figure 6). They appear frequently around the anus or on the tail as hard lumps of varying size. However, they can be found almost anywhere on the skin. They are especially common in grey horses.

The tumours are slow-growing and usually non-invasive. In some circumstances they can become malignant and spread to other body organs.

Treatment
Small tumours or groups of tumours that do not appear to have spread or involve other structures (such as the muscular ring of the anus) can be removed surgically or by freezing (cryosurgery). Large or multiple tumours are very difficult to remove. If there is evidence of metastasis (spread to other parts of the body) surgery is not worthwhile. A promising new treatment involves the use of cimetidine, but a 4-month course is required and is very expensive.

Mosquito bites
Cause
Bites from mosquitoes, particularly after spring and summer rains.

Description

Horses may have many small skin swellings in areas not protected by rugs.

Treatment and control

Rugs and hoods will provide some relief and residual insecticides (pyrethroid types) can be sprayed on the horse. The swellings normally disappear in 2–3 days.

Neurofibroma

Description

Neurofibromas are tumours found on the face, usually on the upper or lower eyelid. It starts as a small swelling under the skin and gradually enlarges. Usually the skin remains intact unless it is bumped or rubbed, in which case the surface of the tumour may become raw and ulcerated.

Treatment

Surgical removal is the best method of treatment, although the tumour often regrows in a different site, necessitating further surgery.

Nodular necrobiosis

Cause

Possibly an allergic reaction.

Description

Small non-painful lumps of up to 1 cm in diameter develop in the saddle area. They often appear in spring and summer and are thought to be an allergic response to insect bites or feed. Pressure from the saddle when the horse is ridden can worsen the condition.

Treatment

Relieving pressure from the saddle is important and sometimes, the lumps will shrink without treatment. Many veterinarians will use oral, injectable or topical corticosteroids; however, there may not be any change after treatment. Changes to the feed often help resolve the condition.

Onchocercal dermatitis

Cause

Larvae of the filarial parasite, Onchocerca.

Description

The dermatitis affects the skin of the underside of the body and occasionally the chest, withers, head and neck (Plate 2, Figure 4). It is caused by larvae (microfilaria) of the parasite migrating through the skin. The adult form lives in ligamentous tissues of the body, especially the neck. It is transmitted by biting midges. Horses become itchy and their hair thins. The skin is scaly and develops lumps, and eventually it begins to weep because of secondary bacterial infection. Later, scabs form over the area. Occasionally the larvae migrate into the surface of the eye.

Onchocercal dermatitis occurs in northern and western Australia. In south-east Australia it only appears in summer and then regresses or becomes less severe in autumn and spring.

Treatment and control

Onchocercal dermatitis can be treated with ivermectin and related compounds.

Photosensitisation

Cause
Sensitisation of unpigmented skin to sunlight following feeding on lush clover or burr medic pastures or St John's wort (see later section on poisonous plants). Chronic liver disease can also lead to photosensitisation.

Description
White skin, particularly on the nose, back or legs, will be swollen and exude serum, later cracking to leave raw bleeding areas that are often covered by scabs and can be confused with rain scald.

Treatment and control
Place the horse in a stable and apply an antibiotic/corticosteroid cream to the affected skin.

Queensland itch

Cause
An allergy to the bite of the sandfly, *Culicoides brevitarsus*.

Description
Queensland itch is very common in northern Australia. When the horse is bitten by the sandfly, small lumps develop in the skin and because they are intensely itchy the horse rubs itself, causing loss of hair and irritating the skin. Serum oozes from the irritated skin and scabs form. In chronic cases the skin becomes thickened and corrugated, especially along the flank, neck, buttocks and tail (Plate 2, Figure 5). The appearance of the skin is distinctive, but other diseases that can look similar include ringworm, tail itch, bites from stable or buffalo flies, and rain scald.

Treatment and control
Provide temporary relief of the skin irritation by applying calamine lotion or anti-inflammatory creams. Your vet may need to give an anti-inflammatory or antihistamine injection.

Paddocked horses on large properties are difficult to treat. Many remedies have been used with variable success. Non-irritant, oil-based compounds may help because they have a sedative effect on skin and also act as a barrier to the fly. Residual insecticides, such as the synthetic pyrethroid-type sprays, can be used, but remember that rain will reduce the effectiveness of sprays and local treatment.

In the long term fly control measures are needed because susceptible horses are usually affected annually. Ideally, keep these horses in insect-proof stables at the times of greatest fly activity (late afternoon and early morning). Rugs with hoods and tail covers will also help.

Rain scald (dermatophilus), mud fever, greasy heel

Cause
The fungus, *Dermatophilus congolensis*.

Description
Rain scald is characterised by matting of the hair with formation of a scab under the hair mat, commonly along the midline of the neck and back, the face, muzzle, legs and on either side of the rib cage (Plate 2, Figure 3). After the scabs fall off the skin underneath is moist and greyish pink. Large areas of skin may be affected. The condition is also called 'greasy heel' or 'mud fever' when present on the legs.

The disease mainly occurs during prolonged rainy weather when the skin becomes excessively wet. The distribution of scabs on the horse follows the pattern of wetting from rainwater or mud.

Rain scald is contagious and spreads rapidly between horses. It can resemble sunburn, photosensitisation or ringworm.

Mud fever or greasy heel occurs behind the pastern in horses that are continually standing or walking in wet grass or mud and is more common in horses with white legs. Mud fever causes cracking of the skin, which can be painful and cause lameness.

Treatment and control
In most cases the disease is self-limiting and will regress after 3–4 weeks. A horse that is severely affected should be housed in a dry stable. Treatment involves iodine-based washes and injectable antibiotics. Consult your veterinarian. Keep the skin dry. Horses that have recurrent bouts of these conditions should be kept out of the rain and damp conditions.

Ringworm
Cause
Fungi of the genus *Trichophyton* or *Microsporum*.

Description
Ringworm is a common condition in horses. Infection of the skin by the fungus causes loss of hair in circular areas that progressively enlarge (Plate 2, Figure 7). Small areas of hair loss, usually up to 3 cm in diameter, can join together forming large lesions. Mostly, the skin is scaly and dry, but can be moist if the hair has matted together with a secondary bacterial infection. The area may be itchy in the early stages.

The most common sites are the girth and head, but the disease can occur on most parts of the body. Hair will begin to regrow at variable intervals, depending on the species of fungus causing the problem. Most cases of ringworm will begin to resolve within 40 days.

Infections usually occur in young horses during the humid months of the year. Ringworm is contagious and is spread between horses by close physical contact and contaminated gear. After being infected, horses seem to develop some immunity against re-infection. Subsequent infections are less severe and regress more rapidly than the original lesions.

Treatment
Many treatments are used for this condition, but most cases will heal eventually, although the affected animal is a source of infection for other horses. Remove all scabs and infected hairs and burn them. Scrub affected areas and the surrounding hair for 1–2 minutes daily for 5 days with an iodine-based wash (e.g. Betadine, Iovone, Vetadine) or 0.1% chloramine (e.g. Halamid, Halasept) or use a specific ringworm cream (e.g. Fungazone).

Thoroughly wash contaminated harness, grooming gear, boots and rugs in one of the compounds listed or fumigate with formaldehyde gas in large plastic bags.

Saddle sores/girth galls
Cause
Girth galls or saddle sores are caused by over-work in poorly fitting harness.

Description
Mild cases have a moist painful area of skin, but severe cases can develop a raw, deep ulcer that is very painful.

Treatment and control

Spelling is essential until the sores heal. Early cases may respond to the application of hot and cold packs followed by application of white lotion or a drying agent. Attention to saddlery is necessary to prevent recurrence.

Sarcoid

Cause

A papilloma virus and certain horses are predisposed to development of sarcoids.

Description

Sarcoids are the most common tumour found in horses. They are skin tumours that can invade the surrounding tissue and cause serious problems, depending on the site. Most parts of the body can be affected, but the usual sites are the skin of the head, neck, chest and legs. They vary greatly in appearance, but there are two main types.

The warty type lies on or above the level of the skin (Plate 2, Figure 8). It grows slowly until rubbed or knocked, after which it can rapidly grow into the granulomatous or 'proud flesh' form.

The granulomatous type starts as a hard nodule in the skin that grows outwards into and above the skin (Plate 2, Figure 9). The tumour may be firmly attached to the skin or attached by a stalk, and can grow up to 30 cm in diameter. The surface of the tumour becomes ulcerated and bleeding, and sometimes fly-blown.

Sarcoids can look similar to other skin tumours as well as to warts and healing wounds. Diagnosis depends on taking a biopsy for pathological examination. In northern Australia sarcoids can be secondarily infected with fungi and habronema (see later), so accurate diagnosis is essential.

Treatment

Any one or more of the following are used:

- ligatures (if the tumour is on a stalk)
- local application of astringents such as podophyllin
- cryosurgery
- surgical removal
- immune stimulation using proprietary products (e.g. Ribgen, Equoid).

Early diagnosis and treatment is vital. The type and position of the sarcoid will determine the method of treatment. Consult your veterinary practitioner because sarcoids can be notoriously difficult to treat successfully. Unsuccessful treatment may stimulate a sarcoid to grow rapidly. Sometimes it is best to leave it untreated and just monitor it. Sarcoids can be infectious to other horses that come into contact with the affected animal.

Spider bites

Cause

The black widow spider, *Ixeuticus robustus*.

Description

Black widow spiders are often found around older stables in funnel-shaped webs. They are nocturnal and a bite causes a very painful swelling, usually on the neck or body.

Treatment and control

Crushed ice should be applied to the swollen area, followed by alcohol rubs and white lotion. Your veterinarian can give further treatment. Stables can be sprayed by commercial pest controllers.

Squamous cell carcinoma

Description

This tumour starts as a depressed, ulcerating sore that gradually enlarges until it becomes raised above the skin surface with a red, raw surface. It can be found on any part of the body, but is usually on the face, penis, prepuce, vulva or eyelids (Plate 2, Figure 10).

Precancerous changes can sometimes be seen as white, slightly raised patches, especially on the eyelids, nose and penis. Diagnosis is based on the appearance of the lesion and biopsy sections taken for microscopic examination. Because there are many conditions that can be confused with squamous cell carcinoma, veterinary diagnosis is essential.

Treatment

Early surgical treatment is essential for even moderate success and often the tumour is inoperable by the time attention is sought.

Stable fly bites

Cause

The stable fly, *Stomoxys calcitrans*.

Description

Stable fly bites can cause intense irritation in some horses. The bites appear as small raised lumps with a central scab (Plate 2, Figure 11).

Treatment and control

The stable fly is most prevalent in summer and autumn. To control fly numbers you should dispose of manure and rotting vegetation in and around the stables, and spray the sunny side of the stable walls with a registered insecticide. Soothing, analgesic creams can be used on badly affected horses.

Summer sores (habronemiasis)

Cause

Larvae of the worm, *Habronema*.

Description

Summer sores can appear anywhere on the body that flies are able to rest and feed, around the eye, face, prepuce and penis, or on any wound (Plate 2, Figure 12). Flies carry the larvae of the Habronema worm onto these areas where they cause a severe inflammatory reaction, sometimes with a yellowish white centre that is 1 mm in diameter. Irritation and biting or rubbing of the area is common.

Often the inflammation will subside in winter and reappear the next summer. Summer sores are most common in very hot, wet areas of Australia, but in some years the condition is widespread wherever large numbers of horses congregate. The lesions can sometimes look similar to proud flesh.

Treatment

Ivermectin, a topical insecticide, is the best form of treatment and for a response to treatment can be diagnostic for the cause. Surgical removal of the lesions is possible, combined with certain insecticides that are applied to the skin or given orally. Fly control is necessay. Veterinary assistance is needed to diagnose and treat habronemiasis.

Swamp cancer, phycomycosis (also known as equine granuloma or 'kunkers')

Cause
Several types of fungi.

Description
Swamp cancer is mainly a problem in northern coastal areas of Australia where the horses have access to low-lying swamps or wet pasture. It occurs mainly on the legs and abdomen of horses, but can be seen occasionally on the head and other parts of the body. Affected areas of skin develop a tumour-like appearance, ulcerating and discharging pus. Inside these growths are characteristic yellow cores, called 'kunkers' (Plate 2, Figure 13) .

The horse often bites and kicks at the growths. Joints, tendons, nerves and blood vessels can be involved with the tumour, making surgical removal difficult. If the nose is affected the horse may have a nasal discharge and respiratory noises because of the growths in the nasal passages. Diagnosis requires a biopsy.

Treatment
Several treatments have been tried, with varying degrees of success. The growths are usually removed surgically, but can also be cauterised under local anaesthesia. Injections of fungicide can be given following removal to increase the recovery rate.

Tail itch (also known as pinworm infestation)

Cause
The pinworm, *Oxyuris equi*.

Description
The adult female pinworm lays eggs on the skin around the tail and anus, which causes irritation and the horse rubs its tail on fences, trees or the stable. Hair may fall out of the tail. Scraping the skin around the tail and anus and examining under a microscope may show the eggs. Pinworm infestation can be confused with Queensland itch.

Treatment
Most broad-spectrum worm drenches will control this parasite.

Ticks

Cause
The cattle tick, bush tick (also known as the New Zealand cattle tick) and paralysis tick can all infect horses. The bush tick is found throughout New Zealand, but is more common in the North Island. In Australia, the cattle tick is limited to Queensland, northern Western Australia and the Northern Territory. The paralysis tick is found along the east coast of Australia. It is usually not present in New Zealand, but has been found on imports into the country.

Description
There is often a severe local reaction to the attached tick with swelling around the site. Paralysis ticks, once adult and engorged, can cause progressive paralysis and death, especially in foals.

Treatment
In foals or adults affected by paralysis tick, antiserum will be successful if given early in the progression of the paralysis. Various chemicals are used to remove ticks. See Appendix 3 for the tick clearance policy for Queensland.

Urticaria

Cause
May be obscure, but insect bites, feed allergies, injection reactions and contact irritants are often incriminated.

Description
There is rapid development and then rapid disappearance of elevated, round, flat-topped plaques on the skin surface, swelling of the nostrils, eyelids, ears and between the legs (Plate 2, Figure 14).

Treatment and control
Urticaria often regresses without treatment, but if it persists for more than a few hours veterinary treatment may be required. If the horse is distressed or has difficulty breathing seek urgent veterinary attention.

Warts (also known as papillomatosis)

Cause
A papilloma virus.

Description
Warts are more common in young horses, appearing on the muzzle, lips, face, eyelids and legs (Plate 2, Figure 15). They are greyish white, cauliflower-like growths with short stalks attaching them to the skin. They vary in size from one millimetre up to several centimetres in diameter. Usually warts are quite harmless, although sometimes they grow very large and can interfere with the eyelids or the lips.

Treatment and control
Although warts will disappear in time and cause no further problems, they may sometimes become infected or fly-blown. If this happens, apply antiseptics and fly-strike preparations. The wart virus is spread by contact between infected and susceptible horses, so isolate infected animals.

Various astringents are used to remove warts, including glacial acetic acid, iodine and podophyllin. Warts can be removed surgically, but often recur. Your veterinary surgeon can make up a crude 'wart vaccine' for your horse, but results with this are variable. Warts will disappear in 3–6 months, regardless of treatment. Some foals have large, congenital papillomas that can be removed by ligating the stalk.

PLANT POISONING

Many plants contain poisonous substances, or are poisonous at particular stages of their growth or at particular times of the year, so all horse owners should be familiar with identifying those plants. A well-fed horse in good condition may occasionally eat limited amounts of a poisonous plant and not suffer any ill-effects, which may lead you to think that the plant is not poisonous. However, on another occasion, the same horse may be hungry and eat a greater amount of the plant and become sick or even die as a result. Most horses will avoid eating poisonous plants, but may do so if they are hungry. A hungry horse with an empty stomach will absorb poisons more quickly and thus is more likely to be affected. Most horses that are poisoned are not well fed and have been forced to eat poisonous plants because other food is not available.

The plants listed in Table 4 have caused cases of poisoning in horses in Australia and New Zealand, or are known to be poisonous even though cases may not have been reported.

If you suspect plant poisoning may have occurred, check the following.

- What signs of illness is the horse showing?
- What plants has the horse been eating?
- What plants are growing in its grazing area?
- Can the horse reach non-pasture plants in a garden, the roadside or windbreak?
- Can these plants, and their toxic properties, be identified?

Seek veterinary help immediately and provide the veterinarian with as many answers as possible to these questions.

Safety precautions

The plants most commonly causing poisoning of horses have been listed, but there are others that could be dangerous, so the following safety precautions must always be observed to minimise the risk.

- Identify the common poisonous plants and be suspicious of unfamiliar plants to which horses might have access.
- Find out the means by which such plants can be removed.
- Do not allow horses access to garden waste or plant poisonous trees near fence lines.
- Seek veterinary advice as soon as poisoning is suspected.

Poisonous plants grouped according to their effects

- Group A: Sudden death or death within a few days.
- Group B: Nervous diseases (e.g. staggers, dummy syndrome).
- Group C: Photosensitisation.
- Group D: Liver damage.
- Group E: Kidney damage.
- Group F: Skin disorders.
- Group G: Heart and lung disorders.
- Group H: Intestinal disorders.

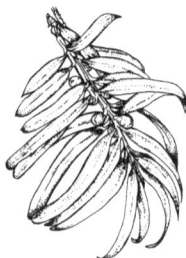

Figure 11. Yew

Group A: Plants causing sudden death or death within a few days

Yew (*Taxus baccata*)

Although yew trees are not common in Australia they can be found near old homesteads and churches in southern Australia. They can also be found in New Zealand. They have leathery dark green leaves that sometimes have yellow markings (Figure 11).

Yew poisoning generally occurs when branches are blown off or clippings are thrown into horse paddocks. A poisoned animal may suddenly drop dead, but in less acute cases the animal will show excitement, trembling and staggering before death. Prevention consists of keeping horses away from yew trees and careful disposal of branches and clippings.

Figure 12. Laburnum

Laburnum spp.

Laburnum is a cultivated plant (Figure 12) and the most likely cause of poisoning is trees that overhang horse paddocks or when clippings have been disposed of carelessly. Horses appear to be more susceptible than cattle. Clinical signs are excitement, convulsions, nausea and finally coma. Prevention consists of keeping horses away from the plant and careful disposal of garden waste.

Figure 13. Hemlock

Hemlock (*Conium maculatum*)

Hemlock is a common weed of roadsides and wasteland (Figure 13). Characteristic signs of hemlock poisoning are trembling, depression, dilated pupils, weak pulse and rapid breathing. Animals that survive more than 8 hours usually recover, although pregnant mares may abort.

Table 4. Plants that are poisonous to horses

Plant	Group	Clinical signs
Aconite (*see* Monkshood)		
Alsike clover	C, D	Liver disease, photosensitisation
Avocado	G	Dullness, colic, heart failure
Birdsfoot trefoil	A	Rapid breathing, collapse, death
Birdsville indigo	B	Nervous signs, behaviour changes, death
Black nightshade	A	Depression, colic, death
Bracken fern	B	Nervous signs, staggers
Buckwheat	C, D	Liver disease, photosensitisation
Burr medic	C, D	Liver disease, photosensitisation
Corn gromwell (*see* White iron weed)		
Castor oil plant	B	Nervous signs, colic, death
Crofton weed	G	Breathing difficulties, coughing
Darling pea	B	Weight loss, staggers, mood changes
Ergot of paspalum	B, H	Tremors, staggers, colic
Fireweed	D, B	Liver disease, nervous disease
Foxglove	A	Diarrhoea, colic, tremors, death
Hemlock	A	Depression, shock, death
Laburnum	A	Excitement, sudden death
Leucaena dermatosis	F	Hair loss, hoof rings
Linseed	A	Rapid breathing, collapse, death
Lupins	B, D	Staggers, convulsions
Marshmallow	B	Staggers, rapid breathing
Monkshood	B	Staggers, trembling
Oak	E	Colic, bloody urine, depression
Oleander	A	Diarrhoea, colic, death
Patterson's curse	D, B	Liver damage, behaviour changes
Poppies	B, F	Staggers, trembling, dermatitis
Ragwort	B	Nervous signs, staggers
Rattlepods	B	Behaviour changes, liver damage
Rubber vine	H	Intestinal damage, death
Russian knapweed	B	Chewing disorders
Ryegrass – Perennial	B	Staggers
Ryegrass – Annual	B	Staggers, convulsion, death
Rhododendron	A	Diarrhoea, colic, collapse, death
Salvation Jane (*see* Patterson's curse)		
Selenosis	F	Loss of hair, gait changes
Sorghum	A	Rapid breathing, collapse, death
Soursob	E	Weakness, incoordination, collapse
St Barnaby's thistle	B	Chewing disorders
St John's wort	C, D	Liver damage, photosensitisation
Storksbill	C, D	Liver damage, photosensitisation
White iron weed	B	Staggers
Wolfsbane (*see* Monkshood)		
Yellow burr weed	B	Staggers, nervous signs
Yew	A	Excitement, sudden death

Figure 14. Black nightshade

Black nightshade (*Solanum nigram*)

This common plant is suspected as a cause of sheep mortalities, but could also pose a risk to horses. It is one of the commonest weeds and grows readily on bare ground in horse yards and paddocks (Figure 14). Poisoning is most likely when hungry stock have access to heavily infested pastures or are restricted to yards containing the weed. Signs of poisoning include depression and trembling, accompanied by loss of appetite, abdominal pain and diarrhoea, and finally coma. The only treatment is to remove stock from the infested area as quickly as possible.

Oleander (*Nerium oleander*)

Oleanders are ornamental garden plants with leathery leaves and bunches of colourful flowers (Figure 15). The whole plant is toxic, but especially the leaves. Stock poisoning is rare because the leaves are unpalatable, although leaves that have wilted after branches have been trimmed will be more palatable. Signs of poisoning, principally diarrhoea, convulsions, loss of appetite and general pain, are noticed within hours of the plant being eaten and death may occur within 12–30 hours. Treatment may not be successful. Prevention consists of ensuring that horses never have access to the plant.

Figure 15. Oleander

Cyanide poisoning: birdsfoot trefoils, linseed, sorghum

Cyanide combines with haemoglobin in the blood stream, destroying its ability to carry oxygen. Symptoms of poisoning usually appear less than one hour after grazing the toxic plants. Breathlessness and an increased rate of respiration also occur. In a severe case, the horse will collapse, froth at the mouth, gasp deeply, and die. In a less severe case, the horse will become drowsy, stagger and twitch, and its breath may smell of bitter almonds. Veterinary treatment in the early stages of cyanide poisoning is usually effective. Three plants are potentially toxic:

- native birdsfoot trefoil and red-flowered birdsfoot trefoil
- linseed
- sorghum.

Figure 16. Native birdsfoot trefoil

Native birdsfoot trefoil and red-flowered birdsfoot trefoil (*Lotus australis* and *Lotus cruentus*): these plants have white, pale-pink or red pea-shaped flowers (Figure 16). They are most toxic when making vigorous, leafy growth, but cases of poisoning are rare.

Linseed (*Linum usitatissimum*): linseed is a field crop, growing up to 1 metre tall, with white or blue flowers (Figure 17). Young or wilted plants and the seeds are the most toxic. Large amounts of the plant or seeds must be eaten before symptoms develop. The seeds are rendered harmless by boiling.

Sorghum spp.: Johnson grass (*Sorghum halepense*) is a tall, summer-growing perennial grass with underground runners (Figure 18). It is most toxic when young and green, but hay containing a lot of this grass can also be toxic. All fodder sorghum species (e.g. Sudan grass) are potentially toxic when young or actively growing (Figure 19). Sorghum toxicity may also cause bladder paralysis with loss of urinary control and scalding.

Figure 17. Linseed

Foxglove (*Digitalis purpurea*)

Foxglove is a biennial plant with characteristic helmet-shaped flowers in a variety of colours including white, yellow, purple or pink. The whole plant is toxic, and does not lose its toxicity on drying or boiling. As such, hay containing foxglove can also be toxic. Symptoms of poisoning include diarrhoea, colic, abdominal pains, an irregular pulse, and sometimes tremors and convulsions. Symptoms may intensify over several days and can lead to death.

Rhododendron species and varieties.

Rhododendrons and Azaleas are not usually eaten so poisoning is rare, but all parts of the plant, including the nectar, are toxic. Symptoms of poisoning are noticed within hours of the plant being eaten, and can include colic, diarrhoea, trembling, a weak pulse, and slow irregular breathing, possibly leading to collaspe or death if significant quantities are consumed.

Figure 18. Johnson grass

Figure 19. Wild sorghum

Figure 20. Ryegrass

Figure 21. Bracken fern

Figure 22. Ragwort

Figure 23. Darling pea

Group B: Plants causing nervous diseases

Ryegrass (*Lolium perenne*)

Perennial ryegrass (Figure 20) forms the basis of most pastures in southern Australia and New Zealand, and although cases of ryegrass staggers are rare in horses, owners should be aware of the risks. The condition occurs most commonly when light autumn rains fall during relatively warm weather. The symptoms in horses are a wide stance and wobbly gait, which may progress to temporary paralysis of the hindquarters. Some horses become recumbent and have to be destroyed. Affected horses should be moved quietly from the paddock, and they should not be ridden or excited.

Bracken fern (*Pteridium aquilinum* var. *esculentum*)

Bracken fern poisoning in the horse causes a staggers syndrome because a thiaminase in the plant destroys the thiamine (vitamin B) in the blood and tissues. Horses must eat bracken for 30–60 days to be affected and typical symptoms of poisoning are lack of coordination, staggering, muscle tremors and recumbency. Horses respond quickly to daily injections of thiamine. Prevention consists of keeping horses away from bracken fern (Figure 21) and ensuring it is not accidentally fed in hay or chaff.

Figure 24. Birdsville indigo

Ragwort (*Senecio jacobaea*)

Horses are very susceptible to the alkaloid toxin in ragwort (Figure 22) and can succumb quickly to nervous abnormalities (see also Group D).

Darling pea (*Swansonia* spp.)

Darling pea grows in the low-rainfall, inland areas of Australia, south of the tropics. Horses are more resistant to poisoning than sheep or cattle, but clinical signs may occur after feeding on the plant for one month. Affected horses lose weight, are depressed, carry their head low, have glazed eyes, lack coordination and are very excitable. Horses can recover if they are removed from the area.

Figure 25. Rattlepod

Birdsville indigo (*Indigofera linnaei*)

Birdsville indigo (Figure 24) grows in semi-arid central Australia and poisoning occurs in dry seasons when grass is limited. Birdsville horse disease is characterised by dullness, weakness, lack of coordination, spasms, behaviour changes and death. Horses can be protected by only grazing them in areas where the plant is not prevalent or by providing a high-quality protein supplement such as lucerne hay.

Rattlepods (*Crotalaria* spp.)

Crotalaria is a tropical or subtropical plant that has pea-shaped yellow flowers with seeds in a narrow pod (Figure 25). Alkaloids in the plant cause degeneration of a variety of tissues. One variety growing in Queensland causes oesophageal ulceration, blockage and death from dehydration. Another variety causes Kimberley horse disease or walkabout disease in which liver disease develops over months, followed by major behavioural changes.

Figure 26. Castor oil plant

Figure 27. St Barnaby's thistle

Figure 28. Marshmallow

Figure 29. Yellow burr weed

Figure 30. Garden poppy

Figure 31. White iron weed

Castor oil plant (*Ricinus communis*)

The castor oil plant is a hardy perennial shrub that grows 2–3 metres high. The branches are hollow, with light-green, glossy leaves often tinged with red (Figure 26), or purple in New Zealand. The only poisonous part of the plant is the seeds, which are extremely toxic. Early symptoms are trembling, sweating and nervous incoordination, followed by a loss of appetite, colic, scouring, and weakness. Death follows if enough seeds are eaten.

Chewing disease

Chewing disease is a nervous condition caused by eating large amounts of St Barnaby's thistle (Figure 27) or Russian knapweed, both of which are found in New South Wales. Symptoms include twitching of the lips, flicking of the tongue, purposeless chewing, with difficulty in eating and swallowing. Affected horses will starve to death unless helped to eat.

St Barnaby's thistle (*Centaurea solstitialis*): an annual herb that grows up to 300 mm. It has yellow flowers and long yellow spines around the seed head. It would only be eaten when summer pasture is dry and scarce and the thistle is green.

Russian knapweed (*Centaurea repens*): a slender, branched, slightly-woody perennial herb with mauve to pale-pink flowers that grows up to 2 m, with dark, almost-black, creeping roots. The seed heads are small and hard. Leaves are silvery-grey when young, turning to dull grey-green when mature.

Marshmallow (*Malva parviflora*)

There is evidence that marshmallow (Figure 28) can cause 'staggers' in horses if they are driven or exercised after consuming the plant over a period of days or weeks. Affected animals move sluggishly, with a stiff action of the hind legs, the back arched and head outstretched. Deterioration leads to rapid shallow breathing and, in severe cases, partial paralysis and death.

Alkaloid poisoning

Many plants contain compounds called alkaloids that affect the nervous system. Symptoms of alkaloid poisoning are incoordination, staggering and trembling. Many alkaloids are also narcotics, so symptoms can include impaired vision, listless behaviour and glazed eyes. The best cure is to remove stock from the toxic pastures.

Monkshood, wolfsbane or aconite (*Aconitum napellus*)

A perennial garden shrub, up to 1 m tall, with bright-blue, helmet-shaped flowers held erect at the top of the stem. All parts of the plant, including seeds, are toxic and will cause rapid death. Destroy any garden clippings containing this plant.

Yellow burr weed (*Amsinckia* spp.)

An annual herb, up to 1.2 m tall, with dark-green leaves covered in short, stiff hairs (Figure 29). The flowers are yellow and bunched together along one side of the upper stem, often resembling a young fern frond. Horses poisoned by this weed often walk around aimlessly. Most cases of poisoning develop only after prolonged grazing.

Poppies

The Mexican poppy (*Argemone ochrolenca*), Californian poppy (*Eschscholtzia californica*) and garden poppy (*Papaver* spp.) are all potentially poisonous if eaten in large amounts. The Mexican poppy is an annual herb, up to 600 mm tall, and is thistle-like, with bluish leaves and stems, pale yellow flowers and yellow sap. The Californian poppy is also an annual herb, resembling the Mexican poppy, except that the leaves are not thistle-like and the flowers are orange-yellow to red. The garden poppy has soft hairs up the stem, and usually has only one yellow, orange or red flower per plant (Figure 30).

White iron weed or corn gromwell (*Lithospermum arvense*)

A small annual herb with narrow grey-green, hairy leaves (Figure 31). The small white flowers are closely attached to a curved stem at the top of the plant. Long-term feeding is necessary before poisoning occurs, which would rarely occur in the grazing situation, but may be likely if grain contaminated with corn gromwell seed is fed for a prolonged period.

Figure 32. Paspalum

Ergot of paspalum (*Paspalum distichum* or other *Paspalum* spp.)

Paspalum is a tufted grass that forms a dense mat under grazing conditions. The seed head consists of 2–5 very thin stalks (Figure 32). The toxic ergot fungus (*Claviceps paspali*) invades the seed head, replacing the seed with a mass of sticky, black material, which eventually becomes hard and seed-like. Horses can take a liking to infected paspalum heads. Most outbreaks occur in late summer–early autumn, when paspalum flowers. Close grazing before this will reduce the likelihood of the grass flowering.

Lupins (*Lupinus* spp.)

Lupins are annual herbs that grow to 1 m high, with leaves made up of 5–10 characteristic finger-like leaflets (Figure 33). The pea-like flowers are blue, yellow or white and the seeds are borne in a pod closely resembling a pea pod. The seeds are pale-coloured, varying from cream to white and are slightly flattened. Lupin poisoning from the alkaloids contained in the seed can occur if large amounts of seed are eaten in the absence of other feeds. This may be a problem if stock graze lupin stubble after harvest where they can eat the seeds off the ground. Feeding of lupins as part of a concentrate diet will not cause disease.

Figure 33. Western Australian blue lupin (*Lupinis cosentini*)

Lupinosis is a separate toxicity associated with a fungus that grows on lupin stubble following rainfall and humid conditions in summer. The best prevention is to remove stock grazing on lupin stubble for one month following rains.

Group C: Plants causing photosensitisation

White horses or those with white patches are particularly vulnerable to photosensitisation. It occurs when animals eating large quantities of plants that contain photodynamic agents (substances activated by light) in the green, rapidly growing plant.

Figure 34. St John's wort

St John's Wort, storksbill, alsike clover, buckwheat, burr medic

Photosensitisation is associated with eating St John's wort (*Hypericum perforatum*), storksbill (*Erodium* spp.), alsike clover (*Trifolium hybridium*), buckwheat (*Fagopyrum esculentum*), and burr medic (*Medicago polymorpha*) (Figures 34–37). Photosensitisation is also seen in horses with liver disorders. Skin lesions occur on white areas of the body, particularly on exposed parts such as ears, eyelids and muzzle. Treatment is usually by applying a sunscreen ointment to the affected areas and protecting the horse from sunlight. The provision of shade is important.

Group D: Plants causing liver damage

Fireweed (*Senecio linearifolius*)

Although rarely reported there is strong evidence to link fireweed (Figure 38) with waratah horse disease. Clinical signs are mainly nervous abnormalities such as compulsive walking, yawning and irritability. There is no specific treatment, but careful feeding may allow some animals to regenerate enough liver tissue to survive.

Figure 35. Storksbill

Ragwort

Diagnosed cases of poisoning by ragwort are very rare. Signs of poisoning do not occur until the plant has been eaten for several weeks or even months. The alkaloid toxin in the plant can result in serious liver damage. Horses show mainly nervous abnormalities such as compulsive walking, yawning and irritability. Weight loss, poor appetite and photosensitisation may also occur. There

Figure 36. Alsike clover

Figure 37. Burr medic

Figure 38. Fireweed

Figure 39. Patterson's curse

Figure 40. Oak leaves

Figure 41. Soursob

is no specific treatment, but animals should be moved off infested areas as soon as possible. Every effort should be made to remove ragwort from the property.

Patterson's curse/Salvation Jane (*Echium plantagineum*)
This plant is becoming an increasingly common weed in southern Australia (Figure 39) and New Zealand, and may be found in paddocks that horses graze or in hay. Signs of poisoning do not occur until after the horse has eaten the plant for several months and continued grazing leads to progressive irreversible liver damage. The signs are similar to ragwort poisoning and there is no specific treatment. An acute liver failure is also seen in young horses after eating large quantities of the plant.

Alsike clover (*Trifolium hybridum*)
Alsike clover, which is a legume, can cause disorders in horses eating either pasture or hay that includes the plant for one year or more (Figure 36). The liver damage leads to loss of condition, behavioural changes and dullness, interspersed with periods of excitement.

Group E: Plants causing kidney damage
Oak (*Quercus* spp.)
Oak poisoning is only likely to occur if large quantities of acorns are consumed. The usual symptoms are dullness, recumbency and eventually death. Eating large quantities of oak leaves (Figure 40) may cause colic and bloody urine.

Soursob (*Oxalis pescaprae*)
Soursob belongs to a group of plants containing oxalates, toxins that can cause low blood levels of calcium and kidney damage. Although soursob (Figure 41) is a common weed it is unlikely it would be eaten in large quantities.

Group F: Plants causing skin disorders
Selenosis
Selenosis is a chronic debilitating disease caused by eating plants that accumulate selenium. It is common on Cape York and near Richmond in Queensland. Affected horses may show loss of hair, particularly the mane and tail, weight loss, gait changes, loss of hooves and hoof cracks. In acute cases, a blind staggers syndrome occur. These horses should be given supplementary feed or moved to other areas. Paddocks with pasture plants containing high levels of selenium should not be grazed by horses.

Leucaena dermatosis (*Leucaena leucocephala*)
This subtropical large shrub or small tree contains a toxic amino acid, mimosine. It must be consumed as a large part of the diet for long periods. Hair loss occurs from the mane and tail and around the coronary band. Hoof changes may include hoof wall rings or laminitis. There is no specific treatment other than reducing access to the plant. The plant can be grazed without harm if it comprises less than 10% of the feed intake.

Group G: Plants causing heart and lung disorders
Avocado (*Persea Americana*)
The avocado tree (Figure 42) contains substances that are toxic for the heart and liver. Grazing the leaves, bark and fruit will cause rapid development of dullness, colic, swelling of the head and neck and difficulty in breathing, symptoms that are related to heart failure. Horses will recover if avocado is removed from their diet and they are treated by a veterinarian.

Rubber vine (*Cryphoshegia grandiflora*)
Rubber vine (Figure 43) is widely distributed through the tropics and the leaves are very toxic for horses. Affected horses may be found dead or be dull with progressive weakness and difficulty in breathing because of heart failure. Horses should not be allowed to graze near the plant because even dried leaves can be very toxic.

Crofton weed (*Ageratina adenophora*)

Crofton weed, a 1–2 m high bush, is common in north-east New South Wales and Queensland, growing on shady slopes and gullies around bushland or rain forest. Poisoning occurs when it dominates the pasture or is selectively grazed, and poisoning can occur after grazing for only 2 months. The plant, which has green trowel-shaped leaves and clusters of white flowers, is more toxic at flowering (Figure 44). Coughing, breathing difficulties, exercise intolerance and in some cases sudden death are related to lung damage. Some damage will be irreversible even if the horse is removed from access to the plant, but veterinary treatment may help some horses. Weed control programs with slashing, spraying and replanting are vital on infested properties.

Figure 42. Avocado

TREATING HORSES

Medications

Many therapeutic products must be administered orally to horses:

- via a syringe as a paste or solution
- mixed in feed
- via a stomach tube

Oral administration can be a cheap and successful method of administering drugs to horses, but errors in dose estimation and poor administration techniques limit its usefulness.

Figure 43. Rubber vine

Pastes and solutions

Worming compounds, vitamins, minerals, tranquillisers and anti-inflammatory agents can all be administered in paste form. The pastes contain a known quantity of compound and the syringes have a graduated scale to help you calculate the correct dose.

Accurate estimation of your horse's weight is essential (see Chapter 5). Inaccurate weight estimation and poor administration will limit the usefulness of the paste.

Set the plunger of the syringe to deliver the correct dose. The syringe may need to be warmed to make the contents easier to expel. Stand on the off-side of the horse, slightly in front, and introduce the syringe into the corner of the mouth. Advance the syringe into the horse's mouth till your fingers touch the lips. Depress the plunger and then remove the syringe, wiping any excess paste off on the horse's tongue and inside of the lips.

Figure 44. Crofton weed

If the paste has been deposited far enough back in the mouth the horse will swallow, but some horses need their head held up to ensure the paste does not drop out. Some horses are adept at letting the paste fall out of the mouth and another dose will need to be given. Some horses are hard to 'paste' and a twitch and quick action are necessary to treat them.

If a powder is dissolved in water or mixed with yoghurt or golden syrup it can be administered into the horse's mouth using a syringe in a similar way to administering a paste.

Figure 45. Pasting
(courtesy of Stuart Myers).

Drugs mixed in feed

Mixing powdered, granulated or liquid drug in the feed is an easier and often cheaper method of treatment than using a paste, but it has a number of disadvantages.

- There is no way of knowing if the horse has eaten all the drug.
- Many horses will not eat feed with an unusual taste or smell.
- Many drugs are less effective when given on a full stomach.
- Many horses eat everything but the drug.

For this method to be successful, the taste must be disguised and the granules, powder or liquid completely incorporated into the feed so it has to be eaten. Molasses is the only agent that can do this reliably. Phenylbutazone powders can be frozen to disguise the taste or smell.

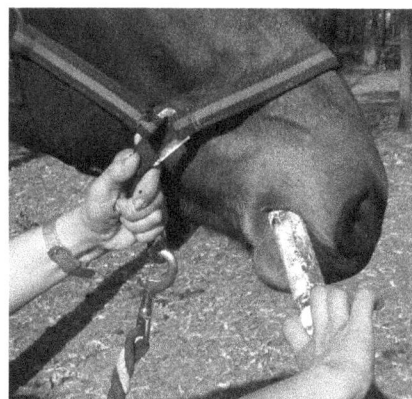

Using a stomach tube

Stomach tubing a horse is the most effective method of giving an oral drug or compound because you know that the drug gets into the horse's stomach. Stomach tubing should only be carried out by your veterinary surgeon. Inexperienced operators can kill the horse by inadvertently inserting the tube into the windpipe instead of the oesophagus, thereby putting the treatment into the lungs, or they can be injured themselves by the horse if it plays up during stomach tubing.

Nebulisation

Nebulization is a relatively new method of getting drugs into a horse's lungs and is mainly used for performance horses. It is the equine equivalent of a 'ventolin puffer'. The required drug is placed in the nebuliser, which is then strapped onto the horse's nose. As the horse breathes in through the nebuliser it carries the drug into the lungs, allowing local action. Nebulization can be carried out by an owner after a vet has prescribed the drugs.

Topical medication

Some drugs can be administered through the skin and because the preparation usually contains a substance that enhances penetration of the drug you must be careful not to allow the medication to contact your own skin. Wear gloves and use an old toothbrush as an applicator.

Injections and vaccination

Types of injections and their preferred sites

There are three types of injections:

- intravenous (into a vein)
- intramuscular (into a muscle)
- subcutaneous (under the skin).

There are different 'preferred' sites for the different types of injections. Intravenous injections should be into the jugular vein in the neck, intramuscular injections can be into the muscle of the neck, rump or brisket, and subcutaneous injections should be under the skin of the neck.

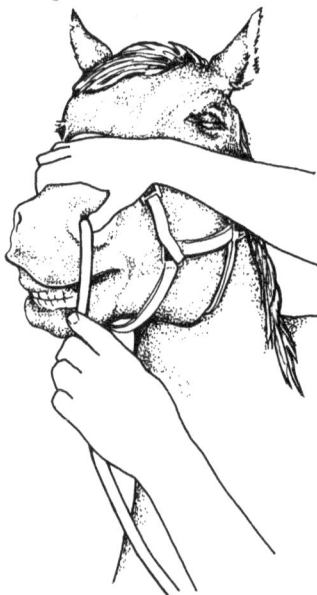

Figure 46. Introducing a stomach tube through the nostril.

Equipment

Use a new, and therefore sharp and sterile, hypodermic needle every time. Use needles that are 40 mm (1.5 inches) long and use the smallest size of needle that will allow the solution to pass through easily; generally 18–21 gauge needles are used for horses.

New syringes are desirable, but not essential. Old syringes can be reused by cleaning and boiling for 10 minutes. The site for the injection should be cleaned with disinfectant, alcohol or cotton wool soaked in methylated spirits (metho). Wait one minute after disinfection before giving the injection.

Technique

Always follow the dosage instructions when loading the syringe. Hold the medication bottle upside down and pass the needle through the rubber cap from below. Inject a quantity of air equivalent to the dose you will be using into the bottle so the solution flows out easily. Pull on the plunger to draw the dose out. Withdraw the needle from the bottle. Expel any air in the top of the syringe before injecting.

Restraint of the horse is essential for successful injection technique. The horse should be held by the halter, but not tied up. It is preferable to have an assistant hold the horse in a stable or corner of a yard while you inject it. Some horses need to be in a crush to be injected. Fractious horses may have to be restrained by twitching or a tinker's grip (see Chapter 3). It may help to cover the horse's eye on the side it is to be injected.

Procedure

Intravenous

Intravenous injections should only be given by veterinarians or a person experienced in the technique. It is easy to inject outside the vein, which may have dire consequences.

Intramuscular

Select your site for injection. If using the neck muscles, it is advantageous to have them relaxed and this can be achieved by bending the neck towards you. Hold the needle hub between your thumb and forefinger and slap the injection site rapidly a few times with the back of your hand. As you move your hand in for the next slap, turn your hand and insert the needle into the muscle up to the hilt.

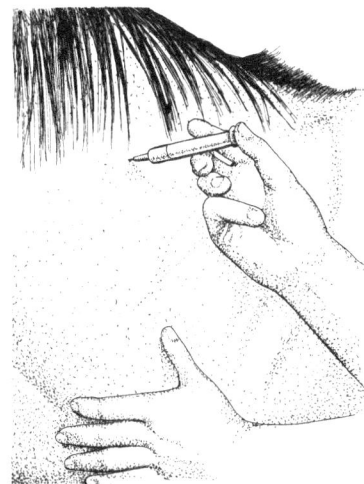

Figure 47. Giving an intramuscular injection.

Observe the needle to see if any blood flows out through it. If that occurs you must remove the needle and re-insert it so that it misses the blood vessel. Attach the syringe and draw back on the plunger slightly at the start of the injection and during its course to check that blood does not come into the needle hub. Some intramuscular preparations, such as penicillin, can have severe, even fatal consequences if they are injected into a vein. When you are satisfied you are not in a vein slowly inject the contents of the syringe. Do not inject more than 20 millilitres at one site.

Subcutaneous

Grasp a fold of skin, pinch it for a few seconds, and push the needle through into the space between the skin and muscle. Inject the contents of the syringe slowly.

Injection reaction

An injection reaction can occur because of poor technique or even with a perfect technique. It will present as a swelling at the injection site that may develop into an abscess or resolve uneventfully. In severe cases the horse will be sick and depressed, the site will be painful and the horse will require veterinary treatment.

Injection reactions tend to be less severe in the brisket or neck, so these are the preferred sites for intramuscular injections. If you have to give several injections over a number of days you should choose a new injection site on the opposite side of the horse each time to prevent muscle soreness. Your injection technique is very important with young horses because it will determine their future response to all injections.

Vaccination

Vaccination is a useful preventative measure against some infectious diseases. Two major horse vaccines are available: tetanus and strangles. A vaccine against equine herpes virus 1 and 4 and salmonellosis is also used in some studs and stables.

The dates of vaccine administration and the vaccine type should be noted on the vaccination certificate or the horse's identity document and signed by the person administering the vaccine.

Tetanus

Tetanus is a fatal disease that occurs as a result of contamination of some part of the body with the tetanus bacteria, which is widespread in the environment and produces a neurotoxin. Certain types of wounds, such as deep punctures, are more likely to lead to tetanus, but any horse that undergoes a surgical procedure or has a wound should receive a tetanus vaccination or antitoxin injection.

Vaccination requires two injections administered one month apart followed by a booster one year later. Another booster is not required for a further four years, but boosters are often given earlier if the horse is wounded. Initial immunity occurs three weeks after the first injection.

An ongoing vaccination program will provide immunity against any future tetanus risk, because the most dangerous wound for tetanus is the wound you are not aware of.

Short-term protection can be provided by a tetanus antitoxin injection, which will provide approximately two weeks' protection by neutralising any toxin produced during that time. When a non-vaccinated horse is at risk of developing tetanus, the best procedure is to commence the vaccination program and inject antitoxin. The antitoxin is usually given subcutaneously, and the vaccine intramuscularly. Tetanus and strangles vaccines are available as a combined product.

Strangles
Vaccination for strangles requires three injections administered two weeks apart, followed by yearly boosters. It is not always 100% effective in preventing the disease, but should reduce the severity and spread of an outbreak. If your horse has contact with many horses from varying backgrounds, vaccination is recommended (see the earlier section 'Respiratory infections').

An injection reaction to the strangles vaccine can occur. Discontinue the course and consult a veterinarian. Horses in high-risk situations can be given two boosters each year and mares can be given a booster just before foaling. Both tetanus and strangles vaccines can be given to foals from 8 to 12 weeks of age.

Bandaging and poulticing
Bandaging
Bandages are applied to protect wounds and prevent swelling, to control bleeding, and to support the structures in the leg. Horse owners must know how to apply a bandage – but how many times have you tried to bandage some part of your horse, only to see your bandage loosen and fall to the ground?

A bandage that it too tight can cause pressure sores, damaged tissue and white hair, so it is important that you make a bandage only as tight as necessary to achieve your purpose. To control bleeding, for instance, a bandage should be put on quite tightly, but after two hours the bleeding should have stopped and the bandage is replaced with one that is less firmly applied. When bandaging for protection there is no need to bandage tightly; for example, a leg bandage should be just a little tighter than if the bandage were gently draped around the leg.

Making sure that a bandage exerts a constant pressure is just as important as making sure that it is not too tight, especially in the case of swelling legs. If the pressure of the bandage is not constant the legs may swell unevenly, with the result that shortly after the bandage is removed the leg will bulge out into the classic 'bandage bow'.

When bandaging leg wounds, always use cotton wool or other padding (i.e. gamgee) under the bandage and cover the wound with a dressing of gauze swabs, or a clean, folded 'Chux' serviette, to prevent the cotton wool from sticking directly to the wound. Always bandage from one joint down to the next and do not skimp on dressings.

If the underlying wound is large or likely to ooze a lot of serum, the bandage should be changed daily. If the wound is small and dry, the bandage can be left on for no longer than four days. Regardless of the wound, check the bandage daily. If the leg swells, this will effectively make the bandage tighter and require more frequent changing.

As is the case with many things related to horses, bandaging is a case of 'practise makes perfect'.

Choice of bandage
Reusable bandages with Velcro tabs are preferred for bandaging tails and for travel bandages, but they do not exert enough pressure to support the leg structures or reduce swelling. For this you need an elastic bandage and there are several types available. Do not wrap these bandages too firmly around the leg or you may create pressure sores or damage the underlying tissue. Many of

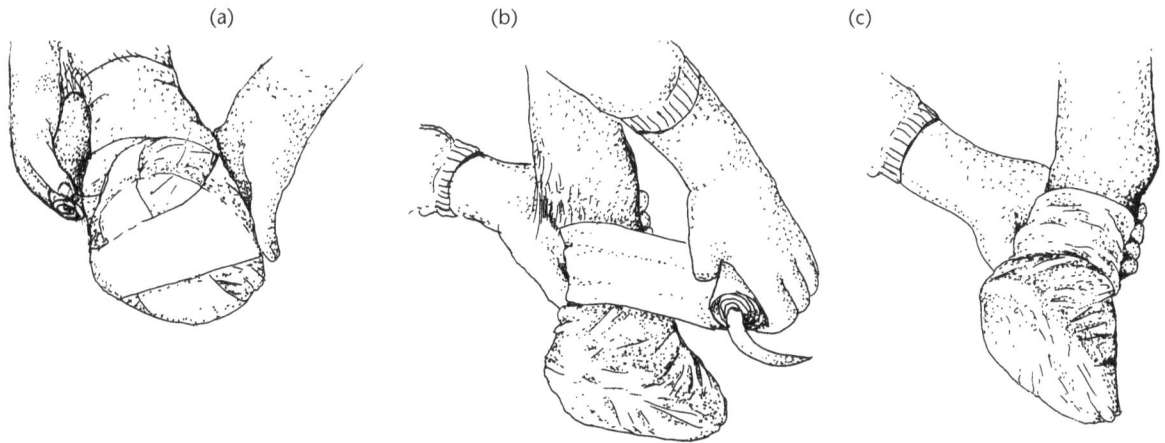

Figure 48. Bandaging an injured foot. (a) Apply a poultice to the foot and cover with a heavy-duty plastic bag. (b) Anchor the bag with a bandage around the pastern. (c) Wind the bandage in a figure-of-8 pattern to cover the foot and anchor it at the pastern.

the new types of elastic bandages are waterproof and self-sealing, but they are expensive. Most so-called disposable bandages can be reused if you are careful in the way you remove them and wash them. Elastoplast® can be used over padding because it will seal well and not slip off, without the risk of too much tension. Otherwise Elastoplast® can be used at the top and bottom of the bandage to overlap the bandage and skin to prevent the bandage from slipping. Crepe bandages are cheap, but are not very useful.

How to bandage an injured foot

Foot bandages are often applied over dressings and are very difficult to keep in place for any length of time. The secret lies in using plenty of padding and applying a firm bandage.

Put the poultice or the dressing on the foot and cover it with a good wad of padding that must be wide enough to be bandaged over completely so that it cannot be pulled out from underneath the foot.

Place a clean, heavy-duty plastic bag over the foot and anchor it by bandaging it firmly around the pastern. Wrap the bandage in a figure-of-eight pattern round and over the foot several times before finishing the bandage up the leg once more. Secure it with Elastoplast® under the heel and toe to prevent wear on the bandage.

How to bandage a leg

The secret to successfully bandaging a leg lies in correct and even tension. Always start with your bandages firmly and evenly rolled up. Gamgee or a similar cotton or foam-backed padding is also essential under the bandage. The bandage should be long enough to be wrapped from just below the knee or hock down the leg below the fetlock and back again in even turns.

There are many different reasons for applying leg bandages, but they are all applied in basically the same way.

A support bandage comprises several layers of cotton wool, then elastic bandage or dressing applied firmly, then the same again for at least three layers. Bandages used for support need to be firmer than those put on for warmth. An exercise bandage should extend down below the fetlock joint without hindering movement. A stable bandage should cover the pastern and finish just above the coronet. Do a

Figure 49. Bandaging a leg. (a) First place a wrap or padding around the leg. (b) Wind the bandage down the leg, overlapping the turns, and then up again the same way. (c) The finished bandage. For travelling bandages, extend the bandage right down to the foot.

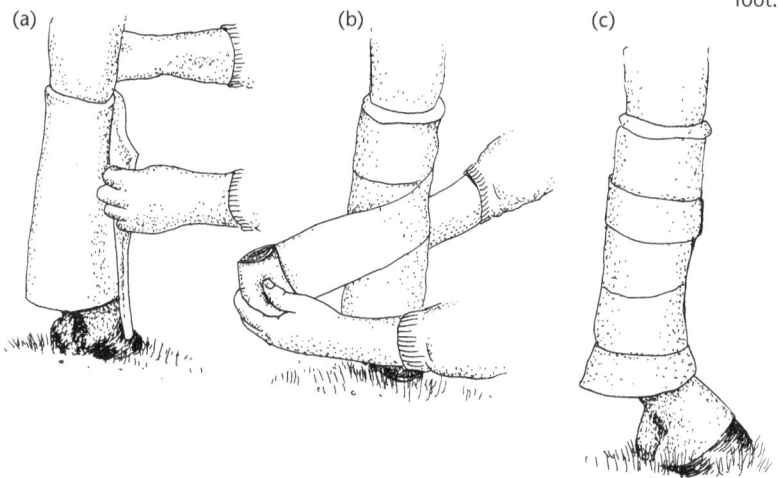

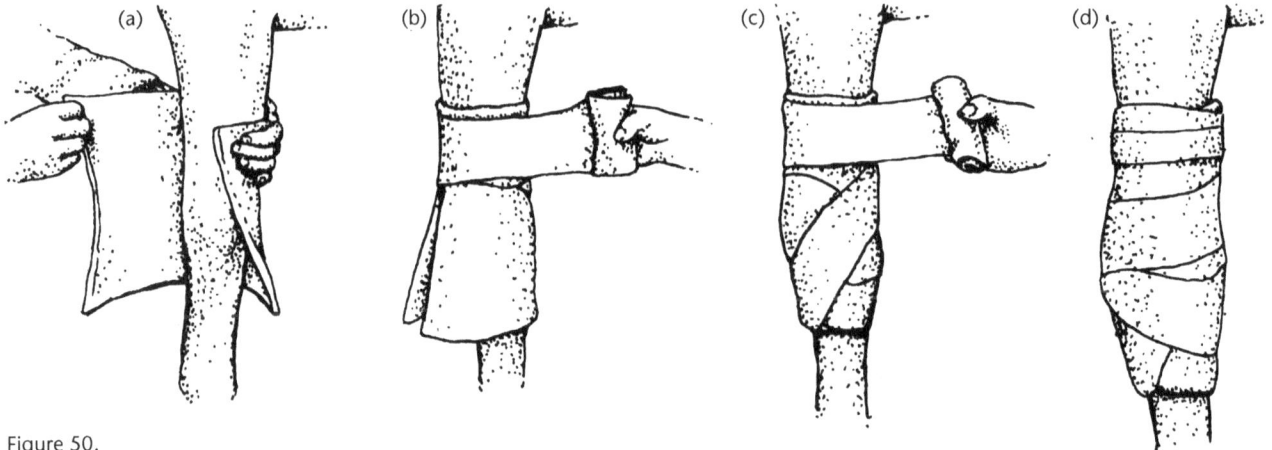

Figure 50. Bandaging a knee. (a) Wrap the knee with padding. (b) Wind bandage above the knee. (c) Wind in a figure-of-8 with a turn at the top and bottom each time. (d) When complete, tie the bandage at the top.

figure-of-8 pattern around the fetlock to support it. When tying the tapes of a bandage make sure they are the same tension as the bandage. In races or competition the bandages should be secured by stitching the tapes.

How to bandage a knee

Bandaging a knee joint is more difficult than bandaging other parts of the leg because of its flexibility. The prominent bones of the knee are particularly vulnerable to pressure. The most sensitive part of the knee, apart from any wounds, is the bone at the back of the knee. If the bandage restricts the circulation here, even a minor injury can become a severe pressure sore. If the knee is to be left bandaged for longer than 24 hours, always slit the bandage behind the knee to prevent a pressure sore developing. Be careful not to cut the underlying skin.

Plenty of padding around the knee is extremely important. The figure-of-8 method of bandaging leaves the back of the knee free from pressure and allows the horse to move reasonably freely.

How to bandage a hock

Hock and high cannon wounds are hard to bandage because of the movement and flexion when the horse walks. You need to apply a well-padded, firm bandage below the hock, then loop the bandage up above the hock in a figure-of-8 for a couple of turns. These loops above the hock must be reasonably loose to allow the gaskin to expand as the horse walks. Secure the bandage with several turns of Elastoplast® above the hock and a turn below, but do not cover the point of the hock with Elastoplast® or the underlying bandage as this will restrict bending of the knee too much. It takes some experience to be able to apply good hock bandages.

Poultices

Poultices are applied to sore areas of a horse's body, mostly the legs, but with a bit of ingenuity they can be applied to other parts. Poultices are either hot or cold.

Cold poultices

Cold poultices are generally used as an emergency treatment for tendon or ligament sprains of the lower leg. Commercial cold packs are available and should be kept in the refrigerator or freezer. Alternatively, ice can be crushed, placed inside plastic wrapping then wrapped around the leg with a stable bandage. Do not apply ice directly to the skin. If you have nothing else available, a packet of frozen peas will substitute.

Hot poultices

Hot poultices are used to protect injured areas, such as the sole of a hoof, and to clean deep contaminated wounds and infected wounds. They can be also used to draw out infections or fluid from a wound. Hot poultices apply warmth to the tissue, stimulating the blood supply and thus increasing the number of white blood cells in the tissue to fight the infection by killing and digesting bacteria.

An old-fashioned poultice is made by mixing boiling water, Epsom salts and bran into a stiff paste. This poultice merely provides warmth and once it cools has no effect. The most used poultice is kaolin, which can be bought under various brand names. It is heated up by inserting the tin in boiling water for approximately 30 minutes (remember to loosen the lid first). The kaolin should not be too hot when you apply it. Test it on your own skin. When it is ready, spread it with a knife or spatula onto brown paper and cover with a strip of gauze so that you can remove the poultice easily.

The most popular commercially available poultice is 'Animalintex', which is a cotton wool and gauze dressing impregnated with bassorin and boric acid. After soaking the poultice in warm water, drain it and apply it to the area to be treated, cover with extra polythene to keep the moisture in and then wrap the poultice in a bandage.

Applying a poultice

When applying a poultice you must ensure that the ingredients are in contact with the wound. Thick layers of poultice material are no more effective than thin ones. All poultices need to be replaced frequently, but poulticing should not be continued for too long. If the edges of the wound become white and soft, it is time to stop.

Although poulticing is helpful, horse owners should remember that it is not an alternative to antibiotic therapy and a vet should be called if things do not improve quickly.

QUARANTINE

Quarantine is where horses (or other animals) are kept separate from other animals for a certain period of time in order is to prevent the spread of infectious disease.

Quarantine is carried out when horses are moved to and from overseas, between States (in some cases) and should be carried out when moving horses to a new property.

Intercountry movement of horses

The quarantine process for horses entering Australia is controlled by the Australian Quarantine and Inspection Service (AQIS), which is a government organisation. In New Zealand, this process is controlled by the Ministry of Agriculture and Forestry (MAF). The aim of the quarantine control is to keep certain diseases that are prevalent in other countries out of Australia and New Zealand.

At the present time, Australia and New Zealand do not have the following equine diseases :

- equine influenza
- African horse sickness
- contagious equine metritis
- dourine
- equine viral encephalomyelitis
- Japanese encephalomyelitis
- surra
- glanders
- equine piroplasmosis.

Horses that are to be brought into Australia or New Zealand have to undergo quarantine at an AQIS or MAF-approved establishment in the country that they are leaving, usually for a period of 14–21 days (depending upon which country they are coming from). Upon arrival in Australia or New Zealand they undergo another period of quarantine, for 14 days or more (again this depends upon where they have come from). During the two quarantine periods the horse must undergo various veterinary tests for diseases. The horse is only allowed to leave quarantine when it is certified as being clear of disease.

Intra-country movement of horses

At present the only restrictions on moving horses within Australia are the requirements set down by the Department of Primary Industries (DPI) with regard to movement of stock from and between cattle tick infected areas for competition or breeding purposes (see Appendix 3).

New property quarantine

Another form of quarantine is when new horses arriving at a property are quarantined before being allowed to mix with the resident horses.

It is recommended that new horses are kept separate from the others for at least one week after they first arrive. This is a good time to worm the new horses. If horses have come from sale yards or other properties where a lot of horses come and go, they may be carrying infectious agents that will not become apparent for several days. Some examples of infectious diseases are :

- strangles
- ringworm
- numerous viruses that cause respiratory disorders.

During this time of quarantine, use separate gear for the horse and wash your hands before handling other horses.

WHERE TO GO FOR MORE INFORMATION.

Go to www.equinecentre.com.au for useful information about horse health.

The NSW Department of Agriculture runs an Australian-wide service using their Wormtest kit. Contacted the department on (02) 4640 6366 or go to www.agric.nsw.gov.au

The Commonwealth Serum Laboratories website (www.csl.com.au) has more information on vaccinations

Go to www.dpi.qld.gov.au for information on stock movements, local diseases and animal management.

The RIRDC website (www.rirdc.gov.au) summarises local research on nutrition and general health relevant to Australian conditions.

The New Zealand Ministry of Agriculture and Forestry website (www.maf.govt.nz) has information on diseases, stock movement and animal welfare.

CHAPTER 7

BREEDING THE HORSE

HORSE REPRODUCTION

The owners and managers of mares and stallions need to understand horse reproduction in order to achieve a successful outcome of a breeding program.

THE MARE'S REPRODUCTIVE SYSTEM

The reproductive organs

The reproductive system of the mare comprises the ovaries, which contain the thousands of follicles from which the ova (eggs) develop; the fallopian tubes (oviducts), where fertilisation of the ovum by a sperm takes place; the uterus, where the foal develops during pregnancy; the cervix, which stops foreign material or bacteria entering the uterus; the vagina and the vulva, the external opening of the reproductive tract.

The oestrus cycle

The basis of mare reproduction is the oestrous cycle, or 'heat', which is the time taken for the development of an egg within an ovarian follicle to the stage that it can be fertilised by a sperm, and for the preparation of the mare's reproductive tract so that fertilisation can occur.

Each stage of the cycle is controlled by substances called hormones, produced in the pituitary gland near the brain (gonadotropins), and in the ovaries and uterus. The level of production of the individual hormones can be affected by external factors, such as day length, temperature and nutrition, and by the presence or absence of other hormones.

What starts the oestrous cycle?

Most (80%) maiden or dry mares are 'seasonally polyoestrous'; that is, they have heat cycles during particular seasons of the year. The mares generally start cycling when daylight hours become longer, temperatures are higher and nutrition improves. The loss of a mare's winter coat usually coincides with hormone activity. The mares continue to cycle until the days start becoming shorter. The remaining 20% of mares cycle the whole year; that is, they are 'polyoestrus' and can conceive at any time of the year if their oestrous cycle is of normal length and they are ovulating. In tropical and subtropical areas, more mares will be polyoestrus than in temperature climates. Mares are in 'anoestrus' when there is no

Figure 1. The mare's reproductive organs (credit Jane Myers).

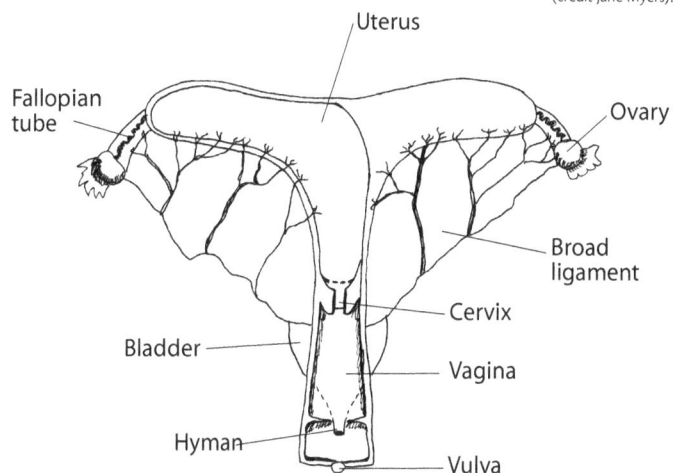

follicle development in the ovary. Anoestrus should not be confused with the 'dioestrus' phase of the cycle.

Mares usually start cycling within a few days of foaling, which means they can be 'on heat' 7–10 days after foaling.

Length of cycle
The length of the oestrous cycle is the time (in days) from the end of heat in one cycle to the end of heat in the next cycle. The day after the mare goes off heat is the first day of the cycle and the last day of the following heat is the last day of the cycle.

The average length of the mare's oestrous cycle is 21 days, which is divided into two phases. The longest phase is called dioestrus ('off heat') and lasts 14–16 days. The length of dioestrus does not vary according to season. However, the length of the remaining phase of the oestrous cycle, oestrus or the 'on heat' period, varies greatly, depending on the season. The average length is 5–7 days, but longer-than-average heats are normal in early spring when the ovaries are not fully active, referred to as a transitional or spring oestrus, and heats lasting up to 60 days have been recorded. The average length of each phase of the oestrous cycle will also vary, often greatly, between individual mares.

Ovulation
Ovulation is the expulsion of a mature egg from a follicle in the ovary into the fallopian tube where the egg survives for up to 12 hours and must be fertilised by a sperm within that time. Ideally, the mare is served 24 hours before ovulation so that the sperm will be already in the fallopian tube when the egg is released. The difficulty is knowing precisely when a mare ovulates: 50% of mares ovulate 24 hours before the end of heat, 30% ovulate 48 hours before the end of heat, some mares may come on heat and do not ovulate, and others ovulate without showing heat.

The percentage of mares ovulating at any time depends on the season. In Australia and New Zealand, the months in which the highest percentage of mares ovulate are December, January and February. Very few mares ovulate in July and August, so the artificially imposed breeding season, starting in late August, does not correlate with nature.

Follicle development
An ovarian follicle is a fluid-filled sac surrounded by a group of cells that protect and nourish the egg. There are thousands of follicles in the ovary, but only one (or sometimes two) begin to develop during each oestrous cycle.

By the 16th day of the cycle, the developing follicle has reached the size at which it secretes enough hormone to bring on the external characteristics of heat. Around day 20, the follicle reaches its full size and ruptures, releasing the egg into the fallopian tube (i.e. ovulation). The ruptured follicle forms the corpus luteum (CL), which secretes the hormones that prepare the uterus for pregnancy.

Follicle testing
On many studs, mares are examined by a veterinarian who will feel the size of the follicle by palpation from the rectum and then advise the stud manager on the best time to have the mare served.

The hormones of the oestrous cycle
On the first day of the cycle, the CL starts producing progesterone, which causes the level of oestrogen to drop, resulting in the mare going 'off heat'. Production of the gonadotropin, luteinising hormone (LH), by the pituitary gland also decreases.

On the 13th day, the uterus starts producing prostaglandin if the mare has not become pregnant. Prostaglandin causes the CL to regress and therefore the level of progesterone decreases.

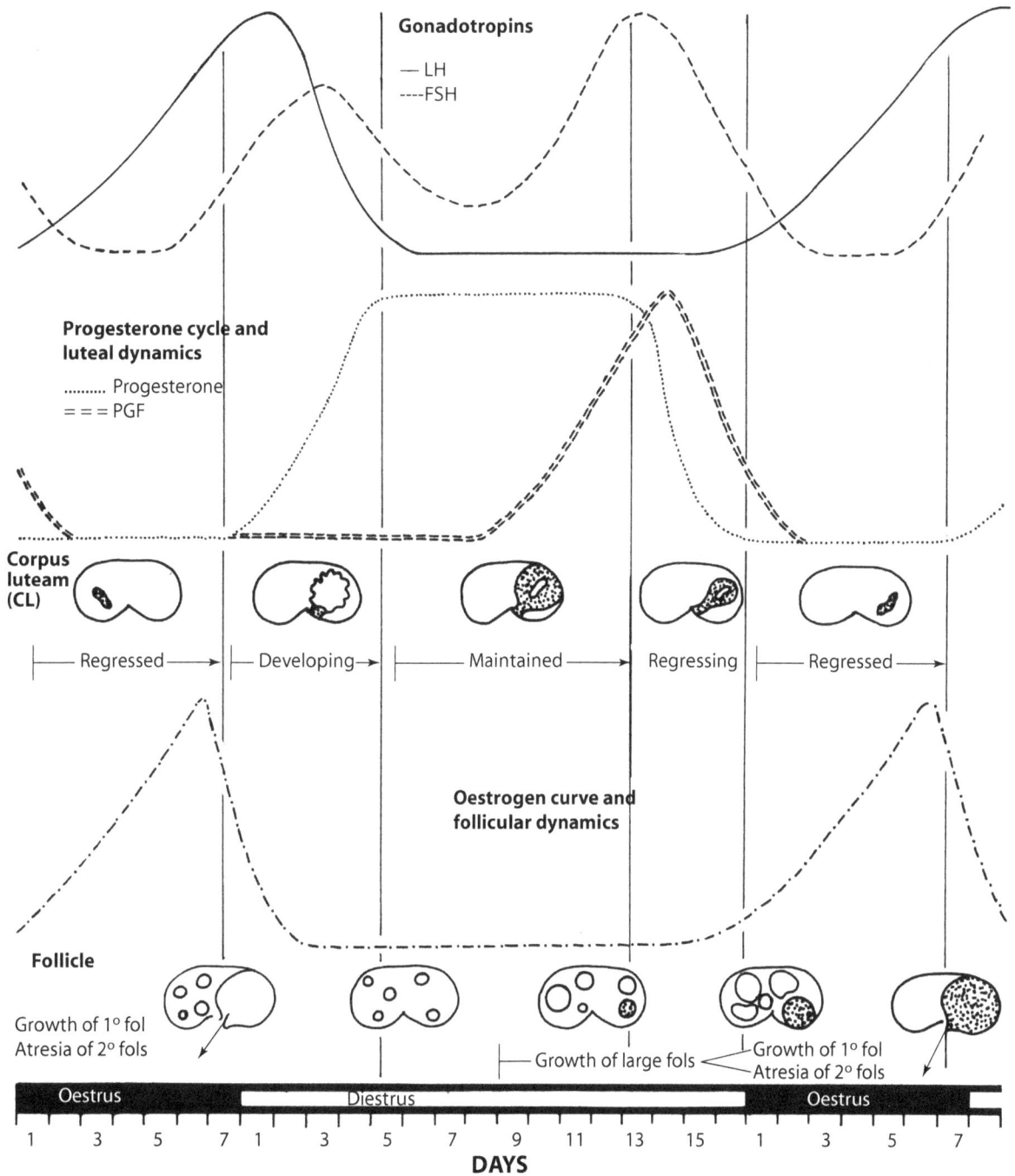

Gonadotropins

— LH
---- FSH

Progesterone cycle and luteal dynamics

.... Progesterone
= = = PGF

Corpus luteam (CL)

|— Regressed —→| |— Developing→| |— Maintained —| |— Regressing —| |— Regressed —→|

Oestrogen curve and follicular dynamics

Follicle

Growth of 1° fol
Atresia of 2° fols

|— Growth of large fols —|

Growth of 1° fol
Atresia of 2° fols

| Oestrus | Diestrus | Oestrus |

1 3 5 7 1 3 5 7 9 11 13 15 1 3 5 7

DAYS

Figure 2. The mare's hormonal and ovarian changes throughout the oestrous cycle. The primary follicle (1°) is the follicle destined to ovulate; the secondary follicles (2°) do not rupture (i.e. they become atretic).

On the 14th day, the low level of progesterone stimulates the pituitary gland to secrete another gonadotropin, follicle-stimulating hormone (FSH), and a follicle begins to develop, producing more oestrogen as it grows.

On the 16th day, the high level of oestrogen causes the signs of heat and stimulates the pituitary gland to secrete LH and decrease production of FSH. Ovulation occurs.

On the 21st day, the CL develops after ovulation and the cycle starts again.

If the mare becomes pregnant, prostaglandin is not be produced on the 13th day, so the CL does not regress and continues to produce progesterone. High levels of progesterone are necessary for the maintenance of pregnancy.

Ampulla gland
Vesicular gland
Prostate gland
Kidneys
Rectum
Bulbourethral gland
Ureter
Bladder
Urethra
Testes
Penis
Scrotum

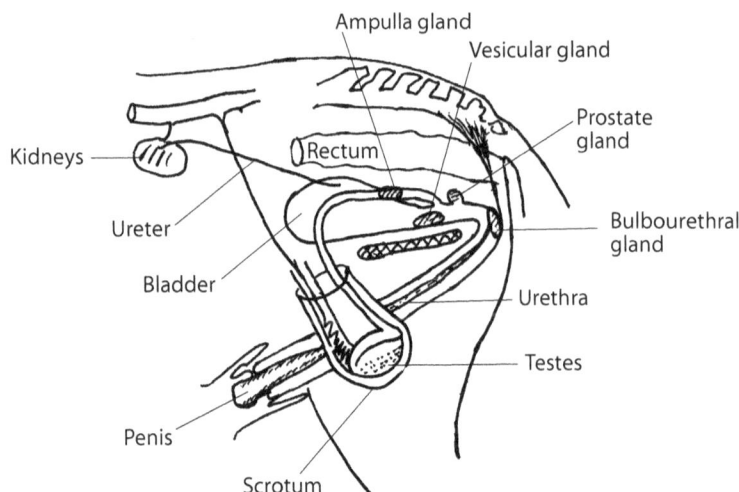

Figure 3. The stallion's reproductive organs (credit Jane Myers).

THE STALLION'S REPRODUCTIVE SYSTEM

The reproductive organs

Scrotum
The scrotum holds the testicles between the thighs and is capable of muscular and vascular adaptation to the environmental temperature to maintain the testicles at a constant temperature that is less than body heat. On hot days the scrotum is suspended lower, on cold days it is drawn up towards the body.

Testes
The testes lie horizontally inside the scrotum, which can predispose them to rotating within the scrotum, twisting the spermatic cord.

Epididymis
The epididymis, which acts as a storage and maturation centre for immature sperm, lies above each testis, loosely attached by its tail.

Spermatic cord
Each spermatic cord contains the vas deferens (the duct that carries the sperm), blood vessels and nerves.

Accessory glands
Seminal vesicles
The seminal vesicles are the largest of the accessory glands and the fluid they secrete comprises almost 60% of the ejaculate. Seminal fluid, which contains protein, acids and sugars, is the transport medium for the sperm. Using the stallion to tease the mare increases the quantity of seminal fluid.

Bulbo-urethral glands
The bulbo-urethral glands secrete between 10% and 20% of the ejaculate, which cleanses the urethra, clearing it for the passage of sperm.

Prostate glands
The prostate glands produce a fluid that comprises between 15% and 30% of the ejaculate and activates the sperm.

Penis
The function of the penis is to convey urine from the bladder to the exterior, and semen from the testes to the female reproductive tract. It is composed of tissue that is highly erectile when filled with blood. When relaxed, the penis is totally enclosed in the prepuce.

Factors affecting the stallion's reproductive system

Season
Stallion fertility is secondary to mare fertility according to some stud masters. Stallions are expected to be fertile and normal all year, every year. However, as with the mare, the fertility in the stallion varies according to the season.

- The sperm count (concentration of sperm per ejaculate) is lower in late autumn and winter because sperm production is almost half that of late spring and summer;

therefore, early in the stud season (August, September) a stallion is only half as fertile as in November and December. The stud manager should not overwork the stallion at this time.

- The total semen volume is lower in winter.
- A stallion is much slower to mate (less interested) in late autumn and winter, and is also more likely not to ejaculate even when he does serve a mare.

Hormones

Two hormones produced in the pituitary gland and one hormone produced in the testes influence sperm production.

Factors affecting male fertility

The fertility of a stallion depends on several factors.

- The total sperm production.
- The percentage of motile normal sperm (a large daily sperm production will be ineffective if the sperm are not motile).
- The size of the testicles (stallions with larger testicles are more fertile than stallions with smaller testicles because they produce a greater number of sperm per day).
- The frequency of ejaculation. If a stallion serves two mares one hour apart, the second mare receives only half the quantity of sperm that the first mare receives. If the stallion has average fertility, the second mare receives enough sperm to enable her to conceive, but if the stallion has poor fertility she may not conceive.
- Ill health or stress. Sperm produced in the testes take almost 60 days to mature and be ready for ejaculation. If a stallion becomes stressed or ill, causing sperm production to either cease or result in deformed, infertile sperm, the effect on fertility will not be seen until 60 days later.
- Handling and care, which will affect the stallion's behaviour and thus desire to mate. Excessive roughness, excessive use as a teaser, isolation from other horses, boredom and stress are all factors that can cause abnormal behaviour and affect the stallion's desire to mate and ejaculate.

A veterinary examination will assess the stallion's fertility, and should be carried out prior to purchase.

Mature stallions with normal fertility can serve two mares each day of the week without lowering their fertility. Younger stallions, aged 2 or 3 years, should be used less; two or three services a week in their first season because their ability to produce sperm is lower than the mature stallion. The number of mares booked to a young stallion is not the criterion. The number of services given per week is the critical factor with regard to fertility and a veterinary examination of mares to determine their suitability for service is more important.

MARE MANAGEMENT

Compared with most other domesticated animals, horse breeding is relatively inefficient. Although results vary between breeds and studs, only 50% of mares sent to stud will produce a live foal. Part of the reason for this low result is that horses are not selected for reproductive performance or fertility, unlike other animals, but for their athletic performance. There are ways to improve the efficiency of your breeding program.

Feeding

Mares should be in good to fat body condition at the beginning of the breeding season (see the section 'Condition scoring' in Chapter 5). Ovarian activity correlates with body condition, but the mare should not be too fat or she may then lose condition when sent to a stud and placed in a new management system. The timing of the loss of the winter coat is influenced by feeding practices and body condition.

Parasite control

Young horses are very susceptible to the effect of intestinal worms and permanent damage can result from heavy burdens, so the parasite control program for mares and foals must be optimal (see Chapter 6).

Routine health care

Mares should receive an annual vaccination booster for strangles/tetanus before they visit a stud or just before they foal. Whenever large numbers of horses congregate there is an increased risk of infections. A booster of strangles/tetanus just before foaling will ensure that the foal receives optimum immune protection from the colostrum.

Brood mares should have their teeth checked annually and have regular hoof care. These aspects are often neglected, resulting in reduced nutrition and mobility of the mares and thus affecting fertility (see Chapter 2).

Getting the mare cycling

Most mares begin to cycle in spring as the daylight hours increase and they lose their winter coat. The peak time for cyclical activity is at the height of summer. Thus in southern Australia and New Zealand the artificially imposed 'birth date' breeding season is nearly two months early, whereas in northern Australia it is better synchronised with the natural breeding season. As day length increases in the later part of the imposed breeding season, cyclical activity becomes more regular in most mares. However, if breeders want early foals (i.e. born in August or September), they must stimulate the mare, using a combination of the following factors, to begin cycling early so that by August or September she is cycling regularly:

- good feeding
- shelter or rugging
- artificial lighting
- hormonal treatment.

Teasing

Unless you are using paddock mating where the stallion does the teasing, you will need a method for determining when a mare is in season.

Some mares will show signs of oestrus behaviour without the presence of a male horse, but most require stimulation via a colt, stallion or gelding (the teaser) to show the signs of oestrus. A teasing set-up is where the teaser and the mare are in adjacent paddocks and the teasing process can occur without the interference of anyone holding the horses. The teaser can also be placed in a small yard adjacent to a paddock of mares, but the fence must be solid so that the teaser, if entire, cannot serve a mare. It is preferable to have a variety of teasing methods because not all mares are suited to the same approach. Some mares need to be twitched to show the signs of oestrus, and mares with foals at foot are especially difficult to assess (e.g. some need the foal nearby while others will only show if the foal is away from them). The key to a good teasing program is knowing the signs of oestrus, having a versatile system and good record keeping.

A mare that is not in oestrus will show a combination of the following:

- rejection of the teaser
- trying to kick the teaser

- clamping the tail down
- putting the ears back.

A mare that is in oestrus will show a combination of the following:

- acceptance of the teaser
- lifting the tail
- winking the vulva (rhythmic pushing out of the clitoris) and urinating
- squatting.

Fertility assessment

A fertility assessment will determine the suitability of a mare for service and evaluate any factors that may contribute to reduced fertility, enabling the stud manager to place the mares in an order of priority for service by the stallion. Such an assessment will ensure that if a mare is served only once in an oestrus period there is a good chance of conception. A detailed veterinary examination of the mare's reproductive tract is not essential for getting a mare pregnant, but it should be done if a mare has not conceived after service in several oestrus periods in order to rectify problems before the next breeding season.

Figure 4. Teasing (courtesy of Glenormiston Campus, University of Melbourne).

The routine veterinary examination has two components:

- an assessment of ovarian follicle development and the time of ovulation
- swabbing or assessment of uterine inflammation and/or infection.

At some studs each mare is swabbed routinely before they are served, but this practice is not essential. It may be better to base the need for swabbing on the presence of a discharge or repeated failure to conceive.

Serving the mare

There are four ways of getting your mare pregnant.

- Hand service.
- Paddock mating.
- Artificial insemination.
- Embryo transfer.

Hand service

Hand serving is the most common method of breeding because it allows the stud master to choose when a mare is bred, protects the mare and stallion from injury, and the results of the breeding can be monitored.

Before hand serving a mare you need to consider the timing of the service relative to the mare's cycle, restraint of the mare and stallion, and safety and hygiene. If the mare has not been follicle tested, it is usual to serve her every second day from the second day she is in season so that live sperm will be in the reproductive tract when the mare ovulates, thus facilitating conception. Some stallions have sperm that lives for only a short time after ejaculation, and in that case the mare will need to be served more often, or the service will need to be scheduled (using follicle testing) closer to the time of ovulation.

Choose an area for the service that is free from obstacles, has a relatively non-slip surface, is dust-free and quiet. If the mare has a foal at foot you will have to confine the foal within eyesight of the mare.

If you need to prevent the mare injuring the stallion, you will have to consider the use of a twitch, serving hobbles, leather serving boots or a sideline. If the stallion has a habit of biting mares, you

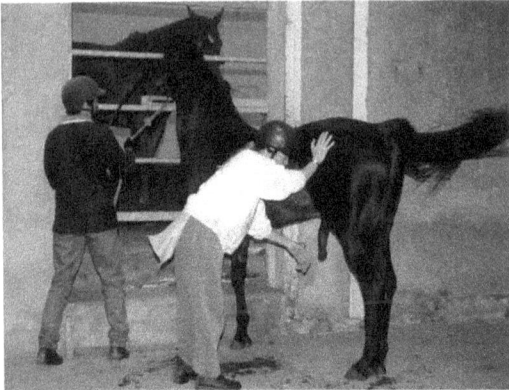

Figure 5. Washing
the stallion
(courtesy of Glenormiston
Campus, University of
Melbourne).

will need to use a leather neck guard to protect the mare. The mare's tail should be bandaged, if it has not already been clipped, and the area around the vulva should be washed with soap and water to clean away dirt and debris that might come into contact with the stallion's penis and thus contaminate the uterus. Dry the area using paper towelling or cotton wool. At some studs the stallion's penis is also washed before service. Detergents can lead to an imbalance of the normal bacteria of the penis, which can lead to infections, so water alone is usually adequate.

The stallion should be trained to approach the mare without rushing to mount and to mount only when he has an erection. Flagging of the tail is a signal for ejaculation, but can sometimes be misleading, so it is useful if someone can assess ejaculation by placing a hand on the base of the penis to feel the urethral pulsations. After the stallion has ejaculated, the handlers must manoeuvre the horses to make sure that they cannot kick each other while the stallion is dismounting.

Figure 6. Serving
the mare
(courtesy of Glenormiston
Campus, University of
Melbourne).

Paddock mating

Paddock mating, where stallions and mares run together in a paddock, is used for horses of lesser value or where the stud does not have the facilities to tease and serve mares. It can be a very effective and economical method of breeding, but foaling rates are usually lower than those achieved using other mating methods because conception cannot be monitored and the fertility of the mares and stallions is usually lower.

Artificial insemination

The use of artificial insemination (AI) of the mare, with either fresh or frozen semen, is increasing because there are many benefits for the owner, mare and stallion.

- It allows one stallion to serve many mares: one AI collection from a stallion contains enough sperm to inseminate at least 10 mares.
- A stallion can 'breed' all year round and still compete in peak condition.
- The risk of injury to the mare and the stallion during breeding is reduced, and when AI is performed properly the risk of transfer of disease is also reduced.
- Transport costs and the risks involved with transport, especially overseas, are reduced. Frozen semen can be used if death or misadventure of the stallion prevents him from being used for breeding. It is thus an 'insurance' for the owner of a stallion.

Figure 7. Artificial
insemination
equipment
(courtesy of Annie Minton).

Nowadays, AI is extensively used in the breeding of Standardbreds and by many equestrian breeders, particularly for Warmbloods.

There are some disadvantages of AI, including:

- the variability in both the freezing of semen from stallions and the response of stallion semen to the four extenders used in the AI industry.
- more intensive management of mares, especially if frozen semen is being used because it has a shorter life span in the mare's reproductive tract.
- the requirement for trained operators (veterinary surgeons or licensed inseminators); people without sufficient training or expertise in the AI technique will produce poor results.

- the lack of recognition of an AI foal in the Thoroughbred industry to date.

Embryo transfer

Embryo transfer is the technique of flushing a 7-day-old embryo from the uterus of a donor mare and placing it into a recipient mare at the same stage of the reproductive cycle. This technique allows the following types of mares to breed:

- those that repeatedly lose pregnancies through abortion or early embryonic loss
- older mares
- those that have difficulty foaling
- those that compete
- young mares (i.e. 2-year-olds).

Figure 8. Collection of the sample for artificial insemination
(courtesy of Glenormiston Campus, University of Melbourne).

Another advantage is that exceptional mares can produce more than one foal each year.

Breed registries are beginning to accept embryo transfer foals because blood typing and DNA can prove the parentage. Embryo transfer is expensive, but for elite bloodlines the costs are not prohibitive. Future developments will enable the importation and exportation of embryos.

Sexually transmitted diseases

There is not an equine equivalent of the human AIDS virus, although equine viral arteritis can be transmitted by carrier stallions and result in illness in the mare after service. However, it is a self-limiting disease and rarely severe. Symptoms are swellings of the legs, fever, lethargy, a mild nasal discharge and dermatitis. Although the virus is present in Australia and New Zealand, particularly in Standardbreds, there have not as yet been any reports of severe clinical disease.

Another viral disease, equine coital exanthema, can cause a short-term problem with blisters and ulcers on the penis and the lips of the vulva. Affected stallions or mares should be withdrawn from breeding for 2–4 weeks while the blister and ulcers heal. Some mares may be left with white scars (see also Chapter 6.)

There are also bacteria that can cause venereal diseases and uterine infections in mares, which will prevent conception. The typical situation is a number of mares with vulval discharges, returning to oestrus after service by one stallion. The stallion will usually not show any signs of infection. Some of these diseases can be very difficult to treat, which is a significant concern for breeders. A veterinarian will diagnose the cause of the venereal infection by taking a uterine swab and then prescribing specific treatment.

Figure 9. Inseminating a mare
(courtesy of Glenormiston Campus, University of Melbourne).

STALLION MANAGEMENT

The principles behind successful management of stallions are based on the purpose for which the horse is to be used: racing, showing, or breeding.

Generally, a stallion can be treated in much the same way as any other horse, particularly if he has not been isolated from other horses for long periods. Horses are social animals and need to have regular group contact in order to establish their social position and form bonds with other horses. Young horses that have always run with colts or geldings can continue to do so, providing the social order of the group is maintained. However, when a stallion is removed from a group and then returns to it, he may attack and injure the other horses in an attempt to regain his previous social position.

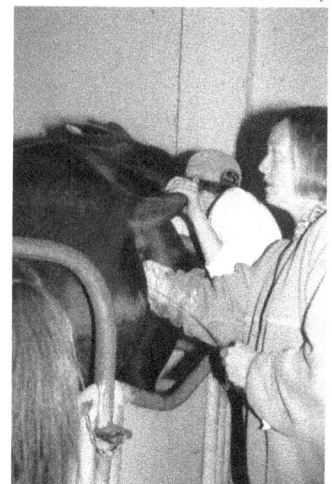

Stallions are usually segregated for ease of inspection by owners of mares, prevention of injury or to control fertility.

Stabling a stallion

It is not necessary to totally isolate a stallion from other horses in the stables, provided the partition walls between the boxes are solid and high enough to prevent the stallion from trying to mount mares over the wall or attacking other colts and geldings.

A floor area of 4.6 × 4.6 metres is adequate for a stallion. The walls must be strong and well constructed. There should not be any overhead obstacles because stallions are more likely than other horses to rear when stabled.

Occasionally a stallion will take a strong dislike to another horse (male or female) and repeatedly try to savage it, resulting in injuries to one or both horses and handlers unless the animals are placed far apart.

The stallion paddock

Apart from issues of safety and 'coat care' there is no reason for a stallion to be stabled at night. The stallion can be kept in a paddock or yard if the fences are extremely secure or, if existing paddock fencing is not adequate for the purpose of housing a stallion, by installing an offset electric wire (see Chapter 8).

Place the stallion yard or paddock in a central area of the stud so that the stallion can observe other horses and activities, which will help prevent him becoming bored and developing unwanted behaviours , such as kicking the stable wall, flank biting, crib-biting, weaving or other more dangerous behaviour (see Chapter 2 for more details about 'vices').

Most stallion paddocks are too small to allow for enough exercise; the paddock should be long enough to allow the stallion to exercise, especially if he is overfed. You will need to collect the droppings at least twice weekly, and regularly slash and renovate the pasture to keep the paddock in good shape (see Chapter 8).

Feeding the stallion

Feed should be carefully regulated according to the exercise and breeding program of the stallion. Most stallions are overfed! In the wild, the stallion is often the thinnest horse in the mob. The fat, glossy-coated stallion may have traditional appeal, but is at a greater risk of infertility, loss of libido and laminitis than a stallion on a carefully controlled feed and exercise program.

If a stallion is in good health and not stressed by abnormal seasons (drought, cold), he can get adequate maintenance feed from most improved pastures. However, a higher level of feeding is needed from the beginning of the breeding season, but need only be at the same level as pregnant mares, allowing for differences in the intake of pasture. Reduce the stallion's feed intake on any day when he is not serving or exercising.

Exercise and handling the stallion

A stallion should have regular exercise and it should become more intensive from early July in readiness for the breeding season. A stallion can be lunged, ridden, walked or allowed to exercise in a paddock. Regular exercise and handling will help develop mutual respect between the stallion and his handler and will also reduce boredom.

When grooming the stallion, keep alert. Although many stallions enjoy being brushed, there will be occasions when it can make the horse kick or bite. Injuries are more likely with inexperienced handlers who cannot anticipate a stallion's reaction. It is recommended that you secure a stallion while grooming it.

Giving a stallion a change of environment and exercise under saddle will often help to manage a difficult animal. The rider has control without being threatened by the animal's natural defensive behaviours of striking, biting and rearing.

Firmness and kindness are good rules to follow and although the handler must always be in control it should not be achieved through severe punishment or isolation.

Remember the following safety rules at all times.

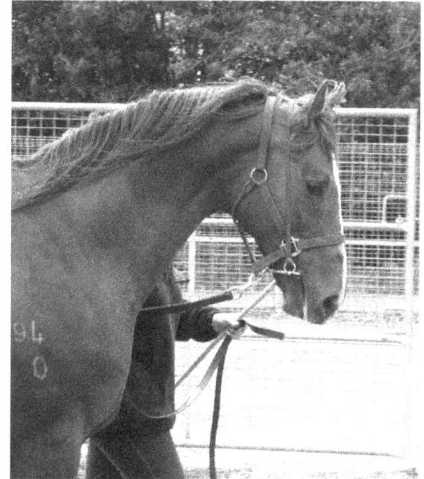

Figure 10. Handling the stallion
(courtesy of Glenormiston Campus, University of Melbourne).

- Stallions can be dangerous animals and only very experienced horse people should handle them.
- One person should handle the stallion on a regular basis in order to develop a relationship of trust.
- Be in control through careful discipline and kindness, not unnecessary punishment.
- Avoid standing directly in front of a stallion; stand beside the shoulder.
- A stallion should have a bit in his mouth whenever being handled.
- The stallion handler should carry a cane or crop.

Parasite control

Parasite control is the same as for any adult horse (see Chapter 6). Collect the droppings from the stallion paddock at least twice weekly.

Routine health care

The health of the stallion during the breeding season is of paramount importance. Lameness or disease can render a stallion either unusable or less fertile. Seek veterinary attention if you suspect the stallion is sick; even a fever of a few days duration can result in a prolonged reduction in semen quality.

Check the stallion's teeth every 6 months, and inspect for lice in early spring because they can be transferred by mares. Give the stallion good foot care, but from a safety point of view it may be better to leave him unshod (some stallions require shoeing during the breeding season). Booster vaccinations for strangles and tetanus should be given before each breeding season.

Fertility assessment and breeding

A stallion's potential fertility should be assessed by a veterinarian before the breeding season begins, particularly if he will be serving many mares. The vet can collect a semen sample and check both it and the stallion for any abnormalities that may reduce fertility. Otherwise, infertility related to semen will not be apparent until the mares start returning to oestrus after service instead of becoming pregnant.

Most young, virile stallions (over 3 years of age) can cover two or three mares each day, seven days a week, but because of the risk of venereal infection and reduced fertility it is recommended to reduce the number of coverings. Follicle testing under veterinary supervision will determine the mare's stage of oestrus and increase the opportunity for conception after one mating.

Not every stallion shows enough sexual interest in mares. Young stallions should not be hurried and can take up to 1 hour to complete a mating. You can try teasing another mare over the fence to excite him, then quickly introduce the mare that is to be covered. Training the young stallion for breeding is a time-consuming business because previous performance training will have quashed his natural instincts and he will need careful retraining.

PREGNANCY AND FOALING

Pregnancy diagnosis

Pregnancy diagnosis is an essential part of an efficient horse breeding program because it will ensure that the mares that are not pregnant are returned to service and those that are pregnant are managed optimally.

All pregnancy tests performed earlier than 45 days after the last service should be repeated, because abortion or loss of the embryo is less likely after 45 days. Confirmation and certification of pregnancy for payment of the stud fee must be done after 45 days and if the pregnancy is to be insured a 63-day test is required.

Twinning is a major problem in horse breeding because most twins are aborted during pregnancy and others are non-viable when born. Twin conceptions can only be differentiated from a normal pregnancy by a manual or ultrasound pregnancy test.

Manual diagnosis

> Warning: Manual pregnancy diagnosis must only be carried out by a veterinary surgeon.

Manual diagnosis is the most common method of determining pregnancy and can be performed from 20 days after service until foaling. As with all pregnancy testing methods the accuracy is lower prior to 45 days because foetal loss can occur between the time of testing and foaling, and operator error is greater. Extreme care must be taken during this procedure because the mare's rectum is very fragile and can easily rupture with rough handling, with resultant fatal consequences.

Restraint of the mare and protection of the operator are essential. If a crush is not available, pregnancy testing can be performed using a twitch and leg ropes or a suitable barrier. Some mares have to be examined in a crush and a suitable one should be built (see Chapter 8).

An estimation of the stage of pregnancy can generally be made and some indication given of the viability of the pregnancy, allowing measures to be taken to prevent abortion or embryo loss.

Ultrasound examination

Ultrasonic pregnancy testing is now widely used by horse veterinarians and is an essential part of the management of most studs. A probe is inserted into the mare's rectum to obtain an image of the uterus on the instrument's screen.

The embryonic foal can be accurately detected from 12 to 14 days after conception, and twins are readily distinguished. Mares are usually scanned between 14 and 22 days after service, so the use of ultrasound pregnancy testing allows earlier return to service of non-pregnant mares, earlier action in cases of twinning and rapid detection of uterine changes that may mimic pregnancy. The scanner can also detect follicle growth, ovulation, and a range of uterine abnormalities. Some specialists can determine the sex and age of the foal. A photograph of the embryonic foal can be produced.

Laboratory tests

Immunological blood test: 'MIP' Test, 'D-Tec' Test

A blood test used to diagnose pregnancy between 45 and 120 days after service has 90% accuracy because it detects a hormone, pregnant mare serum gonadotropin (PMSG), that is produced by the pregnant uterus after day 38.

False-positive tests can occur if there is loss of the embryo after PMSG production has commenced because PMSG is still produced up to day 150, even without a foetus. This test can be performed by your veterinarian, a private laboratory or, in Australia, a Department of Agriculture laboratory.

Mouse blood test: 'AZ Test'

As with the MIP and D-Tec tests, the AZ test detects PMSG, so it has similar limitations, but is considered slightly more accurate. Mare serum is injected into immature mice and 48 hours later the mice are examined for changes to the reproductive tract caused by PMSG.

Cuboni test

The Cuboni test uses a series of chemical reactions to detect oestrogen in the mare's urine after day 150 of pregnancy.

Early embryonic loss

Most losses occur early in pregnancy and 15–20% of mares that conceive will lose the embryo before day 50. If the embryonic loss is detected the mare can be served again later in the breeding season and this is an important reason for repeating the pregnancy tests. An ultrasound examination can detect mares that are in danger of early embryonic loss, enabling treatment or further monitoring. Some mares, particularly older mares and mares with uterine inflammation, have an increased risk of early embryonic loss. Stress and disease also may increase the rate of early embryonic loss.

Abortion

Abortion is the abnormal expulsion of a foetus any time from the first month to full term. Up to 30% of mares that conceive will lose the foetus before foaling. Most of these losses occur in the first 35 days of pregnancy and the embryo is resorbed, after which the mare may come back into heat season, at a longer interval after the last oestrus.

There are procedures that should be followed when a mare aborts. All other mares in the paddock or yard where the abortion occurred should be moved to an isolated, empty paddock. The foetus and the membranes should be collected in a strong plastic bag and taken to a veterinary laboratory for examination. If that is not possible, your vet should be called to do a post-mortem of the foetus and take tissue samples, as well as a cervical swab from the mare, for laboratory examination. To minimise the spread of contagious abortion, the affected mare should be kept isolated until the results of the laboratory test are known, generally in 2 weeks. The area should be disinfected with lime or a similar preparation. Any people who handled the mare or the aborted foetus should thoroughly wash their hands, disinfect their boots and change their clothes before handling other horses, especially pregnant mares.

Some mares will show signs of impending abortion, such as bagging up or having a vulval discharge, other mares will abort without warning. It may be possible to delay an imminent abortion, so consult your veterinarian.

Viral abortion

Equine herpes virus (EHV) infection may cause abortion, stillbirths or deaths of foals within three days of birth. Viral abortion occurs in late pregnancy. The mare will show either mild symptoms or none at all. All horses are susceptible to EHV and the disease is spread by coughing, sneezing or leaving nasal discharge in the environment. The recently aborted mare and the site of abortion are highly infectious. After abortion caused by EHV, the mare develops an immunity and is unlikely to abort again from this cause.

Weanlings and yearlings are very susceptible to EHV and could become future carriers, so they should be kept away from the pregnant mares. Some mares may also become carriers.

To prevent outbreaks of viral abortion, pregnant mares should be run in small groups to minimise cross-infection. Reduce the movement of mares between properties. Isolate new mares for 7–14 days after arrival at a new property with horses. Mares on properties known to be infected should be left there or if moved, they should be isolated on arrival at the next property until they have foaled. Some studs vaccinate against EHV.

Bacterial abortion

With hygiene and good management, some forms of bacterial abortions can be prevented. Bacteria are easier to identify, do not spread as readily and respond more quickly to treatment with antibiotics.

Twinning abortion

In horse breeding, twins are not good news; 2% of abortions in Australian Thoroughbreds are caused by twinning and up to 10% of conceptions may be twins. Usually, one foetus develops more rapidly, progressively assuming the major portion of the maternal blood supply and because of the consequent lack of blood the other foetus dies. The death of one foetus usually results in the abortion of both. Abortion because of twinning is most common between the fifth and ninth months. Occasionally live twins are born, but most are premature and are not likely to survive.

Mares with a history of producing twins should have an ultrasound pregnancy test between 12 and 25 days after the last service, preferably between 14 and 16 days, and then one embryo can be destroyed so that the other can continue to develop as normal. Sometimes one twin will fail to develop and be resorbed without any veterinary intervention.

Hormonal abortion

Mares that have a history of aborting that is unrelated to infection or other known causes are thought to have a hormonal imbalance. Progesterone, oral or injected, may help the mare maintain the pregnancy to term. Orally administered progesterone is more effective than injections, but must be given daily to the at-risk mares.

Stress-related abortion

The stress of long-distance transport, feed shortages or illness in late pregnancy may predispose the mare to abortion.

Early pregnancy

There is no particular need for special care in early pregnancy other than that the mare is fed to maintain good to fat body condition and stressful incidents are minimised. However, if the mare is too fat she may have trouble foaling and then lose condition, which may stop her coming into oestrus.

Late pregnancy

A mare's nutritional requirements increase by 20–30% during the last three months of pregnancy (see Chapter 5). Rapid growth of the foetal foal leads to particular increases in the requirements for protein and calcium. Make sure the mare is in good body condition, but is not too fat.

The mare should be wormed on or shortly after the day of foaling to stop contamination of the paddock. The mare should be watched for signs of impending foaling and checked that she is not running milk; if this occurs for too long prior to birth the foal will not get the valuable colostrum. Some mares have mild colic in late pregnancy, brought about by movements of the foal or compression of the intestinal tract. Some mares may also have swelling of the hind legs or under the belly from impaired blood flow caused by the swollen uterus. These problems may require veterinary attention.

It is also important to check prior to foaling whether the mare has had a Caslick's operation. The procedure involves suturing the lips at the top of the mare's vulva together to minimise the risk of uterine infections caused by aspiration of air. If the operation has been performed, it is necessary to have a vet surgically re-open the vulva just prior to foaling to prevent the foal tearing the mare's vulva.

It is also recommended to vaccinate the mare against tetanus and strangles approximately one month before foaling. At some studs the mares will also be given a vaccination against Salmonella to protect the foal. The foal will receive antibodies to these diseases in the colostrum

Signs of impending foaling
The length of the mare's pregnancy (gestation) is quite variable, usually 330–350 days, but shorter and longer pregnancies are common. It is not unusual for some mares to have a gestation longer than 12 months. Early season foalings (August, September) are often a few days longer than later foalings. Most mares show all the signs leading up to foaling, but some mares may show only a few of them or even none of them. It is harder to pick the onset of foaling in maiden mares.

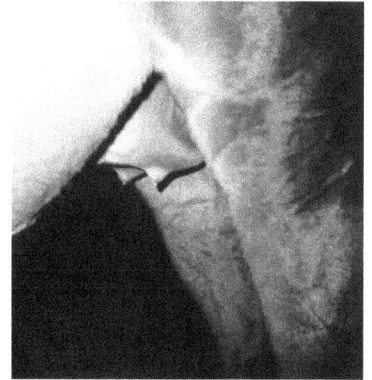
Figure 11. Waxing
(courtesy of Glenormiston Campus, University of Melbourne).

Udder growth and secretion of colostrum, known as bagging up, occur during the last 2–6 weeks of pregnancy. A noticeable swelling of the udder may occur well before foaling in older mares that have foaled before, but will occur much closer to foaling in young or maiden mares.

Although bagging up occurs some weeks before foaling, the teats themselves do not usually fill up until 2–4 days prior to foaling. Approximately 48 hours before foaling, the mare will show a drop of sticky yellow substance on the end of the teats ('waxing'). At about the same time the mare's vulva loses its normal firm tone and appears flaccid.

The muscles of the croup, particularly around the tail, start to relax around one week before foaling; they feel soft to touch rather than firm, and tend to fall away in appearance.

Most mares become a little restless and leave the other horses 12–24 hours before foaling. They may get up and down and walk around immediately prior to foaling.

Foaling
To understand the process of foaling and be able to deal with any problems, it is important to know what is normal and what is abnormal. Although there is no substitute for experience, instructional videos of foaling mares are now available.

Foaling generally takes place at night, with 80–90% of mares foaling between 10 pm and 4am. It is quite normal for mares to foal in the open, but some studs prefer to have them foal inside for ease of observation. If foaling inside, ensure the mare has a large box, at least 3.5 metres square, so that she can get away from the walls.

The stages of foaling
Stage 1
The first stage generally lasts between 1 and 2 hours. The mare seems restless, walks around, gets up and down, swishes her tail and may urinate frequently. The water bag may or may not appear at the vulva. Rupture of the water bag by the mare's early contractions causes the release of straw-coloured fluid from the vagina and is the beginning of stage 2.

Stage 2
The second stage is the actual birth and takes 10–20 minutes. Allow this period of time to elapse before interfering in any way. Almost all mares foal lying down; however, on rare occasions a mare may foal while standing. During this stage there are some very forceful contractions. The first things to appear at the vulva are the foal's front feet covered in a whitish sac, the amnion.

It is usual for one foot to appear before the other. The amnion may be ruptured before it appears at the vulva or it will rupture soon after. As the mare continues to strain, often quite violently, the foal's nose appears, usually at about the same time as the knees. Once the foal's chest is through, the mare often has a short rest before completing delivery and expelling the pelvis of the foal. If

Figure 12. Foaling
(courtesy of Glenormiston Campus, University of Melbourne).

the foaling takes more than 15 minutes, be prepared to assist or call the vet urgently. A mild shoulder lock can cause both feet to appear together and this is simply corrected by pulling on the leading foot, then alternatively on the trailing foot. Pressure should be steady and in unison with contractions.

During the delivery the mare may take short rests, during which she will sit or stand up, then lie down and resume straining. These breaks give her time to regain her strength, and also to reposition the foal if it is not correctly presented.

When the foal has been almost completely expelled, both the mare and foal will rest, with the umbilical cord intact and often with the foal's hind feet still in the mare's vagina. Do not disturb them during this period, because there is transfer of blood from the mare's circulation to the foal.

The complete expulsion of the foal signals the start of stage 3.

Figure 13. Placenta (courtesy of Glenormiston Campus, University of Melbourne).

Stage 3

The final stage of foaling involves the mare standing up and in doing so breaking the umbilical cord at almost 5 cm from the foal's navel. At the same time the foal makes struggling efforts to get to its feet. This is an important time for the development of the maternal bond and the mare and foal should be left alone, despite the temptation to assist the foal.

Usually the foal has managed to stand within an hour and will attempt to suckle. Just as the foal's first attempts to stand appear futile, so do its first attempts to suckle, but the foal is usually suckling within 2 hours.

The end of stage 3 is when the mare passes the afterbirth (placenta), which should occur within 2 or 3 hours of foaling. Check that all the afterbirth has been passed (it looks like a pullover with the ends of the long sleeves sewn up). Spread it out on the ground to check that the entire membrane is present.

For a normal foaling there is very little the owner needs to do. The natural process is efficient and the mare should be left to get on with the job without interference. It is wise, however, to disinfect the umbilical stump of the foal with iodine as this is a potential entry point for bacteria.

Induction of foaling

A veterinarian can use drugs to induce foaling if the mare has an injury or illness, is overdue to foal, continuation of the pregnancy would threaten the mare or the foal, or if it is known beforehand that veterinary assistance will be required at the foaling. Ask your vet for advice.

Problems with foaling

Because foaling is very quick and forceful, if problems arise they can worsen quickly too. Seek immediate veterinary assistance as soon as you suspect that something is wrong. While waiting for the veterinarian's arrival, get the mare up and keep her walking around because this will make it difficult for her to strain.

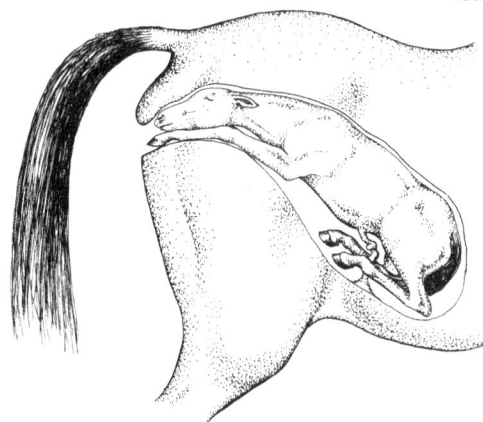

Figure 14. Normal presentation of a foal.

Foal presented wrongly

The usual problems of incorrect presentation are:

- one or both front legs bent back (in which case there is only the head or the head and one leg appearing at the vulva).
- the head is bent back (only the front legs appear).

- a breech birth with the hind legs appearing at the vulva (in this case the soles of the foal's feet face upward rather than towards the ground as in a normal birth).
- a breech birth with the hind legs extending forward (in this case there is persistent straining by the mare, with nothing appearing through the vulva).

If you are uncertain whether the foal is presented normally, gently pass a hand inside the mare's vulva where you should be able to identify the foal's muzzle lying on top of its two front legs. Before doing this, clean the mare's vulva and anus with disinfectant, then disinfect and thoroughly soap your arm with a bland soap in order to provide good lubrication.

Mares can unexpectedly kick out during foaling so it is desirable to have a barrier between you and the mare while you are checking on the foal.

Foal requires traction

The foal may present normally, but does not appear to be coming out, which may occur with young mares or with large foals. Firm traction needs to be applied to the foal's front legs. The direction of pull should be down towards the mare's hocks. If ropes are to be tied around the foal's legs they should be placed above the fetlocks with a half hitch around the pastern. Keep one leg in front of the other leg so that the shoulders are slightly twisted and uneven, thus reducing the foal's diameter.

Foal's feet in the wrong position

If the mare's anus appears to be bulging before the foal's feet appear through the vulva, they are being pushed up into the roof of the vagina and potentially could penetrate into the rectum of the mare. You need to put your hand into the mare's vagina, after thorough disinfection and lubrication with soap, and guide the foal's feet into the correct position.

Post-foaling problems in the mare

Retained afterbirth

There are two potential complications of a retained afterbirth. The first is the high likelihood of a uterine infection and the consequent difficulty of getting the mare pregnant again. The second is the possibility of laminitis, which is much less common but disastrous when it does occur.

If the mare has not passed the afterbirth within 4 to 6 hours call a veterinarian. Do not try to manually remove it because this can cause serious damage to the lining of the mare's uterus.

Tears to the vulva

Check that the foal's legs did not tear the vulva. Tears will need suturing to prevent scarring and a poor seal to the vagina. A rectovaginal fistula is a severe form of tearing between the vagina and the rectum and must be treated or the mare will be infertile.

Discharges

A normal mare will have a discharge for a few days, but check that this does not increase in amount or become purulent (yellowy green indicating pus).

Colic

If a mare shows signs of colic post-foaling seek urgent veterinary attention.

Feeding the wet mare

A wet mare will have double the feed requirements of a dry mare, with protein and calcium being especially important for good growth of the foal. Further details on feeding mares can be found in Chapter 5.

FOAL MANAGEMENT

Newborn foals

Foals are born without any immunity to disease and although colostrum provides them with temporary protection, newborn foals can suffer many disorders that are serious and rapidly life threatening. Seek veterinary attention immediately.

Just after birth

If you are observing the birth of a foal, leave the mare and foal alone once the foal is expelled so that the mare does not jump up and break the foal's umbilical cord prematurely. Ideally, the mare should rest for 5–10 minutes, allowing blood to be pumped into the foal. All you need to do is to check that the foal's nose is clear of the afterbirth and it has started breathing. After the umbilical cord has broken, you can apply some iodine to the foal's umbilical stump.

The normal foal will get to its feet within 1 hour of birth. If the foal has not got to its feet after 4 hours you should seek veterinary assistance. The normal foal will attempt to drink from the mare soon after getting up and on average it will have had a drink within the first 2 hours after birth.

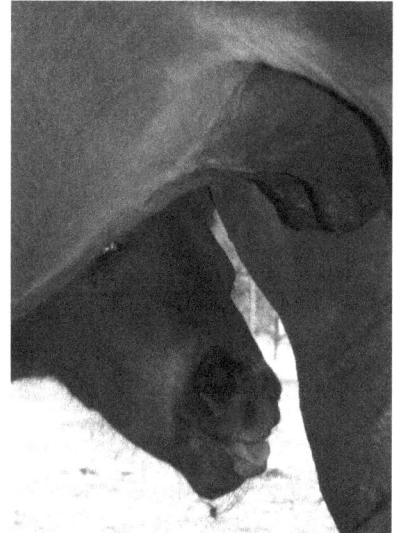

Figure 15.
Searching for the teat
(courtesy of Annie Minton).

Some mares, especially maiden mares or those with sensitive teats, do not co-operate with the foal in its search for its first drink, and keep turning around to nuzzle the foal. These foals can become weaker and confused. If a foal has not had a drink within 6 hours of birth you will need to get the foal suckling or milk the mare and feed the foal. You may have to look closely to see if the foal is getting a drink because it may have its head under the mare's belly but not be drinking. If the foal has not had a drink, the mare's teats will be full, may have wax on the end or be running milk. They do not have the shiny appearance of a sucked teat.

Any foal that is depressed, has laboured breathing or is unable to stand should be examined by a veterinarian as a matter of urgency.

Giving a foal its first drink

If the foal is not suckling within 2 hours of birth, it needs some assistance because it must get a drink within the first 12 hours of life.

Trying to get a reluctant foal to drink from the mare for the first time can be an extremely frustrating experience. The foal can have its head in the right spot, nuzzle the teats, or suck on your finger, but will not suck on the teat even if you put it in its mouth. You have two options if your foal has not drunk and won't drink when you first help it. You can persevere in the hope that eventually it will get the hang of it, or you can milk the mare and have the foal stomach tubed by a veterinarian, or feed it from a bottle. Whichever way it gets it, that first drink of milk seems to give the foal some energy and direction. From then on it knows where the milk supply is. It is vital that a foal gets at least 2 litres of colostrum in the first 12 hours of life (see later section 'Colostrum'), so do not delay in giving the foal its first drink.

Figure 16.
Collecting milk from the mare
(courtesy Glenormiston Campus, University of Melbourne).

If you have to milk the mare, you may need to sedate her so that you can to do the job properly. Your vet can then stomach tube the foal and give it a substantial drink or you could try to feed it from a bottle, which is time-consuming and less successful.

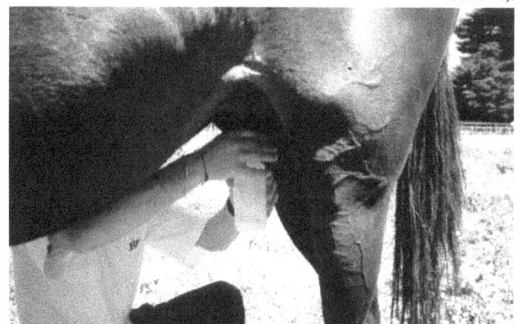

Ensuring immunity from disease

The foal depends on the antibodies it receives in the colostrum to give it immunity against diseases. A foal can be deprived of adequate colostrum because it is born prematurely, the mare does not have a good supply of colostrum, the mare has run milk prior to foaling and lost the colostrum, the foal does not drink enough on its first suckle, or a variety of other reasons. Your veterinarian can do a blood test to determine if the foal has adequate immunity and in these circumstances the use of intravenous plasma, which has been collected from another horse, is a good defence measure against infection. There are kits available for testing the immune status of all foals at 24 hours of age and their use, and subsequent supplementation of foals that have a reduced immunity, appears to have led to a reduction in the incidence of disease in newborn foals.

Sick foals

A number of serious diseases in young foals all present with similar symptoms, with the infected foal usually being described as 'sleepy'. The first sign may be that the mare is bagged up and running milk because the foal has not been drinking. You may notice the foal is spending more time lying down or is lethargic and it will usually have a reduced suck reflex. Many of these foals have a fever, but some have a reduced body temperature. Depending upon which body system is involved, the foal may show other signs such as:

- lameness and joint swelling from joint infections
- diarrhoea
- rapid breathing because of lung infections
- straining and failing to pass urine because of a ruptured bladder
- colic.

All these conditions are life threatening and need urgent veterinary treatment.

Septicaemia

Septicaemia (blood poisoning) is an infection affecting several organ systems and also being circulated in the blood. These foals need intensive treatment and often die despite therapy.

Lameness and joint swelling

The most serious cause of lameness in young foals is infection of the joints, which can be fatal or result in permanent damage to the joints and possible euthanasia of the foal. Any lame foal should be considered to have a joint infection unless proven otherwise by a veterinary examination, so seek help whenever a newborn foal is lame.

Diarrhoea

Diarrhoea is a major killer of newborn foals and you should treat all cases seriously (see Chapter 6). Foal heat diarrhoea is a common and less serious form that occurs when the mare comes back into oestrus after foaling. The cause is unknown, but may be related to changes in the mare's milk because of the change in hormones or to the fact that foals start eating hard feed or the mare's droppings around that time. Provided the foal is bright, alert, keeps drinking and does not become dehydrated, no treatment is necessary for foal heat diarrhoea. Some foals will scald the hair off their rear end and it can help to protect the area with petroleum jelly.

In other types of diarrhoea, dehydration occurs rapidly and will cause death in many cases. You can recognise dehydration from the following signs

- the foal is lethargic
- the mare has bagged up because the foal is not drinking
- the skin tenting test (see 'Health assessment' in Chapter 6).

Seek veterinary attention for any foal that has diarrhoea and is depressed or lethargic. Some cases of diarrhoea are an intolerance of the mare's milk and the mare and foal must be placed in separate, adjacent boxes to prevent the foal drinking. The mare is then milked out and the foal fed for a short period on a commercial electrolyte mixture.

Rapid breathing

Fever, premature birth or infection can cause rapid breathing, so your veterinarian must check the foal to determine the cause and institute appropriate treatment.

Straining and failing to pass urine

Rupture of the bladder is a common problem in newborn foals, especially colts, and signs will develop over the first few days of the foal's life. The foal will appear lethargic with a bloated abdomen. Surgery is needed to correct the problem.

Colic

A range of problems, including retained meconium and intestinal disasters such as an intussusception, can cause colic in a newborn foal. Seek urgent veterinary attention.

Other problems

Tendon trouble

Newborn foals often have weak flexor tendons in one or more legs, which means the fetlocks almost contact the ground, or they have slightly contracted tendons, which cause the legs to be very upright. In either case, most affected foals will assume a normal conformation in a few days. In severe cases, and even in moderate ones, it is important to restrict the exercise of the mare and foal to prevent extra stress on the legs. Ask your veterinarian for advice in cases that are so severe that the foal knuckles over or its fetlocks are contacting the ground.

Retained meconium

Meconium is the term for the first droppings of a foal, which are very hard, dark-brown pellets. Some foals, especially colts, have difficulty in passing the meconium and you will see them straining with the tail up for extended periods. These foals can even be colicky and most of them will stop sucking for a period of time. The foal should pass its meconium in the first 24 hours of life and you will often see it in the paddock. Afterwards its faeces become much softer and yellow because of the milk diet.

If your foal retains the meconium for any period and appears uncomfortable, seek veterinary help. Most cases are resolved quickly by an enema, but it should not be given by an inexperienced operator as there is a danger of damaging the foal's rectum. Do not place anything into the rectum in an attempt to remove the meconium.

The premature foal

Any foal born of a pregnancy less than 320 days has a dramatically reduced chance of survival because many of the foal's body systems, including the respiratory and immune systems, mature just before a full-term gestation. Premature foals will be small, weak and slow to stand, have thin silky skin and not suckle well. They may survive for a few days with intensive care, but the long-term prospects are very poor.

Some foals born after a normal gestation or between 320 and 345 days will show signs of prematurity, but have a better survival rate. They require good nursing care and will develop into normal foals after an initial difficult period.

The older foal

The older foal is not as vulnerable as its newborn counterpart to problems and diseases, but they can still have a significant effect on the growth of the foal and even lead to death. The most important problems encountered in this age group include:

- parasites
- diarrhoea

- rattles (i.e. severe lung infection)
- angular limb deformities
- joint abnormalities related to growth
- strangles.

Parasite control and vaccination

The foal will need to be wormed at 8 weeks of age, but on studs it is usual to batch the foals that are aged between 5 and 8 weeks and worm them as a group. Worm infestation of a young foal can reduce growth or lead to irreversible intestinal problems, so a worming program for young foals and their dams is very important. Foals can be given their first strangles and tetanus vaccination at the same time (at 8 weeks) and they will need one extra tetanus dose and two extra strangles doses to complete the course. Early vaccination of foals (before they are 6 weeks of age) may not produce the desired response.

Hoof care

Care of horses' feet from a very early age is important in avoiding or correcting conformation faults of the lower limbs.

There is a simple five-step check of the feet that should be followed when a foal is born.

1. Check the foal's feet and legs within 2 days of birth.
2. If the legs and hooves are abnormal, confine the mare and foal.
3. If there is a noticeable fault, seek veterinary and/or a farrier's advice.
4. Trim, shoe or have a veterinarian surgically correct faulty feet and legs.
5. Trim the feet every 2–6 weeks.

Correcting angular limb deformities

Foals can be born with or develop angular limb deformities, usually because of asymmetric bone growth. Valgus deformities of the knee and fetlock, in which the leg bends out from the knee, are the most common, but varus deformities, in which the leg bends in at the fetlock, also occur.

Corrective hoof trimming can rectify many of the angular limb deformities that fast-growing foals can develop, so you should seek veterinary/farriery advice as soon as you notice them. It is important to restrict the exercise of affected foals and to perform corrective hoof trimming early. Trimming of the foal's hooves needs to be repeated every few days and the bearing angle of the hoof changed so that the leg is straightened gradually. The outside of the hoof is trimmed when the leg turns out and the inside of the hoof when it turns in.

It is not good management to delay treatment, because the foal's rapid growth will reduce the time available to correct the deformity and the chances of successful treatment are likewise reduced. Fetlock deformities require earlier attention than knee problems because growth ceases much earlier in the fetlock. Glue-on shoes or acrylic compounds such as 'Equilox' are used to provide extensions to the hoof and compliment the corrective trimming. Surgery will treat those deformities that do not respond to confinement and hoof trimming.

Caring for orphan foals

If a mare dies, rejects her foal, or is incapable of producing milk, the foal must be hand reared with a milk substitute or foster-mothered by another mare or even a milking goat.

If the foal is orphaned at birth, it must receive colostrum or a substitute within the first 24 hours of life.

Colostrum

Colostrum is the first milk of the mare. It has high levels of protein and vitamin A and contains the antibodies that temporarily protect the foal against certain infections. Foals that do not receive colostrum at birth can be affected by scouring and other intestinal disorders. A foal can only absorb colostrum within the first 12 hours of its life.

Collecting colostrum

Colostrum can be obtained from another mare within 2 days of her foaling. It is usually a yellowy-orange colour, but white colostrum is still effective. It is preferable to take it from a mare that is a good milk producer, so that her own foal is not deprived. The colostrum could also be taken from the foal's own mother if she dies during or after the birth. Cow's colostrum is not suitable for a foal although if fed in large quantities it will be better than no colostrum at all.

To hand-milk a mare, place your thumb and first two fingers as high up on the teat as possible and squeeze the teat downwards. Two thin streams of milk will squirt from the teat at a broad angle: a wide container, such as a large plastic bucket, will be required to collect the milk. Try to avoid dragging on the teat too much as you will eventually cause painful abrasions. Massage the milk down from the udder to the teat (see Figure 16).

Alternatively, a human breast pump can be used to draw the milk from the mare's teat by suction. Breast pumps can be purchased from a pharmacy.

Hygiene is extremely important. The udder and your hands should be washed with pure soap and rinsed in warm water before milking. All utensils should be cleaned and sterilised as if caring for a human baby.

Frozen colostrum

It is a good idea to collect some colostrum from your mare or mares each year and freeze it. Frozen colostrum can be stored for up to one year in 500 mL amounts. Colostrum should not be refrozen.

Thaw the stored colostrum, put it in a bottle and warm it by submerging the bottle in a hot water bath until it reaches blood temperature (38° Celsius), and feed it to the orphan foal. The hole in the teat needs to be large to allow the thick colostrum to flow properly. Large, black rubber, calf teats are available from stock agents. Alternatively, the foal can be given colostrum by stomach tube. Give colostrum as many times as possible on the first day.

Colostrum substitute (plasma)

If an orphan foal does not receive colostrum, it can be given the necessary antibodies using plasma that has been collected from the blood of another horse by your vet. Plasma can be stored frozen and be given either intravenously or by stomach tube during the first 12 hours of the foal's life. Alternatively, fresh blood can be collected from a gelding (mares may have undesirable antibodies from previous pregnancies) and the plasma given as soon as it has been separated from the red blood cells.

Hand rearing

Providing a suitable environment

A foal that is to be hand reared should be kept in a warm, dry place sheltered from the wind. A loose box with clean bedding is ideal. As the orphan foal will be more susceptible to disease, the box should be thoroughly cleaned.

The foal needs regular exercise, especially as it gets older and stronger. Milk contains very little vitamin D, so the foal needs regular time in the sunlight, at least 3 hours each day, to produce its own vitamin D, which is essential for correct bone development.

Bottle feeding

To teach a foal to drink from a bottle, first squirt a little milk into its mouth so that it learns the taste. Co-operative foals will then begin to suck the teat immediately. Make sure it does not pull the teat off the bottle.

If the foal keeps moving away from the bottle, back it into a corner and try to get it to suck your milk-smeared fingers. Gradually exchange the teat for your fingers. In time, even the most reluctant foal will become hungry enough to begin sucking. If the foal is very weak, however, your vet may need to give it several feeds by stomach tube.

Bucket feeding

Although a bottle can be used to feed a newborn foal, it soon becomes tedious and it is desirable to teach the foal to drink from a bucket once it is 2 weeks old.

A wide, shallow container should be used initially; a normal bucket will enclose the foal's head and possibly cause it to panic. Half-size plastic buckets are suitable. Put either your fingers or a teat in the foal's mouth and, as it begins to suck, slowly lower your hand into the bucket of milk. You may need to push the foal's head down to show it the bucket. After some time you should be able to remove your fingers or the teat and the foal will keep drinking. At worst, a foal may take a whole day to learn to drink from a bucket.

The formula feed

Once colostrum has been fed, the orphan must receive a formula feed, made from either cow's milk or a commercial milk substitute (milk replacer).

Commercial mare's milk replacers are available, but are expensive. Milk preparations for human babies can be used, but they are also expensive. Calf milk replacers are suitable with some additional minerals and energy. Do not use milk replacers that contain antibiotics because they can cause the foal to scour.

Cow's milk contains more fat and less sugar than mare's milk and milk from a Jersey cow contains more fat than milk from a Friesian cow, so use a Friesian's milk.

Formulas for feeding to an orphan foal

Mix 1

- 375 ml cow's milk
- 150 ml lime water
- 20 g lactose or glucose
- brown sugar, molasses or honey

Mix 2

- 50g Denkavite (calf milk replacer)
- 150 ml lime water
- 25 g glucose
- 305 ml water

Lime water can be purchased or prepared at home by adding 50 grams of hydrated garden lime (brickie's lime) to a 10 litre bucket of water. Allow this mixture to settle overnight then pour off the limewater.

Suitable complete milk replacers include:

- Barastoc General Purpose Milk Replacer
- Barastoc Ascend Ezy Mix Milk Replacer

Follow the feeding instructions on the label.

Initially, the feeding mixture should be warmed to 38°C before it is fed to the foal, but the temperature can be gradually reduced to room temperature over the first week.

Amounts to feed initially

Mares produce between 3% and 4% of their body weight in milk per day in the first 2 months of lactation and then 2–3% of their body weight after that time, so the foal needs plenty of milk replacer. A pony foal should not be fed as much a light horse foal or a draft horse foal. Small ponies, such as Shetlands, with birth weights around 18 kg will need up to 180 mL per feed during the first few days. Larger ponies with birth weights around 25 kg can initially be fed 250 mL per feed. Light horses such as Thoroughbreds and Standardbreds with birth weights around 50 kg should be fed up to 500 mL per feed and heavy horses such as Clydesdales, with birth weights around 80 kg, should be fed up to 800 mL per feed during the first few days.

Feeding program for an orphan foal

It is very important not to overfeed the orphan foal because this may cause digestive upsets and scouring. Underfeeding can also cause problems because the foal will not grow as it should. It is therefore important to plan a feeding program to suit the foal. It is also important to remember that any changes to the feeding program or diet should be made gradually.

The following feeding program has been devised for light horses. It is a guide only. Alterations to suit other horses can be made by feeding half the amount for larger ponies, one-third of the amount for smaller ponies and half the amount again for heavy horses.

In the first three days

It is preferable that a foal has its first feed 2–4 hours after birth. Feed 500 mL every 2 hours, beginning early in the morning (6 am) until late at night (midnight), giving a total of 10–12 feeds each day.

The next four days

For the remainder of the first week, reduce the number of feeds gradually to 8 per day and increase the amount fed on each occasion to 600 mL. Feed every 2–3 hours.

Second and third weeks

During this period reduce the number of feeds to 6 per day and feed one litre (1000 mL) each time. Feed regularly, approximately every 4 hours. Have fresh clean water available to the foal at all times and teach the foal to drink from a bucket.

Fourth week

If necessary, increase the amount fed to 1.5 L at each of the 5 feeds. The foal will begin to eat solid feed and it should be allowed to eat as much pasture and supplements as other foals being reared on their dams.

Second month

Reduce the number of feeds to 4 per day and feed up to 3 L at each feed, depending on appetite.

From third month to weaning

Reduce the number of feeds to three per day, given in the morning, midday and at night from a bucket. Up to 3–4 L can be fed at each session. The foal can be weaned when it is 3–5 months of age and regularly eating solid food.

Feeding solid food

Once the foal learns to drink the milk replacer the next step is to introduce solid food into the diet. Lucerne hay and grass can be introduced within the first two weeks of the foal's life. A high-protein creep feed with at least 16% protein and fortified with minerals and vitamins should also be fed with lucerne chaff. Even though the hay can be fed ad lib, the hard feed should be restricted to 0.5 kg per month of age up until weaning. After that time it can be fed at 1% of the weanling's bodyweight.

A companion

A young orphan foal should have a constant animal companion. If it has only humans for company it will lose respect for them, become spoilt and develop handling problems. Do not hesitate to discipline the foal if it begins to fight and kick.

A quiet sheep or goat can be kept with the foal as a playmate. The playmate should not be much bigger than the foal because otherwise it may bully the foal and the foal will then develop an unnaturally submissive character.

Calves are not desirable as playmates because they tend to chew the foal's tail.

Foster mothering

A foster mother is a much easier way of rearing an orphan foal. As well as avoiding the chore of regularly feeding the foal, a foster mother will care for the foal naturally. However, it is uncommon to have both an orphan foal and a mare that has lost its foal and cooperation between breeders is often required. A foal from one stud may have to be mothered up with a mare from another stud.

Another reason foster mothering is uncommon is that a successful mothering depends on the individual characteristics of both the mare and the foal. The foal should not be frightened of the mare and should be immediately willing to suck the mare. Unfortunately, foals that have been drinking from a bucket soon lose the instinct to suck.

A suitable mare is often hard to find. She should have suckled a foal because a mare that has a stillborn foal is more reluctant to become a foster mother and she should have lost her foal only recently. The greater period she spends without a foal, the greater the chance that she will reject a strange foal and the greater the likelihood that her milk supply will have dried up. A mare will become dry 2 or 3 days after sucking stops.

Methods of fostering

A mare recognises her foal by sight when it is at a distance. When it is close to her she recognises it mainly by smell. The mare's sense of smell can be confused by smearing a strong smelling ointment over her nose.

The smell of the dead foal can be given to the orphan in a number of ways. The fresh afterbirth can be rubbed over the orphan foal or if that is not available, rub the orphan with the dead foal's meconium. If the dead foal did not die of a contagious disease, it can be skinned and its hide tied over the orphan to make it look and smell like the mare's foal. Remove the hide after two days.

If the meconium or the hide cannot be obtained, take some milk from the mare and rub it over the head, neck, back and tail of the orphan. Feed some of the mare's milk to the orphan also, which will help to give it an odour that is familiar to the mare.

Before introducing the foal and the mare, make sure that the foal is hungry and the mare's udder is full. The foal should be hungry if it has not eaten for 3 hours and the mare's udder will be full 3 hours after milking.

The mare and the foal should be introduced in a foaling box or a small yard. When they meet, allow the mare plenty of time to smell and examine the foal. One person should then back the mare into a corner while another guides the foal to the udder.

Allow the foal to suck for a brief period before letting the mare again examine the foal. If the mare accepts the foal, observe the mare from a distance for a few hours in case she later rejects the foal.

If the mare rejects the foal, she may need to be restrained with a twitch or subdued with a tranquilliser administered by a veterinarian. Patience is needed, but there is little chance of acceptance if the mare still rejects the foal after 10 hours.

Handling foals

The ideal time to handle a foal is soon after birth. If the foal becomes comfortable with humans very early in life, its early education is simpler, easier and less stressful for all concerned. Facilities such as stables and yards are very useful when handling young horses. If a foal spends time in such an area with its mother, it will be less timid about having people so close.

To catch a foal, enlist the help of another person to manoeuvre the mare so that she wedges the foal in a corner, thus providing you with access to the foal, but keeping it close to its mother.

Gently approach the foal, touch it and quietly move it by placing an arm around its breast and another around its rump. Some people hold the foal's tail up instead of putting their arm around the rump. Do whatever is easiest for you. With careful, gentle handling the foal will become relaxed and comfortable around humans. Each time you handle the foal rub it gently all over until it is not worried about being touched anywhere on its body. A sign that the foal is accepting your touch is if it starts to make a chewing motion with its mouth. Initially, as soon as the foal begins to relax, end the lesson so that the foal is rewarded for relaxing. Do not make the lesson more than a few minutes long, as a foal's attention span is short.

Training the foal to lead

Once the foal has had a few lessons in being handled (i.e. rubbed, scratched and touched over most of the body) and is relaxed, a headstall can be put on. It is usually best to secure the neck strap before fastening the noseband. Initially the feeling of the noseband startles the foal in this very sensitive area. Give the foal time to get used to the feel of the headstall before attaching a lead rope.

Have your helper lead the mare around the stable or yard. The foal will follow its mother and when it feels the restriction of the lead and headcollar, it will resist. The handler should keep within a rope length of the foal and not apply too much pressure until the foal accepts the pressure. Gently work the lead rope so that the foal 'gives' with its head. Within a few minutes it should be walking beside or behind its mother with some continuous pressure on its lead. Be careful because a foal can panic and flip over if too much pressure is applied.

Once the foal is accepting the pressure in the stable or yard, the mare can be led out and the foal allowed to follow. The foal will probably go through the same reaction as when it first felt the pressure on its head. Again you need to work the rope gently, never jerk or pull, until the foal accepts the pressure. This procedure should be repeated every day for a few days, with lessons of approximately 10 minutes in length.

The next lesson should be teaching the foal to lead away from its mother. Do not expect the young foal to go out of sight of its mother (or for her to loose sight of her foal). An arm around the hindquarters can be used to encourage forward movement or a breeching rope can be used (a rope is passed around the foal's hindquarters, crossed over its back and passed through the headstall. If the foal stops the rope can be pulled, putting pressure on the foal to move forward).

Training the foal to load

The foal can be taught to load onto a float soon after it has been taught to lead (in which case the dam may also need to be loaded or at least be nearby) and then repeated when the foal is weaned. With a young foal an arm can be placed around the rump or a breeching rope used to guide it in to the float. Stand the foal in the float for a few minutes, and then back it out gently, placing your hand along the foal's rib cage to prevent it backing off the edge of the ramp. Enlist the help of another person if necessary. Force should not be used and will not be necessary if the foal was taught well initially to lead.

Handling the legs of the foal

The last of the early lessons is the handling of the legs and hooves. Once again, work in an enclosed space and with an assistant. Placing the foal against a wall serves two purposes: it

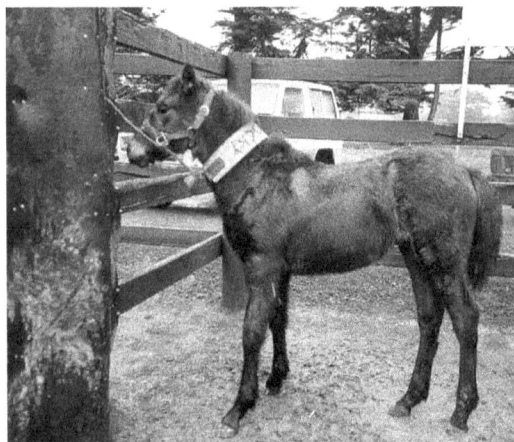

Figure 17. Teaching the foal to tie-up (courtesy of Glenormiston Campus, University of Melbourne).

prevents the foal from swinging away, and as it is not used to balancing on three legs, the wall will act as support for it when it leans. Remember that it requires significant trust for a foal to let you handle its legs. Be gentle, support the foal and it will learn to trust.

Stand at the side of the foal just in front of the shoulder, facing the hindquarters. Start by rubbing the wither and gradually rub down over the foreleg. Most young horses will accept this, the few who are going to be ticklish will warn you by pulling the leg in underneath them and showing signs of anxiety. If the foal does this keep touching the leg, but do not move further down until it has accepted you touching the upper leg.

Continue rubbing down the lower leg. The usual reaction to being touched down close to the foot is to lift it. As it does so, hold the leg off the ground if the foal does not argue too vigorously. Talk quietly, then let the leg gently drop. Go through the whole procedure a few times.

To pick up a hind foot stand level with the foal's barrel facing the hindquarters. Place your nearest hand on the hindquarters and rub; feel the tail, rub the buttocks and check the foal's reaction. If it is calm, rub the hand quietly but firmly down to the hock; again check its reaction and if all is well move down to the fetlock. When you and the foal are ready ease the hind fetlock forward and off the ground a few inches. It may help to lean against the foal to shift its weight off the leg you are trying to lift. Make sure you repeat every thing equally on both sides.

These lessons should be continued on a daily basis, until you have a well-mannered little horse. If the foal has accepted the earlier lessons well, the leg handling should not be a problem, but sometimes a foal will be nervous or be aggressive about having its legs handled.

A short piece of polypipe can be helpful. Used as a long 'hand', you can quietly apply it to the ticklish bits of the foal without fear getting hurt. Another way of getting the young horse to accept having its legs touched is to gently apply water via a hose until the resistance ceases. Start with a trickle and gradually increase the pressure as the foal accepts the previous level of pressure.

Training the foal to tie up

At some time during any horse's life it will have to be tied up, so it must be educated as early as possible in its life to do this calmly. Most people prefer to train by gradually tightening and loosening pressure using a long lead passed through a ring in a wall or post, or by taking a one-turn twist around a post, until the foal accepts that it cannot get away.

The equipment needed is the foal's own headcollar, a rope without a clip and a neck strap, which can be made from a folded hessian chaff bag with a hole at either end. Tie the rope to one of the holes and place the folded bag around the neck (accustom the foal to the bag first), then thread the rope through the other hole. This technique means that the bag cannot tighten around the neck. The collar spreads the pressure over a large area, which is much kinder when teaching a foal to tie up. The rope is then threaded through the ring at the back of the headcollar to keep the head straight.

When the foal finds it is tied up it will usually struggle, which is to be expected, but its final action is usually to sag on its front legs and then jump forward. This eases the pressure instantly, and the handler should immediately sooth the foal and offer a tit bit, so that the giving in is rewarded. Over time increase the time that the foal spends tied up from a few minutes to perhaps half an hour or so. Never leave the foal alone when it is tied up.

Headcollars and foals

It is better not to leave a headcollar on a foal because even in a safe environment there are so many ways in which an inquisitive youngster can get caught up. Also, a foal's head grows very quickly and a headcollar can become too tight. It is not uncommon to see a horse's head with a dent in the nose caused by leaving a small headcollar on a growing horse.

The weanling

A horse that is between weaning and breaking in/starting stage should be handled on a regular basis, not necessarily every day or even every week, but the young horse should have periods where it is handled and all of the previous lessons reinforced, and some new ones are taught. As well as having the early lessons repeated, it can be taught to yield to pressure, which will form the basis of all its future training (see 'Training horses' in Chapter 3).

Weaning foals

Plan the weaning of a foal in advance. There are few other stages in a horse's life when it will be under as great a stress, and procedures that will reduce the stress at weaning can be adopted. The foal is stressed because:

* it loses the protection and company of its mother
* a familiar and valuable part of its diet is replaced by unfamiliar feeds
* it is placed in unfamiliar surroundings such as stables or yards
* it comes into close contact with humans for an extended period of time.

This stress can increase susceptibility to illnesses such as diarrhoea and can result in weight loss or growth setbacks. The foal's stress can, however, be reduced, at least to some degree, by its owner/handler. For instance, handling foals regularly from the age of 2 or 3 days old until weaning will reduce handling stress. Creep feeding foals for at least 4 weeks before weaning will reduce the stress caused by a change of diet. If creep feeding is not practised, feed the mares and foals a supplement similar to that which the foals will get when weaned. The stress of a change in surroundings may be reduced if the mare and foal are kept together for at least 24 hours in the stable or yard in which the weaned foal will be left.

Removal of the mare from the foal and the stress associated with it cannot be avoided, but the weaning process can be staggered (e.g. removing the foal from the mare for 1 hour the first day, 2 hours the next day, and so on). If you do not wish to stable the weanlings to handle them you can remove one mare from the paddock each day until all the mares have gone, leaving the weanlings in the paddock. The mares should be removed far enough from the foals that they cannot hear each other calling out. This is a good method because the foal stays in a familiar environment.

As well as considering factors that will reduce stress at weaning, consider whether the facilities for weaning, and after weaning, are adequate. Weaning can take place in a well-built (metal or wood) yard or stable, which should be at least 1.5 metres high and 4 metres square. If the foal is weaned in a stable that has a divided door, make sure both the top and bottom doors are kept closed so the weanling does not jump out. However, do not put a single foal in a dark box as this can be a traumatic experience.

Try to wean at least two foals together so that they can be company for each other. Weaning two foals in the same small yard or stable often helps to quieten them down. Use an older horse in a nearby yard as company if only one foal is being weaned.

Using an older horse for company can also help prevent injuries to the weanlings when they are let out into a large paddock after handling. An older horse will be more sensible and the weanling will accept it as the leader. Accidents such as weanlings running through fences can often be avoided if an older horse is present. It is very important, however, that the paddock has fencing that is visible.

The age to wean

The decision about the age at which foals are weaned should be made after considering the condition of the mares, the amount of paddock feed, the extent of supplementary feeding needed, and other annual routines. For instance, on most Thoroughbred studs, weaning takes place as a routine after the yearling sales when foals are 6–7 months old. Do not be afraid to wean early, from 4 months, because if you feed the foal correctly it may grow faster than if it stayed on the mare (see Chapter 5).

If a mare can maintain her condition and feed the foal with very little supplementary feeding, there is no need to wean her foal until it is 8 or 9 months old. However, weaning should take place earlier than this if the mare starts losing condition and needs large amounts of supplements to maintain condition or cannot feed the foal correctly. In the late pregnancy the growth of a new foal puts extra demands on the mare, so wean before then.

Under these circumstances, both mare and foal will do better by being separately fed supplements formulated for their particular needs. For example, a foal needs a higher percentage of protein in its supplement than does a mare, and the mare's milk will not supply all the foal's nutritional needs from about 3 months of age onwards.

The weaning process
If the foal is to be weaned in a yard, the day before put the mare and foal in that yard and the next morning remove the mare out of sight and hearing of the foal. Start handling the weanling on the same day.

Handling the weanling
The weanling will be more able to concentrate on the handler once the mare is out of sight and hearing. Catch the weanling, handle it quietly over its whole body and teach it to lead if it has not been taught previously. The more time spent on handling, the greater will be the response. At least two 30-minute sessions will be needed each day.

Each day repeat the previous day's work and introduce new lessons. After one week, the weanling should face up when approached, lead at the walk and trot, lead into a float, tie up without pulling back, let you pick up each leg and trim its hooves, and stand quietly while being drenched.

Always remember that each weanling will be different, and that the time needed to achieve a result will vary. Results will never be achieved if you lose your temper and frighten or hurt the weanling.

Feeding the weanling
The weanling needs time to adjust to the new diet it will receive when taken from its mother and from pasture. Continue the same feed that the foal received in the creep, or shared with the mare. An extra source of roughage, such as high-protein lucerne hay, must also be available.

Do not feed too much. Small amounts that are completely eaten are preferable to large amounts that are left to become stale. Extra supplements should be introduced over at least 7 days (see 'Feeding weanlings' in Chapter 5).

After weaning
The weanling can be turned out to graze after it has been handled for 7 days. Paddocks for weanlings should be the safest on the property because they have the greatest potential for accidents. Loose wire, fallen limbs, baling twine etc. can all cause injury or death.

Fillies should be separated from colts. Weanlings should not live alone, so if other youngsters are not available an older mare or gelding can be used as a companion. As mentioned before, even with a group of weanlings an older horse can be very useful. The number of weanlings per paddock should be restricted to as few as is practicable in order to decrease the risk of injury.

Castration and branding
The weanlings can be branded, and colts castrated if necessary, once they have been handled.

Vaccination
Tetanus and strangles vaccination should be given if not already done. However, it is preferable to vaccinate one week after the stress of weaning to ensure a good response to the vaccination.

Records
Keep a record for each weanling of the drenching and vaccination dates, and injuries and their treatments. Health programs must be monitored if they are to be effective.

WHERE TO GO FOR MORE INFORMATION

The Glenormiston website http://horses.landfood.unimelb.edu.au/ (go to the Links and Resources page).

Go to www.equinecentre.com.au for lots of useful articles about reproduction and breeding.

There are many articles on the subject of orphan foals on the internet including http://ianrwww.unl.edu/pubs/Animals/g1237.htm

CHAPTER 8

HORSE FACILITIES

DESIGNING THE PROPERTY

Correct location of fences, lanes, buildings and stockyards on a horse property is essential for its efficient operation and management. A well-planned property will save money by reducing labour and running costs, and will be a more valuable piece of real estate. Small horse properties especially need careful planning if they are not to succumb to erosion, compaction and weeds.

Creating your plan

When developing your plan, visit as many properties as possible to get ideas and contact the relevant government agencies (see the end of the chapter for more details). There are courses on 'Whole farm planning' available at TAFE colleges.

Aerial survey

Make an accurate and detailed plan before starting any major changes or additions. A good starting point is an aerial photograph, such as those produced by the Crown Lands and Survey Department or one of the private companies that now provide this service.

A series of plans can be made for the various stages of property development. You can include:

- natural features such as ridges, valleys and trees
- water supply
- paddocks
- fences, gates and laneways
- buildings: stables, yards and riding arena.

General principles of horse property design

Buildings

The ideal location for the buildings is in the centre of the property to allow easy access to all parts of the property, but that must be balanced against the extra cost of providing access and supplying power, water and telephone. You should also consider drainage and wind protection.

In practical terms, the buildings will usually be within the front half of the property and close to an existing residence for security reasons.

Figure 1. A horse property (courtesy of Gary Blake).

All-weather access

You will need all-weather, multi-vehicle access to the buildings and around the property. When planning the layout of the property, consider connecting all paddocks to the stockyards or stables with laneways sited on ridgelines where the ground will be dry. Do not site your laneways over ground that is too wet or too rocky.

Security

Horses are valuable animals, so reduce the opportunity for theft. Have only one access road into the property for vehicles and ensure that it passes near the residence. If there are other access points, they should be securely locked when not in use, unless this constitutes a fire hazard.

TREES

Trees are important for shade and shelter from sun and wind. They also attract birds and create an attractive environment. Plan for shelterbelts near yards, foaling paddocks, houses and buildings, as well as along roadways. If there are double fences between paddocks, the area between them is excellent for a shelterbelt, which will also help the horses to see the fence as a physical barrier. (See the end of the chapter for more information.)

WATER SUPPLY

One horse can drink up to 19 000 litres of water per year, depending on its environment and the work it does. An adequate water supply, in terms of quantity and quality, must take into account evaporation, seepage and drought.

Reticulated water system

A reticulated water supply is better than a dam because the water is kept cleaner and fresher. A reticulated (or even 'mains' pressure) water supply usually relies on a float-valve to control the water-flow into the trough. Troughs are usually formed from concrete or plastic and range from small 'automatic drinkers' (usually in a stable) or post-mounted plastic troughs to large multi-animal troughs. The float-valve must be protected from the horses 'playing' with it and damaging it or causing water to flood out of the trough. Cover the float or buy a trough with a fully enclosed float-lever. Alternatively an 'armless' type of float valve can be used, attached near the bottom of the trough with a cord.

Set the troughs on a level sand base where the ground will be dry in winter and/or provide a solid surface around the trough so that the horses do not get bogged. The pipeline into the trough must be sufficiently far below the surface of the ground to be protected from hooves and of a sufficient diameter that the trough refills in an acceptable time. Typically, 32 mm is the starting point in a non-mains pressure reticulation system.

Dams

Dams can be an economical method of water collection, but should be fenced off and the water reticulated. Dams are a common method of watering stock, but are not necessarily the best method. On small properties dams significantly reduce the grazing area.

Over time, horses and other stock will 'walk' soil from the banks down into the dam, causing the water to become muddy as the dam silts up. Trees that are close to the dam add leaves, which as they break down can lead to the build-up of potentially toxic contaminants. Horse manure can wash into the dam, adding to the droppings of ducks and other birds, which can cause severe bacterial contamination of the water. A level of 50 colonies of Escherichia coli per 100 mL of water is recommended as a guide to the safe upper limit of bacterial infection of water for livestock.

In addition, if the dam becomes stagnant (such as in a drought) the water can become contaminated by the toxic 'blue-green' algae.

If water is drawn from the top 200 mm of the dam, it is more likely to be contaminated with potentially harmful micro-organisms and if taken from near the bottom of the dam, it will be colder and much lower in oxygen because of the micro-organisms that use oxygen to break down the organic matter at the bottom of the dam.

As well as bacteriological testing, water can be tested for impurities. Horses can drink 'hard' water (usually caused by large amounts of calcium), once they are accustomed to it. Magnesium salts are the first to reduce palatability (and therefore the amount of water consumed), so the recommended maximum allowable amount of magnesium is 200 mg/L. Two other factors that affect water quality and can be tested are pH (acidity and alkalinity) and electrical conductivity (salinity).

PADDOCKS

How many paddocks?

Properties fall into two categories: small acreages, which can be anything from a house block to four hectares, and large properties (usually studs). On a large property, the number and size of paddocks will depend on the number of each class of horse, and of other livestock. The classes on a typical horse stud include wet mares, dry mares, yearlings, weanlings, stallions and spelling horses. As well as providing for the different classes, you need to provide for groups within the classes. Some spare paddocks to allow for grazing rotation will also be needed. The ideal number of paddocks on a large property is at least 30 (although this may be difficult to achieve initially), depending on the relative size and carrying capacity of the property.

On small properties a balance must be struck between having enough paddocks to rotate grazing and not having too many small paddocks, which are more susceptible to compaction. Small properties can be harder to manage because it is easier to overstock. The minimum size of paddock for a horse is 04 hectare (1 hectare is the ideal).

What is the best shape for paddocks?

Square paddocks are the most economical to fence, but may not fit the shape of the land. Rectangular paddocks are the best for exercise for the horses, and are often easier to work with in terms of erecting temporary electric fences; however, if they get too narrow, the horses can cause a lot of soil damage when 'running the fences'. Ridges and valleys can be the starting points for fencing when planning subdivisions. Badly planned subdivisions can be the start of problems such as soil erosion.

Pasture management

Horse owners from many countries envy Australians and New Zealanders because we can usually graze horses on pasture all year round, but many Australian horse owners do not make full use of pasture and instead copy the overseas practice of relying heavily on expensive hard feeds.

The advantages of pastures for horses have long been recognised in New Zealand and this is a common explanation for the success of New Zealand horses in Australia. Well-managed pasture is an excellent feed that is a cheap, convenient and balanced for most horses. Hard feed need only be given to certain horses when pasture is scarce or if the horse is in heavy work, experiencing rapid growth or lactating. You should monitor your horse's condition to assess the need for supplementary feed (see Chapter 5 for information on condition scoring and weight estimation).

Suitable pasture species

Most pastures in southern Australia and New Zealand contain mostly ryegrass or cocksfoot, which grow well and are quite suitable for horses, although they are not as palatable other less productive species such as demeter fescue. Under heavy grazing, horses may eat out a paddock of demeter fescue. The most suitable pasture for horses is a grass/clover mixture.

There is some disagreement about the suitability of clover-dominant pasture for horses. The main criticism is related to weight gain (see Chapter 5). One variety of subterranean clover, called Yarloop, should never be included in a pasture for horses because it contains sufficient oestrogens to affect reproductive performance. The other varieties of sub-clover do not have this effect.

A typical horse pasture will contain mostly ryegrass and/or cocksfoot, 20–30% clover and some demeter fescue. Other species may be included depending on climate and soil type.

There is not a typical horse pasture for northern Australia. In the more temperate areas winter crops such as barley and pasture grasses such as ryegrass and cocksfoot can be grown to overcome winter/spring feed shortages. In the more tropical areas, horses at pasture may need supplementation if weight loss is greater than acceptable. Some introduced tropical pasture grass species can be hazardous for horses to graze, leading to a condition known as 'big head', so native pasture is preferable (see Chapter 5 for more detail).

Horse owners need to be aware of poisonous plants and take steps to remove or control them (see Chapter 6).

Your local office of the Department or Ministry of Agriculture can help with pasture recommendations and arrange for soil tests to be carried out. The soil test can also be used to formulate fertiliser recommendations. Most soils in Australia and New Zealand require regular applications of superphosphate and sometimes lime. Lime adds calcium to the pasture, which will, in turn, help to build strong bones in the horses. In general, lucerne and clover contain more calcium than grasses. Lucerne can be added to the pasture sowing mix at the rate of 0.5–1 kg/hectare.

The psychological and physical value of grazing is immense and horses should be allowed to graze good-quality pasture. However, if you have a horse that is likely to get too fat and subsequently founder, it may be necessary to put the horse in a yard with hay (grass hay, not lucerne), rather than allow it to graze when the grass is green and rich (see 'Laminitis' in Chapter 6).

Seasonal variations in pasture quality
Both the quality and quantity of pasture varies markedly throughout the year and this will influence the need for supplementary feeding. Young, growing horses and lactating mares have the highest nutritional requirements and are most susceptible to variations in pasture.

In summer, pastures dry off and the protein levels drop, so protein supplements may be needed, whereas lush spring pastures are low in calcium and fibre, requiring hay and calcium supplements. Extra roughage may be needed in the cold months.

Good drainage
A wet, poorly drained paddock will have reduced pasture growth and the movement of horses around the paddock will destroy the pasture and create bare areas that are sites for weed growth. During extremely wet periods remove horses from the wettest paddocks.

Weed control
If parts of the pasture become dominated by weeds, they should be eliminated using herbicides. Uncontrolled weeds will quickly dominate the pasture plants and dangerous weeds, such as Patterson's curse, must also be controlled.

It is best to spray in June or July (southern Australia and New Zealand) or November to April (northern Australia) when the weeds are very small. Make sure you spray along the fence line and under trees where nettles often grow thickly.

Grazing management

When grazing pasture, horses will dung in one area of the paddock and eat in another. Horses will not eat around their own dung (which may be a natural worm prevention strategy), so their grazing behaviour leads to increasingly large areas where the grass is long and rank ('roughs') and areas that are overgrazed ('lawns').

Over time, the roughs will become dominated by weeds and the whole paddock has an imbalance of nutrients (i.e. the paddock is 'horse sick').

One or more of the following management practices can be used to prevent the deterioration of pasture grazed by horses.

Figure 2. 'Horse sick' paddock (courtesy of Jane Myers).

Rotational grazing

Sheep and cattle will graze the areas that are avoided by horses. Sheep will tackle most weeds found in horse paddocks and although neither cattle nor horses will eat pasture around their own droppings, they will eat pasture around each other's droppings.

Horses graze with their teeth; that is, they bite the grass off. Cattle tear the grass off with their tongues. Therefore, horses can graze close to the ground, whereas cattle need longer grass. Because of this, it is preferable to graze cattle in front of horses in a cattle–horse rotation. If horses are grazed first, however, the cattle will eat down the long grass in the 'roughs'.

Alternating cattle or sheep with horses is also a worm prevention strategy because, with one exception, the intestinal worms of horses cannot survive in cattle or sheep.

In larger paddocks (more than 10 hectares) it is probably more practical to graze the horses with the sheep or cattle, but the horses may chase the other livestock. Very occasionally horses will attack calves, lambs or sheep and some horses will allow calves to chew their tails.

If there are only horses on the property, having several small paddocks rather than one large one will allow for paddock rotation, which will improve pasture growth and assist with parasite control.

Harrowing

Whenever a paddock is rested it should be harrowed to break up the dung pats and expose the worm larvae. The best time to harrow paddocks is in the hot, dry months (summer in southern Australia and New Zealand, winter in northern Australia). If the weather is warm and moist, the larvae are not killed and the paddock will need to be rested for at least four months before allowing horses back on it. Harrowing returns to the dung to the soil as a fertiliser. The end result is increased grazing area in the paddock due to more even grazing.

Mowing or slashing

If cattle or sheep are not available to eat down the rank growth, then the paddock should be mowed or slashed periodically, particularly during the spring period of rapid pasture growth. Ideally, the pasture should be mowed to a height of 8-10 cm, which keeps the pasture short over the whole paddock, encouraging the horse to graze the whole area and preventing dominance by broadleaf weeds and thistles. However, some broadleaf weeds such as cape weed, which grows outward rather than upward, may not be effectively controlled by mowing.

A further advantage of mowing is that it helps to remove old, tough, dry grass and encourages the growth of young green grass. Young immature plants have more leaves than stems and are more nutritious because there are more nutrients in the leaves than in the stems of plants.

Manure removal

Removal of manure will help control parasites and increase pasture availability, thus allowing an increased stocking rate. It is essential for very small paddocks or those that cannot be rested and

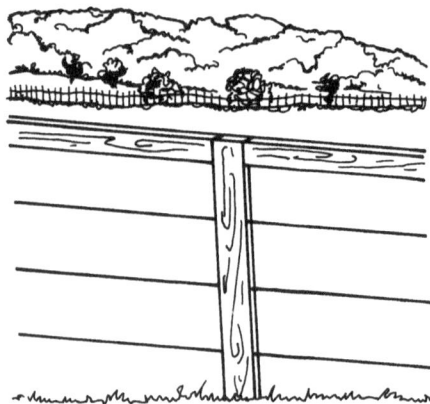

Figure 3. Horses need highly visible fencing.

harrowed. However, removing manure is labour intensive. On a large property it can be done with machinery and an operator.

FENCES AND GATES

Safety, cost and appearance are important considerations when designing the fencing for horses because many horses are injured by fences and gates. Unfortunately, there is not a perfect fence for horses. Even the costly timber post-and-rail fencing so favoured by some can cause horrendous injuries if horses charge through it.

Fence height

The recommended height for a horse fence is 1.2–1.4 m. Fences for yards must be higher, approximately 1.7–1.8 m. The spacing between the rails, wires or pipes should be 200–300 mm so that the horse cannot get its head through the fence. The bottom rail, wire or pipe should be 300 mm from the ground to reduce hoof injuries. Any higher than 300 mm, sheep will get through the fence and foals may roll underneath if they lie against it.

Fence visibility

The colour of the fence is not important because a horse can see a dark fence just as easily as a white fence. The main consideration is that the fence presents a visible barrier to the horse. Obviously, a fence with widely spaced posts and only a few wires will be difficult to see, so a white PVC sighting wire or rail, or coloured discs attached to the wires, will make it more visible.

Types of fence

Post-and-rail fences

The most expensive fence is the traditional timber post-and-rail because it is costly to erect and maintain. It can be painted white or stained black with creosote, but a white fence needs to be painted more often.

The boards of a post-and-rail fence should be placed on the inside of the posts to prevent the horses knocking their hips or other body parts on the post, but this is not possible if only one fence separates the paddocks.

Mesh fences

The gaps in mesh need to be small enough that a horse's leg cannot fit through. The stronger variety of chicken wire mesh can be used with plain wires above it (e.g. in foal paddocks). Ringlock fencing (including dog fence) is dangerous because a hoof will fit through the mesh. If you use or have that type of mesh you will need to have an offset electric wire (see later) to keep the horses away from the fence.

Cable fences

Cables makes a very strong, safe fence for horses if they are well strained (usually with 'turnbuckles' at one end) and the connections are covered. A top rail of either wood or pipe is needed.

Flexi safety fences

Flexi safety fencing looks like post-and-rail fence, but is safer and cheaper. It comprises PVC panels

Figure 4. A post-and-rail fence, 1.3 m high
(courtesy Jane Myers).

moulded over two strands of wire that are 10 cm apart. These fence panels can be strained up to the usual posts and can comprise all or part of a fence (e.g. three panels or just a top panel) to form a strong flexible barrier. It usually has a long life.

Wire fences

Wire fences are the most common type of fence for horses in Australia and New Zealand because they are comparatively cheap to erect, maintenance costs are low and they are relatively safe if built properly.

Barbed wire should not be used for fences for horses. It does not stop the horse from leaning on the fence (only electric wiring will do that) and it is expensive, harmful and dangerous to handle. It must never be electrified.

Figure 5. A pipe fence
(courtesy of Stuart Myers).

Wire

Use high-tensile plain wires, such as 2.80 mm (11 gram) or 2.50 mm (12.5 gram), which are strong and reasonably priced. Soft wires, such as 4.00 mm (8 gram) or 3.55 mm (9 gram) wires, can be used but they are more expensive, generally not as strong as high-tensile wire and they stretch and sag over time, which may cause horses to become entangled in the fence, unless they are regularly restrained.

Droppers

Droppers keep the wires evenly separated over a long span of fence (see Figure 6), thus reducing the number of posts required. They can be made from wood, steel or plastic.

Wooden droppers made from peppermint gum are becoming less common, probably because of their weight and wind resistance. Sawn hardwood or treated-pine droppers are available, either as plain boards or with pre-cut holes or slots. Wooden droppers are usually attached with soft wire, staples or pre-formed wire clips. Ironbark droppers are commonly used for electric fences. Pre-formed light-gauge galvanised steel droppers are available with slots or hooks that allow them to be attached to the wire at predetermined heights. They have a life expectancy of more than 30 years. Black or white plastic droppers are easy to use and cost-effective.

Erecting a wire fence

The cost of wires and posts varies greatly so compare prices before buying. The length of a coil of wire will vary, so it is the cost per kilometre that is important, not the cost per coil. High-tensile wire is cheaper than soft wire on a per-kilometre basis.

Good tools make the job of erecting a fence much easier, so buy a good pair of fencing pliers, a wire spinner to feed out the wire and a fence strainer that shows the amount of tension.

Use the wire spinner on a trailer when running out plain wire. Tie one end of the wire to one end post and then drive to the other end of the fence to minimises the damage to the zinc galvanising and lengthen the life of the wire. Strain the wire to the recommended load using a load-measuring strainer. Once strained, the wire can be stapled to timber posts or tied on with wire if steel posts are used.

It is poor practice to drag the wire over the ground or thread it through holes in the posts. Not only is the zinc coating scraped off, but holes can harbour water, increasing the chance of corrosion. Holes are also funnels for heat during fires and will promote burning of the fence. For the same reasons, do not use thread-on droppers; their use also needlessly lengthens the construction time.

Using a staple driver to fix the wire will save damage to your thumbs from using a hammer. A fencing salesperson will advise you on the

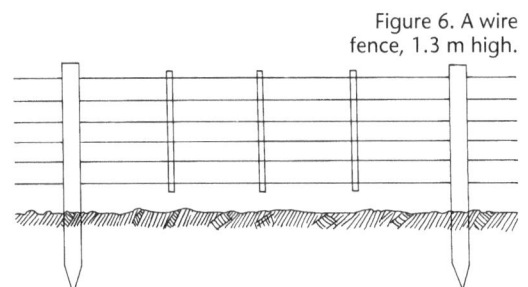

Figure 6. A wire fence, 1.3 m high.

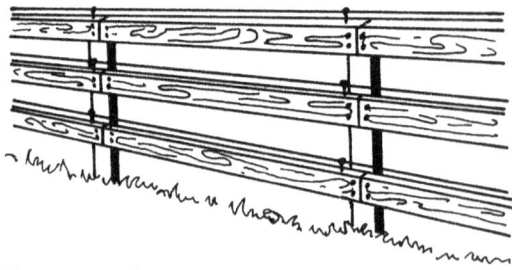

Figure 7. An electric wire on top of wooden rails.

right staple for the job. Angle the staples downwards and at 45° to the vertical. Stagger them as you work down the post to avoid splitting the wood. Do not knock the staples in all the way because then the wires cannot later be tightened.

If possible fix the wires to the northern or western side of the post. In the event of a grass fire, the northerly wind will tend to fan the heat away from the wire, increasing the possibility of it being salvaged.

Electric fences

For many years, rumours about electric fences have circulated through the horse industry, but most are unfounded; for instance, mares will not abort after touching an electric fence. Electric fences pose no more danger (and probably much less) to horses than traditional styles of fencing.

Electric wires can be set on top, or close to the top, of wooden rails of fences or in yards to protect them from being chewed, leant on or rubbed against by the horses.

Figure 8. An offset electric wire using poly pipe outriggers (courtesy of Jane Myers).

Introducing horses to paddocks that have been traditionally fenced for sheep and cattle is often of great concern to horse owners because fencing that incorporates barbed wire or mesh is often associated with injuries to horses. Poor visibility of the fence may also be a factor. For these reasons, combined with the high cost of the traditional types of horse fences that form a strong physical barrier, there is increased interest in using the existing fences with the addition of electric wires. Horses will be much less likely to touch a fence protected by an electric wire, let alone lean on it, and therefore there is less chance of the horses damaging either themselves or the fence.

An offset electric wire

The cheapest and quickest method of making fences horse-proof is the addition of a single electric wire, held 300–380 mm away from the fence by outriggers in the form of wire loops, fibreglass rods, poly pipe (PVC) lengths, or hardwood brackets. Some outriggers can be attached to the wires of the fence, while others are drilled into, nailed or bolted to the fence posts.

The outriggers can be fitted to the fence at any height to suit the size and breed of horses on the property. Some owners prefer to fit the outriggers at full fence height (approx. 1.3 m), while others have found a wire placed 0.8–1 metre above the ground to be equally effective. Those who are likely to run cattle or sheep in the same paddock should place the offset electrified wire at an effective height for those animals, which is usually lower than that for horses. It may be necessary to have two wires at different heights.

Figure 9. Wire outrigger can be attached to either a post or the wire.

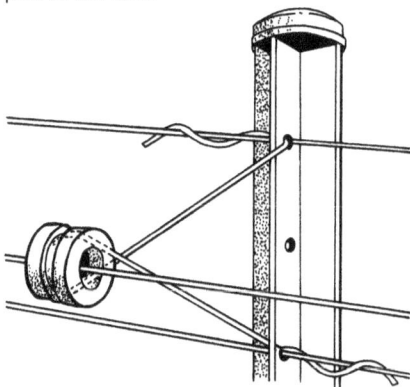

Insulators

The efficiency of an electric wire depends mainly on its ability to carry the voltage over the length of the fence. Any large leakage of current can make the fence ineffective. There are porcelain or plastic insulators available, so ask the advice of your fencing salesperson. Metal outriggers must have an insulator. Cheap, effective insulators can be made by cutting lengths of 37 mm polythene pipe. Hardwood outriggers, made from ironbark, are good natural insulators if they have not been treated with salt-based preservatives.

Live wires and earth wires

The effectiveness of a single electric wire depends on an electrical circuit being made when the horse comes in contact with the wire. The horse may not get a 'kick' from the fence if the soil is very dry, or if the earthing system at the energiser is insufficient. In the drier inland areas, it may be necessary to incorporate at least one earth-return wire with a live wire to get satisfactory results from a permanent electric fence.

Testers

An electric fence tester is essential, and if you can afford it a digital readout tester is the best because it will show you how much voltage is in your fence-line when tracing partial short-circuits. When designing your electric fence layout make sure that you incorporate a number of cut-out switches so that you can isolate sections for working on or for fault-finding. Some testers have an in-built arrow system that shows you in which direction the fault is from the tester. Others have a device for stopping and starting the current at the point where you are on the fence line so that you can get through the fence or repair it.

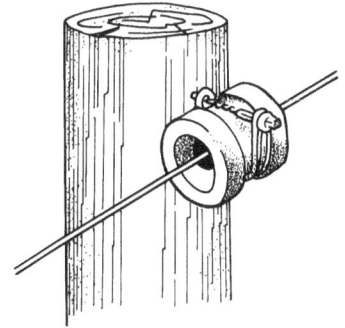
Figure 10. Porcelain insulator.

Introducing horses to electric fencing

Although it is true that electric fences are not dangerous to your horses, you must be careful when introducing a horse to them. Horses quickly learn to avoid an electric fence, but it is wise to avoid introducing a group of horses, particularly weanlings, to electrified wires. A horse that is new to the property and not familiar with electric fences should be put in a paddock with horses of a similar age that are used to this type of fence.

Posts

The diameter of a post has very little effect on its stability, but the method of insertion does. A driven post is more stable than a post set in an over-size hole by ramming.

Posts should be set in the ground to a depth of 0.75–0.9 m, the lesser depth being for heavy soils. The further the post is in the ground the less likely it will be pushed over.

Wooden posts

Treated wooden posts outlast any non-treated posts. Copper chrome arsenate (CCA), which is used to treat pine posts, gives the longest life expectancy, though it is not as fire-resistant as creosote, which also prolongs the life of wood posts. CCA colours the post green, whereas creosote gives a dark-brown oily finish.

Ironbark, red gum and grey box are the hardwoods that make the best untreated posts. However, the life expectancy of untreated wooden posts varies greatly, depending on the soil type and environmental factors. Any wooden post will burn in a hot bushfire, regardless of the preservative treatment.

Figure 11. Plastic insulators.

Split posts are hard and durable, although their life expectancy does vary between the species. Most will last 20–30 years. Round posts can be treated under pressure with preservatives. The most popular species of eucalypt for fencing is black peppermint (E. amygdalina), which also splits well. Browntop stringybark (E. oblique) and whitetop stringybark (E. delagatenis) are also popular. Eucalyptus sideroxylon, commonly called ironbark and used for posts and droppers for electric fences, is marketed as Insultimber™ or Lecwood™. It has good insulation characteristics and does not need extra insulators when used in electric fencing.

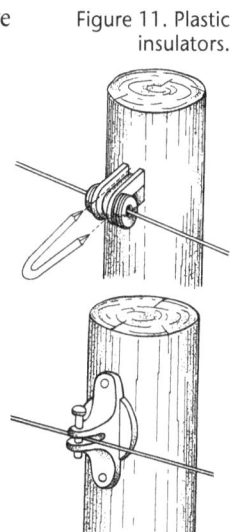

Radiata pine can be used for most types of fence. It is normally treated under pressure with preservative salts, giving a minimum life expectancy of 35–40 years. It is lightweight, cheap and long-barbed staples and insulators can be attached easily.

Steel star posts

Steel posts are fire-resistant and are recommended in areas where bushfires are likely. They are also good for temporary fencing because they are light and easy to install. However, for safety reasons, it is essential to cap the posts with either steel-pipe or plastic caps and offset electric wires should be used to keep horses away from this type of fence.

Concrete posts

Only pre-stressed, reinforced concrete posts are recommended, but their high cost often does not warrant their use except in regions affected by termites (white ants). They have been known to explode in bushfires and even a small crack allows water to corrode the reinforcing, causing early failure.

Fibreglass posts

Fibreglass posts are used for electric fencing because they are self-insulating. The wire is attached to them with stainless steel clips. Fibreglass posts can bend without breaking and are stabilised for ultraviolet light. They are available in a range of lengths, with 8 mm diameter posts being the most suitable for temporary electric fencing and 10 mm diameter posts for permanent fences. Temporary fibreglass and plastic posts (tread-in types) are used for temporary electric fencing.

End assemblies

End assemblies are placed at each end of a fence line and each side of a gate. They are the foundations of a fence; a good end assembly is essential for a good fence. The end assembly must be 100% effective or the fence will fall over.

Steel end assemblies are available and they are preferred in areas of high fire risk. They are more expensive than wooden ends, but are simpler to erect.

There are a few styles of timber end assemblies, but the most successful is the horizontal stay 'H' assembly. Ideally the length of the top stay rail should be 2.25–2.5 times the height of the posts, with the diagonal wires tight and wound around the base of the post at the furthest end of the fence line.

Gates and gateways

Horses tend to stand and paw at gates so they must be safe. Gates should be meshed so that either a horse cannot get its foot through the mesh or the gaps should be large enough for a horse to remove its foot easily.

Locate gateways in the paddock corner that is nearest to the yards to reduce travelling time. Ideally, the gateway should be located one fence panel (horizontal 'stay' length) back from the corner, especially if the fences are electrified, to allow for ease of moving horses in and out while other horses are present. The gateway should be wide enough for a vehicle to pass through, open both ways, be approximately 15 cm above the surface on level ground and be close to the post, with the closing latch or chain placed near the top of the gate.

A common type of accident occurs when a horse rubs its head on the post holding the latch. Its head slides down between the post and the gate, jamming the neck and instead of lifting its head to release itself, the horse pulls straight back, causing injury to the side of the face that often results in paralysis, or even a broken neck. Make sure there is no gap for the horse to be able to get its head stuck.

The latching chain and hook should be placed so that they are unlikely to catch a horse as it walks through the gateway, nor cause injury to a horse standing next to the gate. The fittings should protrude as little as possible.

BUILDINGS

Yards, crushes and riding arenas

Yards are useful, indeed necessary, for the many tasks involved in horse management (e.g. teasing, holding horses for close inspection, feeding, training or breaking). The design of any set of horse yards must be based on two main factors, efficiency and safety. Consider the number of horses using the yard at any one time and the reasons for putting them in the yards. In addition to being well drained, the surface of the yard needs to suit the purpose for which it is to be used. Sawdust, sand and fine gravel are three possibilities. Yards in which horses are to be worked are usually best surfaced in sawdust or sand.

Holding yards

Many horse owners need yards for holding large groups of horses and the design usually includes a teasing yard, crush and a race system.

The yards must be well drained. Do not build yards in an area that will become boggy after rain. Surface the yards with river gravel on a hard base, which will remain firm when wet.

Figure 12. Safe fencing for yards.

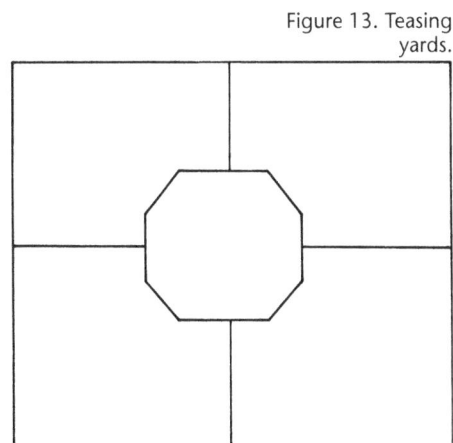

Strong fences are also important. Many studs use post-and-rail fences, which are excellent for retaining the horses, but they tend to splinter and cause injury if horses run into them. Galvanised piping is excellent for fencing yards, but do not use it for the crush. Horses that are restricted on both sides may get their legs tangled in open-sided races or crushes. Weldmesh™ (with a grid that is small enough that the horse cannot get its leg through) or chain mesh are excellent as yard fences if they have adequate support. Weldmesh™ should be topped with a timber baulk rail and at the base the mesh should be either buried or secured tightly with high-tensile wire. Alternatively, use roll-top panels of Weldmesh™. Chain mesh (or even chicken wire) needs much more support, but has the advantage of being more flexible and ideal for young stock. A combination of a top row of piping, securing the mesh, and high-tensile wire in the middle and at the base will keep this type of yard fence in strong condition.

Make sure that gateways are free of any projections and are wide enough to prevent horses pushing through. Gates must swing both ways and should lie flat against the fence when open. When closed there should be no gaps between the fence and the gate.

Teasing yards
The best design for teasing yards is one in which there does not have to be any handling of the mares. The teaser should be kept in a small yard that is fully enclosed so that he cannot injure himself if he strikes at the mares. The teaser yard should be in a central position in the holding yards so that a number of mares can be teased at one time. Separating large groups of mares into smaller groups has the advantage of letting shy or timid mares approach the teaser without fear of attack from other mares. Laneways from the yard to the paddocks should also allow for minimum handling of the mares.

Handling yards
Round yards

Figure 13. Teasing yards.

A round yard can be used for handling, dressage or intensive training. The diameter of a training yard should be approximately twice that of a round yard used solely for handling.

It is best to build your round yard from hardwood, and cover it with a roof if possible. A suggested size for handling yards is 11.5 m in diameter with walls that are 2 m high.

There are many uses for a covered round yard. Mares and foals can spend their first day 'outside', especially in inclement weather, it can be used for post-surgical turnouts, turnouts for horses off the track before they go to paddocks, stallions can be exercised in them, and if tanbark is used as the surface material the yard makes an excellent breeding shed etc.

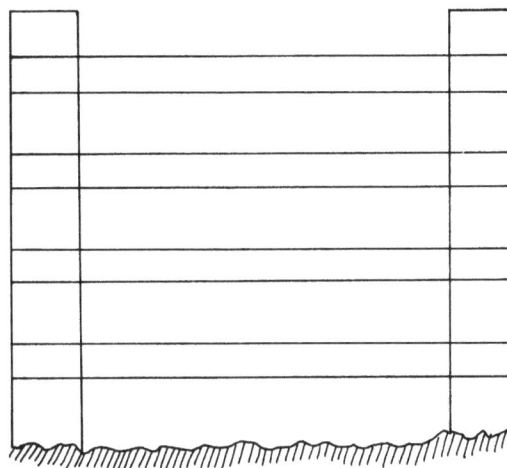

Figure 14. Covered
round yard
(courtesy of Annie Minton).

The walls can either be solid, which some trainers prefer because it prevents the horse from being distracted, or post-and-rail. The height of the walls depends on the type of horses being worked. To be safe, the top, baulk rail should be at approximately waist level when the rider is mounted. Make sure the posts are the same height as the top rail as projections can be dangerous. On a post-and-rail yard the bottom rail should be no less than 0.25 m from the ground to reduce the risk of leg injuries if horses roll close to the fence. Rails must be on the inside to prevent riders (and horses) from banging into the posts.

A novel and less expensive method of constructing a round yard is with old car tyres. These are stacked around a stay pole and filled with sand. They are a good buffer, solid and will last indefinitely if the yard is roofed.

Figure 15. Open
round yard
(courtesy of Gary Blake).

Solid yards

A solid yard can be any shape you think is appropriate, including round. Solid timber sides or post-and-rail fencing is preferable. As with the round yard, the ground surface should be soft to prevent injury to the horse and have good drainage.

Cover at least one post with rubber matting and use this as the training tying post. The panels on either side of the tying post should be close-boarded with no gaps between the rails to prevent leg injuries.

Crushes

A crush is a narrow yard that can safely confine a horse when needed, mainly for veterinary examination of breeding mares or for handling and/or restraint.

Labour costs often become an important factor when many horses have to be handled. On many larger properties horses are now moved from yard to yard as would cattle. With this type of set-up, make sure that the crush is incorporated into the design of the yards and is easily accessible. The site for the crush should be well drained and, where possible, both the crush and the surrounding area should have a base of concrete or hard gravel. The crush must be solidly fixed to the ground and not be able to be moved by the horse.

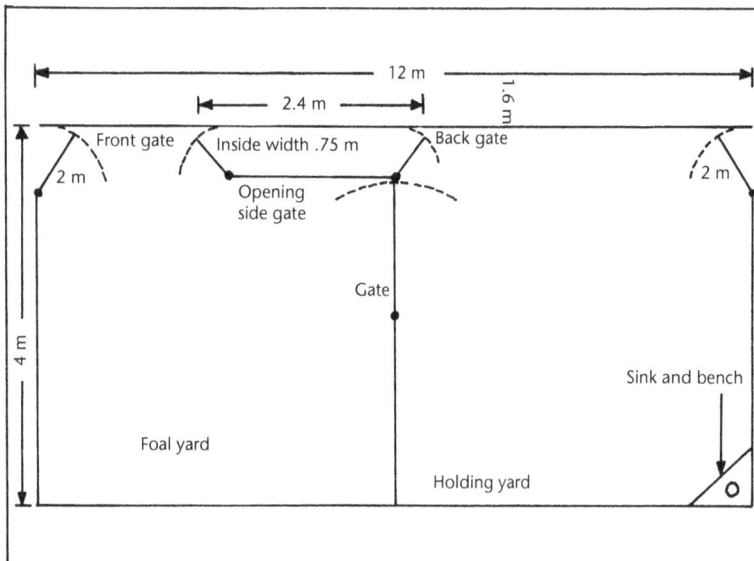

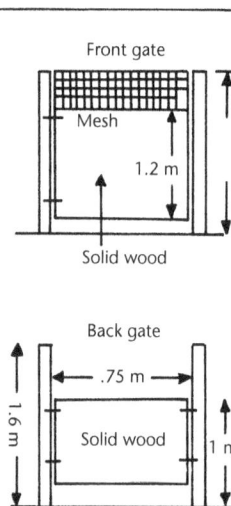

Figure 16. Design for a multipurpose crush that can easily be incorporated into existing yards. It is suitable for breeding examinations by a veterinarian if the mares can be led into the crush. It can also be used as a handling crush if the horse can be led into it.

Figure 17. Side, back and front views of a multipurpose crush. The design of both sides is the same so that almost any section of the horse's body can be treated or handled. The side gates open completely for quick release of the horse in an emergency.

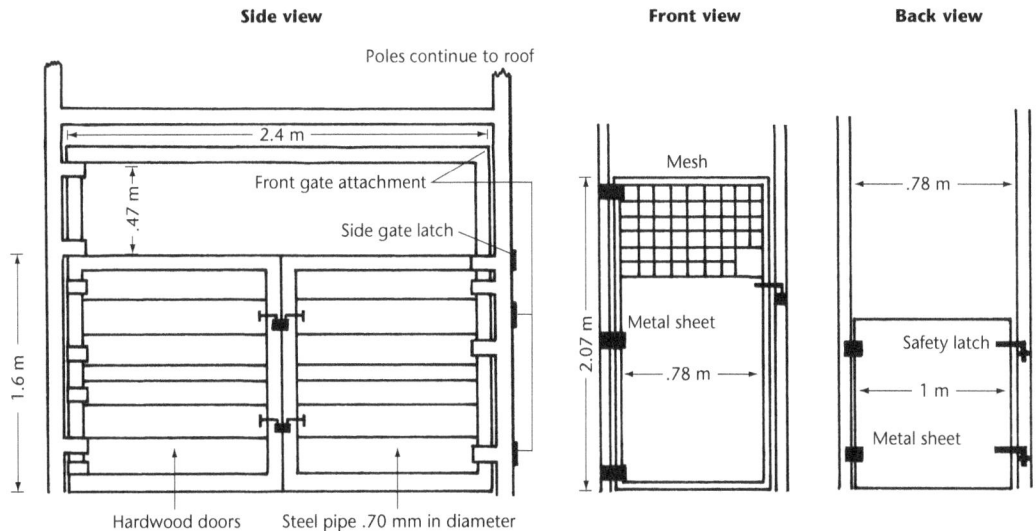

Side view

Poles continue to roof

2.4 m

Front gate attachment

Side gate latch

.47 m

1.6 m

Hardwood doors   Steel pipe .70 mm in diameter

Front view

Mesh

Metal sheet

2.07 m

.78 m

Back view

.78 m

Safety latch

1 m

Metal sheet

The crush must be safe for both horses and handlers, so it must not have any projections or dangerous sections that may cause injury. Crushes should not have any overhead bars.

The dimensions of the crush will depend on the size of the horses for which it is most commonly used. The dimensions of the crushes shown in Figures 16–18 are for horses that are 14 hh (hands high) or taller.

A crush for breeding or veterinary examinations

The crush should have:

- clean water supply nearby, and hot water readily available
- a low back gate so that the vet's arm will not be crushed if the mare 'squats' down.
- a work bench within reach when standing behind the mare
- a nearby power source
- lighting, but able to be darkened for eye examinations/reading of instruments etc.
- provision for the vet's car to park nearby so that equipment is easily accessible
- a nearby safe foal-holding pen to keep the foal in sight of the mare
- if the crush is not inside a building, then at least a roof over the whole crush so that it is possible to work when it is raining
- provision for fitting weighing scales inside the crush

A crush for handling or restraint

A crush that is used for restraining horses rather than for breeding examinations can be simpler in design. It needs high back and front gates so that the horse cannot jump out. The sides should be either completely solid to the ground or made from rails with a bottom rail no less than 150 mm from the ground. The gaps between the rails should be wide enough that if the horse puts a leg between the rails it can withdraw it without injury. Solid sides are usually better.

Using crushes for previously unhandled horses is dangerous. Unhandled horses enclosed in a crush will panic and often injure themselves. Small yards with high post-and-rail fences are safer until the horse has been handled.

Figure 18. A crush for breeding or veterinary examinations.

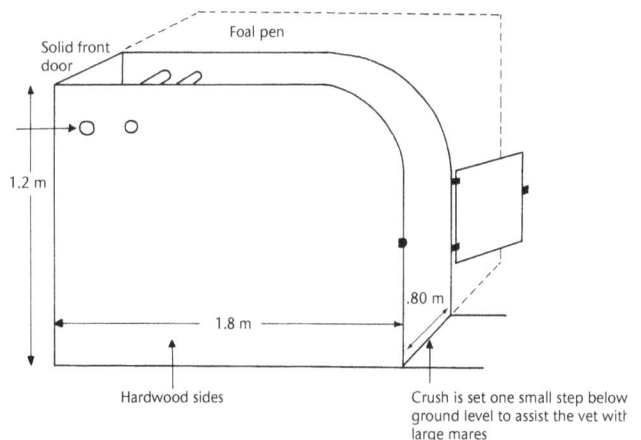

Foal pen

Solid front door

1.2 m

1.8 m

.80 m

Hardwood sides

Crush is set one small step below ground level to assist the vet with large mares

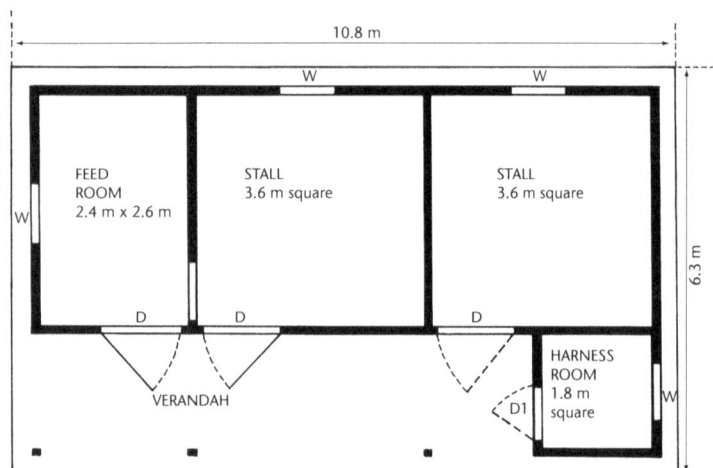

Figure 19. Plan for a stable for two horses.

Riding arenas

The ideal situation is to have your own riding arena, but check with your local council prior to proceeding because considerable earthworks are often required, which could have an effect on water catchments or protected trees or have other environmental implications. Neighbours should also be consulted. Council regulations may limit the arena to 'personal use only', prohibiting its use for commercial purposes such as coaching.

A standard dressage arena is 20 × 60 m, but 20 × 40 m is an option if space (and money) is limited. A number of surfaces are available, ranging from shredded rubber compounds to sand and/or hardwood sawdust. The selection of surface will depend on:

- availability of funds
- access to the base material
- exposure of the arena to wind and sunlight. High prevailing winds require heavier materials, whereas high levels of sunlight and heat (e.g. in Queensland) will exclude rubber composite surfaces that are more prone to UV degradation.
- availability of water. Dust can be a major problem with some surfaces, which may alienate your otherwise compliant neighbours, and the arena will need to be watered during the drier months.

Installing an arena is expensive. The total cost will vary according to the extent of the earthworks, surface preparation and the type of surface required. A 20 × 60 m arena may cost at least $25,000 for professional installation. Beware of 'do-it-yourself' arenas. Many do not drain correctly, have dangerous soft patches and need constant maintenance. Use a construction company that specializes in riding arenas to guarantee both the surface and the base.

Stables, exercise yards and paddock shelters

Stables

Most horses that are being prepared for showing or for sale, or being trained for competition, may need to be stabled. Sick horses may also need to be housed. Stables that are well-planned and well-designed make horse care much easier and more enjoyable. The floor plan for a stable for two horses shown in Figure 19 includes two box stalls, a feed room, and a tack and harness room. This design can be extended by adding more stalls. The floor area of each stall needs to be at least 3.6 × 3.6 m (i.e. at least 12 square metres for horses or 9 square metres for ponies).

Figure 20. Barn-style stables (courtesy of Annie Minton).

There are many manufacturers that will build the stables to your specifications (Figure 20) or you can build them yourself if you have the skills and the time. Another option is to have a shed erected (or buy a kit shed and erect it yourself) and then build the stables inside the shed (Figure 21).

Walls

It is also possible to buy prefabricated sections that will turn your shed into stables or you can make the interior walls yourself. These walls can be made of various materials such as concrete blocks or timber.

Concrete blocks eliminate the need for lining because they can withstand rough treatment. Alternatively the walls can be lined with rubber belting/matting to minimise injury to the legs of horses that kick out.

Any tin/Colorbond® exterior walls must be lined with timber kick boards. These should be 152 × 25 mm planks laid to a height of 1.3–1.5 m, otherwise a horse can kick through the exterior wall and injure itself.

The inside walls of the stall should be smooth and free of sharp corners or protruding objects. They should be flush with the floor so the horse cannot catch its feet under the lining.

If possible have a partition between two of the boxes that can be removed if more space is needed, such as when a mare and foal need to be stabled.

Figure 21. Stable and tie up stalls in an L-shaped, open-fronted shed (courtesy of Stuart Myers).

Doors
Doors should either open outward or preferably slide. Sliding doors must be suspended from the top. It is safer if sliding doors are fitted to the inside of the stall, to prevent the possibility of a leg being pushed through any gap between wall and an outside-hung door.

Floors and bedding
The floors of the stables can be dirt, sand, brick laid on sand (but do not fill the joins with concrete) or concrete. A good arrangement is to have the aisles concreted and leave the stalls with earth floors. When having concrete laid make sure it has a rough finish because smooth concrete is dangerously slippery for shod horses. Concrete stable floors are hard to keep dry so they should have a slight slope to allow water and urine to drain away. Rubber matting laid over a concrete or brick floor in the stalls will significantly reduce the amount of bedding required.

Some form of bedding is essential, no matter what the floor type. The main types of bedding are straw, hard or soft-wood shavings, or sawdust. Straw is easier to handle, but is not as absorbent as shavings or sawdust and must be changed more often. Make sure that shavings are not dusty. Pine sawdust has a pleasant aroma, and a pressure-dried version (although more costly) will absorb more moisture. Shredded paper has also been used with some success. When cleaning the stall, daily remove all dung and bedding, sweep the box clean and sprinkle slaked lime on the damp floor to reduce odours.

Feeding and watering
Automatic watering systems are best for supplying drinking water, but are expensive; placing a bucket of water in a corner of the stall is a cheaper alternative that will also allow you to monitor the horse's water consumption. Make sure the horse cannot tip the bucket over, by placing it inside an old tyre for example.

Figure 22. Automatic watering system (courtesy of Annie Minton).

Feeders should have rounded corners and smooth edges for easy cleaning and safety, and should be deep enough to prevent feed being nosed out by the horse and wasted. The triangular feeders that fit into a corner are excellent. Be aware that some horses will rub on the feeders and dung into them if they are at a convenient height and location.

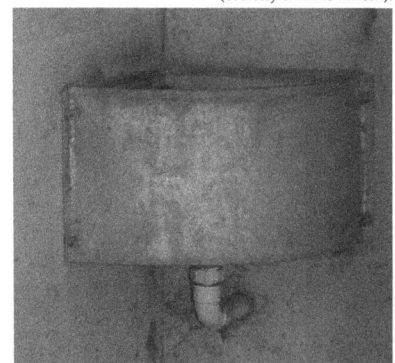

Feed storage
Storage bins for feed must be located in a convenient position and must be accessible for removing the feed and replenishing the bin when it is empty. The bins must also be vermin- and waterproof, otherwise mice, rats, insects and birds will eat the feed and can bring disease, and water will soil the feed and make it unpalatable.

The amount of storage space needed will depend on the number of horses to be fed, their feed requirements and how often the feed supply is to be replenished. Do not have feed bins that are so small they must be refilled every few days. Neither have the bins so big that if you fill them up the feed becomes stale before you use it. Each feed bin should be emptied completely before being replenished so that feed at the bottom of the bin does not become stale. Alternatively, remove the old feed, put in the fresh feed and then place the old feed back on the top so that it is fed first.

Exercise yards

Exercise yards are necessary for stabled horses that are not turned out into a paddock every day. If possible build the yards so that the horse has access from the stable.

An exercise yard needs to be a minimum of 40 square metres per adult horse. The minimum width is the width of a stable and the length can be any distance that can be accommodated by the property. Although square areas require less fencing, horses like to run along fence lines, so for maximum use of a given area, several long narrow runs are best.

Plain board fencing is the most common form of fencing for exercise runs, and electric wires can be incorporated; however, in a very narrow run (i.e. 3.6 m wide) the electric outriggers will reduce the space further and make it dangerous for the horses to exercise themselves. Welded pipe is another suitable option for construction and tends to be safer if horses are on either side of a fence (such as runs attached directly to the stables). With this type of fence, electric wires are not needed and the horses can interact safely with one another over the fence.

Long, narrow runs should be laid out across a slope to minimise soil erosion. Generally, try to avoid slopes that are greater than 3%. The exercise yards should be located on reasonably level, relatively stone-free, well-drained soil. Footing is improved by spreading sand at least 5 cm deep over the existing soil. Sand should be added regularly as the previous applications become mixed with soil. Sand will also reduce the dust, mud and soil erosion that result from confinement of a horse in these areas.

Paddock shelters

Paddocked horses need shelter from hot sun and strong winds. Trees and hedges do the job admirably, but as horses are destructive, especially in small paddocks, vegetation may have to be protected from gnawing teeth.

If a shed or shelter is to be provided, an open-sided design is more acceptable to the horses, preventing them from feeling enclosed.

Horses tend to use a shelter more for shade and to escape flies in hot weather more than they will use it for sheltering from cold weather. Often the horses spend their time standing in front of the wall out of the wind and tend to ignore wet weather, unless it is driving rain.

MANURE MANAGEMENT

Figure 23. Paddock shelter (courtesy of Jane Myers).

Carefully collected animal manure was once the main source of fertiliser for crops. Today, most horse owners usually do not have enough land for crop production to use the manure their animals produce. Consequently, this valuable by-product is often considered as a waste or, at best, a disposal nuisance. Yet all livestock manure can be a valuable commodity if composted and spread on the paddocks, or made available to gardeners, landscapers and other plant growers in a useable condition.

Manure accumulation and composition

Each year a 500 kg horse can generate 8–10 tonnes of manure. If the number of horses in Victoria alone is estimated to be approximately 300 000, there is certainly a lot of horse manure being produced each year.

The composition of manure varies according to the type and quantity of bedding used, the age and function of the animal, the type of feed and how the manure is stored. Typically, 1 tonne of fresh horse manure, with bedding, would have a nutrient composition of almost 5 kg of nitrogen (as N), 2 kg of phosphorus (as P205) and 6 kg of potassium (as K20). Nearly half the quantity of these nutrients would be available to a crop during a growing season with a Spring application. The remaining nutrients will provide fertiliser in subsequent years.

Manure also contains other valuable trace elements. It improves soil texture and soil moisture retention, thereby reducing the need for irrigation.

Manure storage

A single horse will produce 300 cubic centimetres of manure each day. Bedding can easily bring the total volume of material that must be managed each day to 500 cubic centimetres per animal.

Although daily removal of manure from the premises is ideal, it is usually impractical and, contrary to popular belief; it does not eliminate fly breeding problems. If manure cannot be removed weekly in warm weather, provision must be made for proper handling and storage together with a plan for effective utilisation.

Decomposition of manure starts as soon as it is passed and the rate of decomposition depends on handling and storage methods. Ideally, horse manure should be kept compact and moist to prevent excessive loss of nutrients. Manure left in a loose heap loses nitrogen rapidly to the atmosphere in the form of ammonia.

A confined storage space of 13 square metres will conveniently hold manure from one horse for a year. Accumulation might be 30–60 cm in depth. Large storage areas should be well constructed and accessible for the use of machinery.

Locate the storage site(s) so that loading and unloading of a vehicle is convenient. Create a positive image by storing and handling manure as neatly and inoffensively as possible. A covered manure pit would be ideal. Grade the surrounding area to keep surface water from running over or through the manure and into streams or other surface water. Covering the manure pile will also prevent liquid leaking into the storage area.

Manure use

Using manure as soon as it is removed from the stable is desirable. If there is a nearby crop or market garden the fresh manure can be tilled in immediately, thereby saving nutrients and alleviating storage problems. Crops grown and harvested annually on 0.5 hectare of land can easily use the nutrients available in the yearly accumulation of manure from a single horse. Fresh manure is best for crops that have a long growing season because some of the nutrients are not as readily available to the plants as the commercial fertiliser nutrients.

Fresh manure should not be spread on pasture grazed by horses because it will increase the risk of parasite infestation. This risk can be reduced, however, if the paddock is spelled for four months after the manure is spread or is grazed by other types of stock for four months.

Manure left exposed will lose nutrients rapidly. Manure left exposed on sloping surfaces is subject to erosion, possibly contributing to water pollution. Turning the manure under the soil immediately after spreading will reduce the loss of valuable nutrients and the risk of erosion.

Never spread manure on sodden earth and never spread or store it on land subject to flooding.

Composting manure

Most horse owners do not have access to enough cropping land to enable them to dispose of their fresh horse manure. One solution is to compost the manure alone or with other materials such as leaves and lawn clippings that will yield organic matter in a form similar to potting soil.

Decomposition of manure and bedding under composting conditions makes the nutrients more available to plants. It changes the organic matter into substances that more readily form humus in the soil. Availability of phosphorous is increased and many weed seeds that might be present are destroyed. All this means that manure that has been composted generally has a better sale value.

Composted manure can be spread on paddocks to return nutrients to the soil that are lost by grazing. To compost manure, you must pile it properly, keep it moist, and turn it over several times for one or two months. Various techniques can improve and hasten the composting process. Processing methods can be kept quite simple, or be quite sophisticated, depending on the desired condition of the end product and the time needed to complete the composting process.

Marketing manure

Correctly composted horse manure can be marketed to home gardeners, nurseries and crop farmers. Nurseries are the most likely customers for large volumes of less-than-completely composted manure and tend to prefer shavings as the bedding type. Home gardeners are a good outlet for smaller quantities of composted or aged manure and many gardeners would welcome bulk delivery of uniformly composted manure at competitive prices.

FIRE PREVENTION

Fire prevention should always be considered in the design of a horse facility. Whether your horses are out to pasture or stabled, there are number of precautions you can take to keep them as safe as possible from fire. A fire evacuation plan must be drawn up, with the most suitable area(s) for access used for horse holding. Properties should be designed so the livestock will be safe under normal circumstances.

Although all trees will burn eventually, they do offer some protection by reducing the wind speed and thus reducing the speed and intensity of the fire. Trees absorb the radiant heat, protecting animals and buildings on the lee side. They also catch airborne cinders, stopping them from flying into other areas and starting spot fires.

Most fires in southern Australia move on a north-west wind, so shelterbelts should be planted at right angles to the wind. They should not be too close to buildings or yards and they should be planted so that 50% of the wind can pass through; solid shelterbelts do not slow the wind as much because the gusts hit them and jump over them whereas permeable belts slow the wind as it goes through.

For a shelterbelt of trees to be useful as fire prevention or retardation, it should be at least 20–25-fold longer than it is high and should consist of flame-resistant varieties of trees. Local conservation departments will provide you with the names of useful shelter belt trees for your district.

Areas around water troughs are usually worn bare by stock, so place the water troughs where that will be of most help in a fire, on the windward side of the stables and other buildings, both protecting them and leaving a clear area in which the stock can stand in dangerous times. Planting fire-resistant shade trees in these areas will encourage stock to rest there on hot days, helping to keep the area free of grass.

Hedges of plants such as privet, boobialla and pittosporum planted no closer to buildings or yards than 6–10 m can act as radiation shields. Most native shrubs, however, are too flammable to put close to buildings, and even neatly trimmed cypress hedges are extremely hazardous because of the dead clippings that accumulate in the centre of them.

One of the traps a horse owner can fall into is to attempt to beautify the area around stables by planting gardens close to walls and fertilising them with straw and manure. Both straw and manure burn strongly if given the chance so if you do use these or similar garden mulches, dig them into the ground so that they cannot burn.

In high fire danger weather make sure the horses can be quickly and easily moved to the safe areas. Stock yards are often unsafe areas, especially if they are timber. More often than not, horses will be better off in a bare area such as a small paddock so that they can move about a little.

Always equip your stable yard with the following:

- an alarm system that is regularly tested
- appropriate fire extinguishers, also regularly tested
- a hose permanently attached to a tap, with sufficient length and pressure to reach all buildings
- prominently displayed 'no smoking signs'.

Flammable material such as petrol, kerosene or paint should never be stored near the stables and never burn refuse close to the stables. Large quantities of hay should not be stored in or adjacent to stables.

If the stable is a barn-type structure, make sure there is more than one exit, and that they are as far apart as possible. Exits must be kept clear. Do not park manure skips or vehicles in the aisles. Make sure that all the stable doors will open and always leave headstalls in a designated location so that you will be able to find them quickly and easily in an emergency, even if it is dark.

Make sure that everyone associated with the stables is instructed in fire drill. Display a notice telling where a telephone is and the telephone number of the fire brigade.

If possible get a representative from your local fire brigade to come out and advise you of potential dangers on your property. Do this before the bushfire season.

Electrical fittings

Many fires are started through electrical faults. It is your duty to reduce the risk of an electrical fire.

An electrical contractor must fit any electrical installations. Make sure that all cables are away from areas within easy reach of the horses and that they are covered in pipe and are waterproof. Switches should also be well out of the reach of the horses.

Do not leave extension cables lying in hay or shavings. All coiled electrical cords generate heat when in use and a fire may be started by a short-circuit. Keep stables free from cobwebs. When cobwebs become covered in dust they make the flash point lower.

Do a monthly check for loose wires in plug tops, and for cracked plugs. Check for signs of frayed, cracked or chewed leads. Also check for signs of overheating on plug tops and socket outlets.

What to do in the event of fire

Do not panic. First, call the fire brigade: if no one knows you are in trouble you will not get the help you need. Give your name, address and phone number.

Second, before you do anything else, dress yourself in fireproof clothing including a woollen jumper, long trousers and boots. Cover your hair, preferably with a woollen cap.

If you are alone, remove all horses before attempting to halt the fire, unless the outbreak is small, or you are there when it starts and you feel confident you can contain it. If you think that you can contain it, attack the fire with the appropriate type of extinguisher (e.g. water or foam for timber fires, and dry powder for electrical fires).

Remove the horses closest to the fire first and take them to the nearest paddock that is not affected by the fire. If a horse is reluctant to leave its stable, a coat, bag or rug over its head may help. If it will not lead out, it may back out.

Never turn all the horses out loose into the stable yard unless there is no alternative because they will greatly hinder the fire brigade when it arrives and the horses could get out on to the road. Close the doors behind the horses because in their panic they may try to run back into the stables.

URBAN AND SMALL HORSE PROPERTIES

These properties can have particular problems, but with careful planning and management horses can be kept successfully.

The urban horse property will have houses nearby and must be kept spotless if the horse owner is not to raise the ire of neighbours. In very small paddocks manure must be removed daily to reduce the worm infestation, reduce the fly population and preserve the grazing area.

Urban horse properties tend to be small, which usually results in a lack of grazing at certain times of the year. Once the grass has been eaten out the constant traffic of the horse's hooves will cause a lot of damage to the paddock. The soil becomes compacted and finally the only plants that will grow are weeds. Before it gets to this stage the horses should be put in yards for part of each day and fed hay. In very wet and very dry periods reduce the time in the paddock further to conserve the grass. With good management the horse can have limited access to grass all year round.

Neighbours may cause problems if they throw garden rubbish into the paddock, thinking that the horse will eat the plants. This can lead to unwanted plants (some poisonous) infesting your paddock. Garden rubbish may also have toxic sprays on it (such as herbicides). Lawn clippings can kill a horse because the horse swallows the fermenting grass without chewing (and therefore mixing it with essential saliva). You will need to educate your neighbours without alienating them. Presents such as composted manure for their garden can keep them sweet!

Fences need to be strong enough to retain the horse. They must also be childproof. Electric fences are ideal, but check your local by-laws before installing them. Adequate warning signs must be located around an electric fence in an urban area. Padlock the gate to prevent unauthorised use or theft of your horse.

WHERE TO GO FOR MORE INFORMATION

Books can be purchased from the Rural Industries Research and Development Corporation (RIRDC). Telephone 02 6272 4819 or email publications@rirdc.gov.au or go to the website www.rirdc.gov.au/eshop

www.cashdown.com.au/fencing.html is a website that has amusing and practical tips for fencing, particularly if you are combining types of livestock. It also has good drawings and instructions for end assemblies.

www.diamondmesh.com.au/linkssafety.html is a website dedicated to the safety (in particular fence safety) of horses.

Go to www.equinecentre.com.au for information about horses and bushfires. This website has a wealth of information on many equine subjects.

www.affa.gov.au is the website of the Australian Federal Department of Agriculture, Fisheries and Forestry.

www.maf.govt.nz is the website of the New Zealand Ministry of Agriculture and Forestry.

www.gov.au is the portal for gaining access to all Australian State and Territory government information. When you have found the webpage of your government, you can get information about the various agencies associated with the land.

CHAPTER 9

TRANSPORTING HORSES

INTRODUCTION

There are many factors to take into consideration before embarking on a journey with a horse because otherwise travelling can be traumatic for both the horse and the handler.

If transport is needed only once or twice each year, commercial carriers provide the safest and cheapest method of travel, even over short distances. If, on the other hand, you plan to be travelling on a regular basis, it is probably more cost effective to have your own outfit.

THE OUTFIT

If you have decided to purchase your own horse transport, you have a choice of either a truck or a car and float. Each has advantages and disadvantages.

Truck

Advantages

- Can carry up to 10 horses, but most trucks carry between 4 and 8 horses.
- Horses tend to travel better in a truck than in a float because the horse usually stands facing the side of the truck and in this position can spread its legs to balance itself.
- Smaller trucks that carry three or four horses can be driven using a car licence (check with your State or local transport department).

Disadvantages

- Trucks vary in size and weight and therefore are affected by different regulations. You will need to contact your State or local transport department to ascertain what size of truck you are able to drive with your current driver's licence.
- Registration and insurance for a truck will be more expensive.
- Older models of trucks may be expensive to maintain and will use a lot of fuel.

Car and float

Advantages

- A special licence is not required other than a full driving licence (check with your State or local transport department about towing on a probationary licence).
- Maintenance, registration and insurance for the towing vehicle are cost-effective because the vehicle is used for purposes other than transporting horses.

Disadvantages

- It is easier to lose control of a car and float combination while driving; however, careful driving plus experience will reduce the risk.

Figure 1. Vehicle
and float
(courtesy of Stuart Myers).

- Some horses are difficult to transport in a float, but not in a truck, usually because of previous careless handling and/or driving.
- Most vehicles can only tow a float carrying two horses; if you need to transport more than two, you will need a much larger towing vehicle.

Regulations

Your outfit must be certified as roadworthy and registered with the relevant transport authority. The registration labels for both the towing vehicle and the float must be prominently displayed.

In Australia, you must become familiar with the laws and regulations regarding towing for both your own State/Territory and the State/Territory you are travelling to/through. These can be obtained from your State transport department.

The weight of a float must be kept in the correct ratio with its towing vehicle to ensure stable and safe operation. By law, the weight of the float when loaded must be within the vehicle manufacturer's towing rating, which you will find in the vehicle's manual, or from a car dealer or the manufacturer. There are also different regulations throughout Australia on weight ratios, so you need to check with the relevant transport department.

As a guideline (these figures will vary according to the construction and brand of float, size of horses carried and the amount of gear/feed etc. that you put in the float), a the laden weight of a single-horse float is approximately 900–1100 kg, that of a two-horse float is approximately 1500–2000 kg or more, and a three-horse float is approximately 3000 kg.

Security and insurance

You will need to insure your float against accidents that may happen on the road (e.g. if your float became detached from your vehicle during the journey and caused injury to a third party). Insurance is usually available through a company that deals with caravans and trailers.

Check that the company that covers your car for roadside assistance also covers your float (otherwise the car will be towed to a garage, but the float will not). See the end of this chapter for more information.

Because floats are very easy to steal, you should also insure against theft. There are also security devices available, such as a cover and lock for the hitch of the float, or wheel clamps. All floats are required to have a 17-character vehicle identification number (VIN), which must be on the trailer plate and is usually also stamped on the float's drawbar. The VIN is used in the recording and tracing of stolen vehicles. You can also mark your float in such a way that it is less attractive to steal; for example, you can paint identifying words or numbers on the roof.

The towing vehicle

It is essential that the weight and power of the towing vehicle is sufficient to tow the float. The vehicle manufacturer provides recommendations in relation to the maximum towing weight and that information will be in the vehicle's manual or can be obtained from the relevant transport authority or from a car dealer. However, it is just one of the factors affecting the weight you can legally tow. You must also consider the rating of your tow bar and the float manufacturer's recommendations for its carrying capacity.

The vehicle can be either manual or automatic, depending on the driver's preference. If an automatic transmission vehicle is used for towing, it is advisable to fit a transmission oil cooler if the vehicle does not already have one. This device assists with the additional heat that is generated from the load placed on the transmission by the attached float.

A four-wheel drive vehicle has a low gearing system that is useful for towing. Using a front-wheel drive vehicle to tow a float without the use of suitable weight distribution devices is not recommended and could have disastrous results.

The float
Types of float
Single floats
Most commercial manufacturers are no longer producing single floats, so you will be limited to second-hand items. Bear in mind that many horses are difficult to load into the small area of a single-horse float.

Figure 2. Dual-axle double and triple float
(courtesy of Jane Myers).

A single-axle, single-horse float is not recommended because they are unstable, especially in the event of a puncture while travelling. Therefore, make sure your float is dual axle (two wheels on each side rather than one).

Some manufacturers do make a three-quarter float, which has dual axles and is designed to carry only one horse. It is wider than a single float, and therefore more stable, and is lighter than a double float.

Double floats
A double float can be used even if you plan to travel with only one horse. However, you will need a powerful towing vehicle, especially if you intend to transport two full-size horses. A double float will either have two forward facing bays or be angle loading, where the horses stand at an angle in the float. Standard double floats usually weigh 700–900 kg unladen and 1500–2000 kg or more when laden. The weight will increase if the float has an extended front.

Triple floats
Triple-horse floats can have either three forward facing bays or be angle loading. An angle-load, triple float may not be able to carry three full-sized horses; it will depend on the float's dimensions. Triple-horse floats require an even more powerful towing vehicle than that required for a double float. Triple floats weigh approximately 1000–1400 kg unladen and approximately 3000 kg when laden.

Hiring or borrowing a float
If you decide to hire or borrow a float you are still bound by the regulations for towing. Floats can usually be hired from petrol service stations or hire companies. You should carry out a thorough safety check each time you use a float. Check that:

- the floor is strong and there are no cracks or holes in the flooring. If the floor is lined and sealed with rubber, check the floor from underneath the float. If the floor is timber, check that it has not rotted.
- the various attachments on the float are well fitted and operating correctly.
- the walls of the float are strong and sturdy.
- there are no projections within the float or outside the float that might damage the horse.
- your vehicle is able to tow the float.
- the braking system of the float is working properly. If the float is fitted with over-ride brakes, check the level of hydraulic brake fluid in the cylinder at the hitch point. If the float's brakes are electric, you can only tow the float if your vehicle is fitted with a brake control unit.

Figure 3. Inside a double float, showing a shoulder partition, rump bars and a door above that operates with gas struts (courtesy of Jane Myers).

Buying a float

New floats are usually offered with many optional extras in addition to the basic design, but manufacturers vary in what they term as standard features and extras, so check thoroughly before buying. It is a good idea to speak to other horse people who already own a float and ask them what they think are useful features.

If purchasing a second-hand float, ensure that it has undergone a thorough safety check and check it yourself, following the guide lines for hiring a float. If you do not have the necessary experience to perform the required checks, obtain assistance from an expert, such as a motor mechanic.

Features of floats
Projections

There should be no sharp projections of any kind on or in a float. Bolts and fixtures should be flush with the wall. Remember that horses can get into unusual positions in a float and any projection is potentially dangerous.

Dimensions and construction

The physical dimensions of the float should be large enough to accommodate the horse(s), which will need plenty of headroom. Full-size horses need the roof to be at least 2.10 m (7 feet).

Materials used for the walls of a float include galvanised steel, painted steel, wood or even polyethylene (a UV-resistant, synthetic material). The floor can be made from steel or wood and the roof is usually made from fibreglass. The most important consideration is that the float is well constructed and safe.

Breast bars

The breast bars are located at the front of the internal compartment of forward facing floats to prevent the horse from falling forward. Some floats have fixed breast bars, but if a horse becomes frightened, or in an accident, the horse can become wedged either under or over the breast bar. Therefore, a removable bar is much safer.

Check that the height of the breast bar is suited to the horse's size. A small pony may get its head caught under the bar if it is too high, or a bigger horse may get its front legs over the bar if too low. Adjustable breast bars are preferable.

The ramp

Entry for the horse is via a ramp that is lowered at the back of the float. The ramp should be covered in rubber or coconut matting to ensure a non-slip surface and wooden slats should be placed at intervals across the ramp.

Figure 4. Ramp into a float showing a rubber covering and slats to reduce slipping (courtesy of Jane Myers).

The ramp should not be too steep. The shorter the ramp the steeper it will be when lowered. A steep ramp tilts the horse upward as it places its front feet on the ramp, making the horse frightened of banging its head on the roof as enters the float. This is often the beginning of the horse becoming a problem loader. Ramps that are steep can be improved by fitting steel props to the underside of the ramp to make it less steep.

The ramp should be spring loaded for safe operation.

Rear safety chain, rump bar or breeching door

The rear safety chain/bar/door prevents the horse from backing out of the float before the handler is ready. It also assists in containing the horse within the float once it has been loaded and before the

ramp goes up. Chains and rump bars need to be adjustable so that small ponies cannot escape underneath them and larger horses cannot flip out backwards. Take care when undoing the bar, chain or breeching door because it will still be fastened to one side of the float and can hit the handler if the horse suddenly jumps backwards. Breeching doors are deeper than a bar or chain, which means that they fit a larger range of horses and ponies without needing adjustment.

Doors
There are doors at the front of the float for the entry and exit of the handler while the ramp is closed. All doors should open outwards.

Partitions
Partitions divide the float into stalls that separate the horses. A solid, full-length central partition (i.e. one that extends from the roof to the floor) should be avoided because horses find it difficult to balance if they cannot spread their legs, particularly when the float is going around a corner. A partition that has a solid top half and either an unfilled or rubber lower half is better and it is useful if the centre partition can be removed completely when required, such as transporting a mare and foal or even a large load of hay.

Figure 5. Inside a triple angle load float with barn doors, large tail lights and a rug rack
(courtesy of Jane Myers).

A very useful addition is a head partition (sometimes called a stallion divider) that extends from the centre pole forward to prevent the horses from biting one another while travelling. It is usually in the form of a grille that can be removed if necessary.

A partition that extends back from the centre pole (a shoulder partition) is another useful addition because it protects the horses from the danger of getting their head stuck on the wrong side of the pole.

Tie rings
There should be a tie ring or loop for each horse, attached to the front wall. The horse should not be tied directly to the ring, but to an attached piece of twine so that in an emergency the twine will break or can be cut. You should always carry a sharp knife in your tool kit.

Most floats also have tie rings/loops fitted to the outside of the float at withers height or higher. Never tie a horse to an unhitched float. Again, tie the horse to a piece of twine rather than directly to the ring or loop. Never tie a horse that 'pulls back' to the outside of a float.

Padding
Padding is always a good idea and the areas that should be padded include the breast bar, partition, rump bar/breeching door and the walls. Padding helps to prevent loss of hair from rubbing and cushions any blows caused by sudden movement of the float and/or the horse.

Flooring
The floor of the float must be strong and, ideally, covered with rough, heavy-duty rubber that is either fixed and sealed or removable (smooth rubber is very slippery when wet). If the floor lining is removable, it must be taken out and the floor cleaned every time the float is used.

Covered rear opening
A cover for the opening above the ramp is essential if you are transporting foals because a foal can turn in a small space and try to escape over the rear door. A cover will also reduce the amount of dust and noise. The cover can be made from a canvas or vinyl tarpaulin, or can be part of various styles of doors.

Figure 6. Roof vents
(courtesy of Jane Myers).

Ventilation

All floats need good ventilation to reduce the incidence of travel sickness. Ventilation can be gaps in the sides of the float, side windows, vents fitted in the roof or front wall of the float or a combination. Make sure that the ventilation system does not funnel the exhaust gases into the float. It is possible to buy an extension for the towing vehicle's exhaust pipe that channels the fumes away from the float.

Windows

Large, front windows are not necessary, but many floats have them. When travelling on narrow roads the oncoming traffic can suddenly appear in the view of the horse on the right-hand side of the float and the horse may panic. Horses can start to panic for many reasons and try to jump out through the window. If your float has such windows, place strips of tape over the glass or even better, have a grille welded to the inside of the window.

Openable windows placed high on the side walls are useful for extra ventilation and for letting light into the float.

Internal light

An internal light is not essential, but is very useful if you have to load horses at night. Always allow the horse time to adjust its eyes to the light before loading.

Jockey wheel

The jockey wheel is a small auxiliary wheel that is used to keep the trailer stable when it is not hitched to the towing vehicle; it is released when towing the float. Older floats usually have a bracket that releases the jockey wheel completely from the float and you can now purchase a bracket that enables the jockey wheel to swing up out of the way, but remain attached to the float, reducing the risk of losing it.

Safety chains

Safety chains are welded onto the drawbar of the float and when the float is hitched to the towing vehicle the chains are attached to the vehicle's tow bar with shackles. The chains keep the float attached to the vehicle in the event of the tow ball detaching during travelling.

Lights

Floats should have clearance lights placed high up on the front of the float, plus reflectors and stop, tail, indicator and number plate lights on the rear. Large tail lights make the float more visible and therefore safer.

Figure 7. Side-windows for light and ventilation
(courtesy of Jane Myers).

Mudguards

Mudguards must not have sharp edges that could injure a horse either while it is tied to the side of the float or during loading.

Wheels and tyres

Light truck tyres are essential for their load-carrying capacity. One spare wheel is essential, but two are better, placed in an accessible location on the float and maintained in good condition.

All tyre manufacturers have recommended inflation pressures based on tyre size and load-carrying capacity, and your tyre dealer will advise you.

If you have a punctured tyre while travelling, a dual-axle float will usually remain stable and it should be possible to change the tyre or wheel without unloading the horses, using a block of wood that has been sloped at each end so that the undamaged wheel can be driven onto it.

Brakes

Horse floats are fitted with either electric or hydraulic over-ride brakes. The size of the float determines the type of brakes required. Floats weighing more than 2000 kg when laden must have an additional break-away system. This system stops the float moving and holds it for at least 15 minutes in the event of the float detaching from the towing vehicle. Smaller floats can also be fitted with a break-away system.

Dual-axle floats weighing more than 2000 kg when laden also must have brakes that operate on all four wheels rather than two, and such brakes are also preferable for floats under this weight. All brakes must be operable from the driver's side of the towing vehicle, except for the over-ride brakes.

In addition, the float is required to have a parking (hand) brake for when it is not attached to the towing vehicle.

Some older floats with hydraulic over-ride brakes have a manual reversing lock (positioned on the hitch) that must be disengaged for travelling. The lock prevents the brakes from engaging when reversing; however, if it is accidentally left on while travelling the brakes will not operate.

Additional features

There are many optional extras for a float, such as rug racks, tack boxes, awnings etc. Your budget and the advice of other horse people will determine which features, if any, you choose.

MAINTENANCE AND PREPARATION OF YOUR OUTFIT

General maintenance

Trucks and towing vehicles should be regularly serviced by a qualified motor mechanic. Floats should also be regularly serviced by either a mechanic or the manufacturer.

Your outfit should also undergo a maintenance check before and after each journey. The float should be thoroughly cleaned out after every use and the floor inspected. Wooden floors show most wear on the edge along the walls and will wear out much more quickly if the horses have shoes with studs or the heels turned down.

Preparation for travel

Brakes, mirrors, fuel and tyres

Adjust the float's brakes so that they slow the float simultaneously with the tow vehicle, ensuring that the vehicle and float act as one unit without either entity pushing or pulling the other to a standstill. Make sure that you can see your horses through the observation window of the float via the rear-view mirror. If not, consider fitting a larger mirror or an additional mirror. Fit larger or extended side mirrors to the towing vehicle so that you can see beyond the float.

Have a full tank of fuel and check the pressure of the tyres on the vehicle and the float before you load the horses. Check the tyres often for irregular wear, bulges, cuts, nails etc.

Pack the float and the vehicle in advance so that as soon as the horses are loaded, you can leave without delay.

Tow bar

Heavy-duty tow bars, and possibly a weight-distributing hitch system, which is a levelling device, are required for towing floats. The tow bar should be the type recommended by the towing vehicle's manufacturer and should be stamped with its maximum towing weight. Never use poor

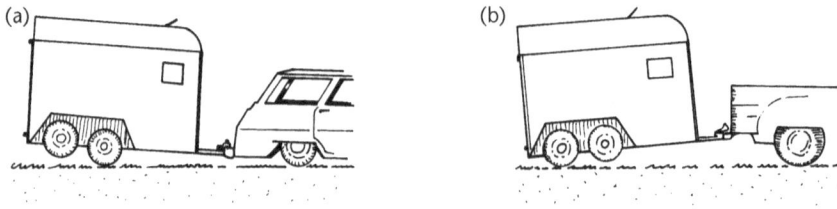

Figure 8. (a) Hitch is too low. (b) Hitch is too high.

quality and/or homemade tow bars. At least twice a year check the tightness of the bolts and studs that attach the tow bar to the vehicle. Tow bars must also be fitted with attachments for the float's safety chains.

Hitching

Check the light globes and wire connections after hitching, each time the float is attached. The indicator lights, reversing lights and brake lights should all be working correctly before you leave on your journey.

Check that the jockey wheel has been released for towing, that the float's hand brake is not on and that the safety chains have been correctly shackled to the towing vehicle.

Matching the hitch heights of the towing vehicle and float is most important to ensure the safe handling of the vehicle and the comfort of the horse (see Figure 8).

Tow height too low

The back of the vehicle drops down under the weight of the float and the float tips forward, causing a number of problems:

- reduced steering control
- headlight focus directed upward because the front of the towing vehicle is raised
- excessive wear on the rear suspension and back tyres of the towing vehicle and the front tyres of the float
- the back of the vehicle or the front of the float or both may scrape on driveways and/or the road
- the float is angled down making the horse feel uncomfortable and unstable.

The towing vehicle must be the right size, in terms of weight and power, and not be exceeding the manufacturer's tow rating. If this is the case the solution is to install a weight-distributing hitch that will ensure that the weight is transferred correctly across the vehicle and float. These hitches can be purchased and fitted at specialist tow bar shops.

Other types of levelling devices are overload springs and equaliser bars, which can also be purchased and fitted at specialist tow bar shops.

Tow height too high

If the tow bar is too high for the float, often the case with light trucks and 4-wheel drives, there is upward strain on the ball and tongue of the tow bar, excessive wear on the rear tyres of the float, and the horse is tipped backward onto its tail and will not travel well. Again, the weight-distributing hitch, together with an adjustable ball mount, will solve the problem.

Travelling gear

The parts of the horse that are most likely to be injured during travelling are the legs, tail and poll, and these can be protected with specially designed accessories.

Boots and bandages

The legs below the knee, and the coronet band in particular, should be protected. Horses can injure themselves by stepping on their own coronets, pasterns and heels, or those of the horse in the adjacent stall. There are various types of travel boots available, ranging from those that cover only the cannon bone, to those covering from the coronet band up to the horse's knees. Special knee and hock boots are also available.

If the weather is hot remember that anything on the legs will increase the overall temperature of the horse.

Choose strong, sturdy boots. They should be fitted to the legs with the fasteners on the outside of the leg, pointing backwards.

Bandages are preferable to cheap, flimsy boots that tend to come off as soon as the horse starts to struggle. Bandages should always be used with over-reach or bell boots and with boots that do not cover the coronet band.

Bandages need to be padded and there is a skill to applying them, so practice before you intend to travel. You may also need to bandage a wound one day.

When applying or removing boots or bandages, do not kneel on the ground beside the horse because if the horse startles, you will not be able to move away quickly and you may be injured.

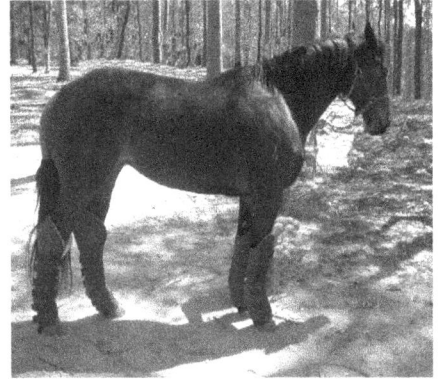

Figure 9. Good quality travel boots (courtesy of Jane Myers).

To apply leg bandages, wrap the padding around the leg, just below the knee, with the overlap to the outside of the leg. Start to bandage down the leg, passing the bandage around the leg from one hand to the other. Keep the pressure even (but not tight) as you work down the leg. Go as far down as the coronet and then bandage back up the leg. The bandage should end just below the knee. Do not fasten the bandage tightly (see bandaging in Chapter 6).

To remove a bandage, unfasten the ties and quickly undo the bandage, without attempting to roll it up as you go. Once the bandage has been removed, you can roll it up, starting with the ties in the middle of the roll.

Tail protection
A tail bandage will prevent the horse damaging its tail by rubbing it on the tailgate. It also prevents damage to the skin under the tail if the horse gets it over the tailgate. It is easier to bandage a damp tail. Do not wet the bandage, because it will shrink as it dries and may cut off the blood circulation to the tail.

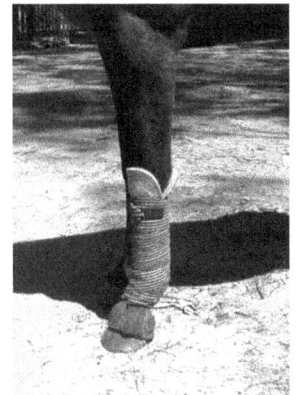

Figure 10. Travel bandage and bell boot (courtesy of Jane Myers).

Position the bandage underneath the thick end of the dock. Wrap firmly and evenly down toward the end of the dock, and then back up again. Tie the fasteners on the outside of the tail approximately half way up. Do not tie the bandage or the fasteners tightly. Never leave the bandage on overnight because the blood supply to the tail is easily damaged by pressure.

To remove the bandage, undo the fasteners and ease the bandage gently from the top of the dock down the tail and then slide it off without trying to unwrap it. If the horse has a very full tail, this might be difficult and you will then have to unwrap the bandage. Remember to be gentle, both when putting the bandage on and pulling it off, because you are working in a very sensitive area of the horse.

Figure 11. Tail bandage (courtesy of Jane Myers).

In some cases a tail bandage may not be sufficient protection because the horse habitually leans on its tail while travelling. The damage to the tail can be extensive. Make sure that the stall in the float is large enough for the horse and if there is room, hang a tyre or tyre inner tube at the back of the stall for the horse to lean on and 'sit in'. Such horses have often been subjected to poor driving technique, which has caused them to lean back as they try to balance.

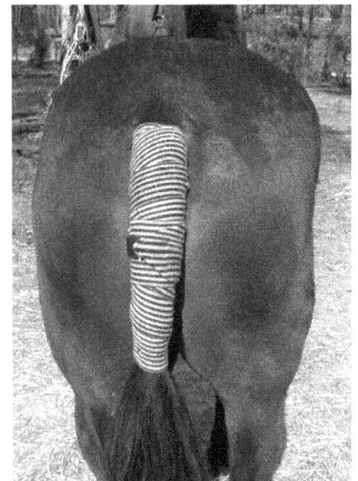

Poll protection
If the horse receives a heavy knock to the poll area (between the ears) by hitting its head on the roof of the float or truck, serious injury or death may occur. If the horse tends to panic while travelling then you should fit it with a poll guard. If you cannot find one (they are rare nowadays), you can improvise by attaching some foam rubber around the headpiece of the headcollar.

Rugs

The number and type of rugs that the horse needs while travelling are both determined by the external and internal temperatures. In most cases it is best not to rug the horse when travelling because the stress of being transported can cause increased heat generation. If there is more than one horse in the truck/float the internal temperature can rise dramatically, depending on the external temperature and the amount of ventilation. Nevertheless, always pack spare rugs and on long journeys make frequent stops to check the horse(s).

If the horse is travelling while rugged, check that the chest straps do not stick into the horse as it leans on the breast bar.

Feed and water

Water

Pack a supply of water from the same source that the horse has at home. Horses can detect changes in water and may refuse to drink water from an alternate supply, even when they are thirsty.

An adult horse needs at least 25 litres of water (5.5 gallons) per day. If the weather is hot you will need to double or triple this amount. A lactating mare needs even more water. If you are going to be away from home for a few days and cannot carry the amount of water required, then you can try to condition your horse to drink water with an additive such as molasses before you leave. The additive will then mask the smell and taste of the water from other sources. Pack plenty of buckets for feed and water

Feed and hay

If your horse is usually fed hard feed, you will have to allow for rest breaks during a long journey so that you can feed the horse. Again it is important that the horse has its usual feed. The digestive system of the horse cannot cope with sudden changes, least of all during a stressful situation such as travelling.

Opinion is divided on whether to feed hay. On a longer journey the horse should have access to hay because the stress of travelling combined with the stress of an empty gut can bring on colic. Ensure the hay is as dust-free as possible by either soaking it beforehand or at least shaking out before feeding it in the float. If a hay net is used, ensure that it is fastened at the top and bottom and positioned high enough that, when empty, it will not become a hazard to the horse.

Other things to pack

The length of the journey determines some of the other things you will need to pack. Some essential items are:

- a block of wood with sloping ends (see previous section)
- a small tool kit, a wheel brace and suitable jack, and a spare set of wheel bearings, grease and seals for the float wheels
- a first aid kit for both horses and humans
- a mobile phone, although in some parts of the country, reception is either non-existent or only very poor
- a spare headcollar, lead rope and twine
- a sharp knife
- a flashlight
- a fire extinguisher
- spare fuel
- reflective triangle and cones in case of a breakdown or accident.

Loading your horse

Frequently practice loading your horse, even if you do not plan to travel very often. You never know when you may have to load your horse in an emergency such as a fire or flood.

In order to be a good loader, a horse must have well-established basics; that is, it moves forwards, backwards and sideways from pressure whenever and wherever asked. Many horses do not have good basics and this is why problems occur with loading. The best time to teach a horse to load is when it is a foal.

There are occasions when a horse must be loaded without prior preparation (e.g. at a sale yard) and you will usually need help from other people. Once you get the horse home you can work on improving its loading so that you can load the horse alone in future. Your goal should be to load your horse(s) by yourself.

With a new horse it is sensible to have extra people around until you have established a routine. Many horses will load for a familiar person, but not for an unfamiliar one, or will walk willingly into a familiar float, but are suspicious of a strange vehicle.

Choose the loading area carefully; load from flat ground and make sure the area is clear of obstacles. If you anticipate problems with loading, it can be useful to park the float next to a wall or fence so that the horse is less able to escape, but ensure the wall or fence is safe or it will cause more problems than it solves. Another option is to reverse the float in to some yards or a gateway with the end of the float as close as possible to the gateposts.

Ensure you have the right equipment. Use a strong headstall and a lead rope with plenty of length. Make sure the front door of the float can be opened quickly and easily if you need to escape. Assuming your horse is good at loading, lead it towards the float, walking beside its shoulder. Let the horse investigate the ramp. Do not start tugging at the horse's head, just let it have a look. Then ask the horse to step up the ramp. When you reach the top of the ramp, you have three options:

- walk in front of the horse into the stall and duck under the breast bar as the horse follows you into the float
- walk up in the adjacent stall (only if it is empty), staying parallel with the horse
- stop at the top of the ramp and direct the rope over the neck of the horse as it walks past you into the float.

A good loader will walk straight into the stall and stand still while you fasten the chain or rump bar. Raise the ramp before tying the horse to the tie ring. The horse must be secured behind before you tie its head because if the horse pulls back and there is nothing behind it, it will pull until it breaks either the lead rope or the twine. The horse will usually bang its head on the roof at the same time, which will frighten it even more. Once it is free the horse will run backwards down the ramp and may injure anyone standing in the way.

Always, even with good loaders, ensure that whoever raises the ramp stands to the side rather than directly behind the float; otherwise if the horse were to kick or rush out backwards (which can sometimes happen regardless of the safety chain etc.), this person could be seriously injured.

A quick release knot should be used to tie up the horse so that it can be quickly freed in an emergency. Always tie the lead rope to a piece of twine that can be cut more easily than the rope in an emergency. It is important that the horses are tied short enough so that they cannot spar with each other or get their heads around the wrong side of the central pole of the float. However, they need to be tied long enough that they can lower the head to clear the airway. The usual length will be approximately 50 cm from the tie ring to the headcollar, but it will vary according to the height of the tie rings in the float. If the float has head dividers (both forward and back) the horse can be given more length.

If you are loading two horses, one option is to load the most sensible/experienced horse first. Often the more timid horse can be coaxed on by the presence of an older or quieter horse. On the other hand, it may discourage an inexperienced horse from loading because it has to walk into a narrow space next to a horse it may not know well. Or you can move the centre partition to give the illusion of more space and load the less experienced horse first. You will have to decide on the day which method is likely to work best. Because of the camber of the road (i.e. it is higher in the middle than at its edges), loading the heavier horse on the right-hand side of the float will give better balance to the float while travelling.

Loading with assistants

If you have to load a horse immediately, without time to train it, you will need help from two experienced, confident people and two lunge lines. Fasten the lunge lines to either side of the float using the ramp fasteners and cross them over behind the horse with someone holding the end of each line. The assistants put some pressure on the lines so that the horse is being 'pushed' into the float. The horse will usually go forward; however, if it is a problem loader and/or very frightened, you will need the help of someone who is experienced with this type of horse. If you lose your temper easily, do not even attempt to load a difficult horse because you will certainly aggravate the situation.

Some helpers may suggest sedation or blindfolding. Sedating horses for transport is not a good idea and should only ever be carried out by a veterinarian under special circumstances (see later). Blindfolding in order to load a horse is very dangerous for both the handlers and the horse because the horse can thrash about and fall.

Loading unassisted

If you are on your own, you can try a 'breeching rope', which is the method often used to teach a foal to lead. It comprises a rope passed behind the horse's hindquarters, knotted on the back and passed through the headcollar. The handler holds this rope together with the lead rope. Each time the horse will not move forward the handler pulls on the breeching rope. Most horses respond well to this method. Once it is loaded, the horse will have to stand still while you go behind it to fasten the safety chain and put up the ramp. Remember, do not tie the head of the horse until it is secure behind. Unless you are experienced do not attempt to load an untrained horse on your own.

Training to load

To teach a horse to load properly and without fuss, you will need a sturdy halter, lead rope and a dressage whip. Using taps from the whip, train the horse to move back from pressure on the halter and to move forward from both pressure on the halter and taps of the whip (see 'Training Horses' in Chapter 3).

Once your horse has established these basic skills you can introduce the float. Do this on a day when you are not planning to travel in the float; it should be a training session only. You now have the training tools to load the horse. Horses that sidle down the side of the ramp can be tapped back into the correct position. Tap and ask the horse to walk up the ramp and if it instead goes backwards keep tapping until it stops going back. It is very important that you stop tapping with the whip when the horse makes a move in the right direction. After a short break ask the horse to walk forward again (with a tap). Remember when training to reward small gains. Do not let the horse turn away from the float. If you persist the horse should eventually move into the float. With practise you should be able to control the horse's movements so that you can ask the horse to walk forward and backward and stop at any point on the ramp or in the float.

Having control over your horse is very important and is dealt with in more detail in Chapter 3. There are also many clinics with horse trainers, advertised in the media, where you can learn these skills, with or without your horse.

Problem loaders

Here are some suggestions for problem loaders, but if you are inexperienced you will not be able to load a difficult horse on your own. You will need the assistance of an experienced horse person to teach you and the horse how to load.

There are many reasons why a horse can be difficult to load and/or travel.

- The horse has been allowed to refuse to load in the past.

- The horse has had bad experiences with floats and travelling and is now frightened.
- The horse and/or handler are inexperienced.

Make sure your horse does not have any reason to be afraid to load. The float should be safe, spacious and solid.

When trying to load a problem horse, keep its head toward the float at all times. Problem loaders may do various things rather than load.

- Rear and twist off to the side when you try to get them to go up the ramp.
- Stand at the base of the ramp and refuse to move.
- Go in to the float or truck, but back off again very fast before you have time to secure them.

Many horses sidle round the ramp, but can still get on. Tap the side of the horse with the whip until it steps sideways and faces into the float again. Do not turn the horse away (or let the horse turn away) and re-present it. Above all, remember you cannot pull a horse into a float; no matter how strong you are, even a small horse is stronger.

Unloading your horse

Never unload onto a slippery surface. Always untie the horse and put the lead rope over its neck before undoing the ramp. Lower the ramp, making sure to stand beside it rather than directly behind it. Once the ramp is down. undo the safety device and allow the horse to slowly back out. A well-trained horse will wait for the command before backing out. If a second person is available, have them stand beside the ramp and place a hand on the horse's hip to prevent the horse from backing off the side of the ramp. That person can then catch the lead rope as the horse emerges.

Never pull on the lead rope as the horse is backing out. This can lead to serious injury if the horse throws its head upward and bangs its head on the roof of the float.

TRANSPORTING BREEDING STOCK

If transporting foals any space above the ramp must be fully enclosed. Foals can quickly become frightened, panic and try to jump out. If possible, transport the mare and foal without the centre partition in place and allow the foal face backwards so that it can nurse whenever it needs to. If this is not possible the foal will need to feed at least every three hours.

When loading a mare and foal it is usually easier to put the foal on first and the mare will follow even if she is not usually a good loader. Her maternal instincts will override her principles! Conversely, even a mare that is normally a good loader may not load unless her foal is either in front or beside her as she walks in to the float.

Transporting pregnant mares is risky, especially in the first 50 days and the last two months of pregnancy. If the mare is a poor traveller the risks are increased further. Mares should not be transported within seven days of foaling and long distance travel with mares in early lactation should also be avoided.

Stallions will usually travel well with other horses as long as they are accustomed to the other horse(s). Partitions and head dividers must be used. Be aware that other horses, especially geldings, may be intimidated by the stallion, which will increase the stress for all the travellers.

TRAVEL SICKNESS

Horses are most often affected by sickness during and after travelling for a long period of time (i.e. for more than 12 hours). However, if a horse is unduly stressed, even a short trip can cause problems.

The possible problems include injuries, colic, laminitis, dehydration and diarrhoea, but the main concern is pleuropneumonia, which can be a fatal illness. Long-distance transport seems to reduce the capacity of the horse's lungs to resist infection and it is thought that stress plays a role, as it does in the cause of the other travel disorders. Pre-travel factors of importance include a pre-existing mild respiratory disease and poor recovery from work just prior to travel. Exhausted and dehydrated horses with increased respiratory rates after exercise have a higher risk of developing travel sickness if transported over long distances immediately after competition.

Types of travel sickness
Colic
Colic can be induced by stress and is often secondary to other problems such as laminitis and diarrhoea. Colic can also occur when the gut has been empty for a long period of time. Horses must be fed at regular intervals on a long journey (see also Chapter 6)

Laminitis
Laminitis can also be induced by stress. Horses with laminitis are reluctant to move, take short shuffling steps and have hot painful hooves. If only the front feet are involved, they will be placed out in front of the horse and the rear legs placed under the horse's body. Even with prompt and vigorous treatment, horses with laminitis can be permanently disabled. A severe case may result in the horse being destroyed. Early veterinary attention is vital (see also Chapter 6)

Dehydration
Severe dehydration may occur in a horse that has sweated profusely during a race or a competition, and is then transported without sufficient time to recover from the exertion. The dehydration is compounded by sweating brought on by the stress of travel.

Sweating during the journey may not be noticed because it dries on the horse's coat. These horses are usually lethargic, not eating and often drink less water than normal. To prevent this situation the horse should be given time to cool down and rehydrate after competing and before travelling. The horse should be closely monitored while travelling. The 'skin-tenting' test (picking up a fold of skin on the neck and checking the time taken to return to its normal position; it should return by the time you count one, two) can assist detection of this problem.

Diarrhoea
Diarrhoea is induced by the stress of travelling long distances and can be life threatening if severe (i.e. watery or projectile; see also Chapter 6).

Pleuropneumonia
Pleuropneumonia is a build-up of fluid in and around the lungs. Early recognition and prompt therapy are vital for a successful outcome.

The horse may be obviously sick at the end of a long journey, but more often the first signs are seen a few days after the trip has ended. Horses developing pleuropneumonia are usually depressed, reluctant to move, not eating and have rapid but shallow respiration with a fever. The presence of chest pain is an important early feature of this condition and an alert owner can detect this from changes in the horse's posture and behaviour. Because of the pain the horse may stand with its forelegs wide apart and its elbows out, or may grunt when disturbed or turned around, especially if pressure is applied to its side. The horse may not be able to lift its head to eat hay from a hayrack or lower its head to graze. It may be restless, paw at the ground and frequently turn its head to look at its side, which can lead to a mistaken diagnosis of colic. A cough may be present, but is always short and suppressed in character because the act of coughing causes pain.

It can be difficult for the veterinarian to non-invasively detect the build-up of fluid in the chest in the early stages of the disease. Although inserting a needle into the chest will reveal fluid accumulation, the technique is not without risk.

The use of ultrasound to examine the chest has recently proved to helpful in making an early diagnosis and thus increasing the success of treatment. This procedure, however, has to be carried out at a specialised equine veterinary hospital. It involves shaving the hair from the chest wall and applying an ultrasound probe to provide a view of the inside of the chest. It can detect fluid accumulation, lung consolidation and lung abscesses, and may assist with the drainage of fluid that is usually necessary as part of the treatment.

Pleuropneumonia is generally a difficult and very expensive condition to treat. Intensive therapy is necessary and some horses that are saved become poor doers and their athletic future is reduced. Deaths and destruction of intractable cases are common.

Preventing travel sickness

Check health status prior to travelling

The majority of horses travel without incident and arrive in good health. However, the environment in which a horse is transported can present a challenge not only to its general health, but to its respiratory health in particular. Although most horses are capable of overcoming the environmental stress, horses with pre-existing respiratory disease are not able to do so as effectively as their healthy counterparts. Research has shown that signs of travel sickness in horses with even a minor pre-existing respiratory disease can become evident within six to eight hours of departure.

Horses with a nasal discharge and swollen glands beneath the jaw, or an acute, unexplained cough or other clear evidence of disease should not be transported. Depression and loss of appetite may also herald the onset of illness, so think carefully about transporting a horse showing these symptoms.

Diseases in horses are not always easy to identify. Check the health indicators: body temperature, heart rate, respiration, hydration, mucous membrane colour and gut sounds (details in Chapter 6). Any competent horseperson is capable of carrying out these simple procedures, but just a simple check of the rectal temperature will indicate if the horse is well. Horses with either an elevated or subnormal rectal temperature and/or any other signs of ill health should not be transported.

Allow horses to cool down before loading

If horses are travelling after strenuous exercise they should not be loaded until they are cool, breathing normally and rehydrated. This may involve walking, grazing, hosing and provision of some feed and water prior to transporting.

Ensure good hygiene

Good hygiene is necessary to prevent transfer of infections between horses, particularly in the case of trucks/floats used to transport more than one horse. Regularly clean and disinfect the vehicle thoroughly and ensure it is dry before loading.

Adequate ventilation

Adequate ventilation is required to clear urine fumes, dust and airborne bacteria from the interior of the float/truck. Ventilators and ventilation spaces should be kept open to allow fresh air into the interior.

Absorbent floor covering

Cover the floor surface with clean straw, wood shavings or other absorbent material to minimise the release of ammonia and other products of the accumulation of urine and faecal matter that is inevitable on a long journey. The absorbent material should be removed and replaced with clean material prior to re-loading after overnight rest periods on long journeys.

Placement of horses

Horses that travel in a truck/float with another horse that has a respiratory infection have more risk of developing travel sickness.

The direction the horse faces during travel appears to be important. Some studies show that horses that face toward the back of float or truck appear to be more stable and less stressed than those facing forward. Rear-facing floats are available overseas, but in Australia and New Zealand this option is only available in multi-horse trucks.

Some studies show that horses that are placed sideways or on an angle in the float/truck appear to travel better, probably because the horse is able to spread its legs out wide to maintain balance. It is when the horse cannot do this that it develops bad behaviours such as 'scrambling'.

Placement of horses so that 'pecking order' effects are minimised will also help reduce the stress of travel for the lower order horses.

Correct tethering

Tether the horse so that it can lower its head to expel the dust and mucous that build up in the respiratory system. (See the previous section on loading)

Plan for changes in temperature

Plan for changes of weather, which can affect the stress level of the horse(s), and be prepared to remove or add rugs en route. A horse generates a lot of body heat in the confined space of a float or truck, so avoid over-rugging, which can lead to excess sweating and predispose to dehydration.

Regular rest breaks

Have a rest break every 4-6 hours for watering, feeding and walking the horse. If it is not safe to unload the horse or the horse is hard to reload, attend to its needs in the float; check the horse's overall appearance, particularly in relation to its temperature; offer a drink of water; check that all leg bandages/boots and tail bandages are still fitted properly. Do not leave horses unattended. Park in a shady position and because the ventilation in a horse float is almost entirely dependent on movement, ensure that maximum ventilation is available to the horses by opening the front door and rear cover. If the weather is hot and you are not unloading the horses, you will have to quickly attend to their needs and your own before moving off again.

Overnight rest breaks if possible

Significant metabolic changes can occur in transported horses within 6 to 12 hours of departure. It is therefore recommended that if horses are to be transported by road for 12 hours or more, an overnight or appropriate rest period should be provided. If you are unable to reach your planned destination it is usually possible to unload your horse at a local show ground, many of which will allow camping and provide yards for a minimal fee or for free.

Regular feed and water

Provide water and hay every 4–6 hours. Use dust-free feed. The hay should be placed so the horse can get access to it, but will not breathe in any dust from the feed. There are many fungal spores in a bale of hay and these present a significant challenge to the respiratory system of horses and humans. To reduce the dissemination of fungal spores from hay, soak the hay for a few hours prior to travelling or at least break open the hay bale and shake the hay with a stable fork outside the float before putting it into hay nets.

Changes in the fluid balance and electrolyte status of horses can occur within a few hours of commencing a road journey, even in cool weather. Horses travelling for 6–8 hours can lose up to 20 kg of bodyweight, even when the air temperature is 10 degrees Celsius. These changes may occur more rapidly in warmer or more humid climates.

Electrolytes

The loss of electrolytes may have an effect on the horse's subsequent performance. Some horse owners offer electrolyte-enriched water to horses in transit. If you wish to do this, you must accustom the horse to the new taste of the water (with the electrolytes added according to the manufacturer's recommendations) a few days prior to departure.

Avoid unnecessary medication

Some horse owners have a policy of transporting all horses in their care under tranquillisers, or following an injection of antibiotics. Such policies should be abandoned; tranquillisers and anti-biotics should be used only when necessary. Some tranquillisers have a profound effect on blood pressure and the corresponding reduction in blood flow to various body organs can be damaging. Inappropriate dosage and timing of administration of some tranquillisers may have other undesirable side effects, including subsequent extreme excitement and frenzy with consequent risk of injury to horses and personnel.

The inappropriate use of antibiotics may suppress the normal bacterial population of the horse and allow the overgrowth of potentially damaging micro-organisms that become resistant to the effect of drug, compromising its effectiveness should treatment later become necessary. Antimicrobial therapy must always be initiated under veterinary supervision and continued for the recommended period at the appropriate treatment intervals.

Laxatives

Horse owners sometimes provide a light laxative diet of bran laced with mineral oils to horses that are scheduled to travel on a long journey. This practice is based on the observation that prolonged confinement of horses can predispose them to intestinal impaction and other forms of colic. However, although colic in horses in transit is a serious problem that can be difficult to control, it occurs in less than 1% of horses that are transported and the administration of large volumes of mineral oil by stomach tube within a few hours of departure may predispose the horse to diarrhoea.

Check health status after arrival

On arrival at their destination, horses should be bright, alert, and have a keen interest in feed and drink within one or two hours, and have a normal body temperature. Depression, a lack of appetite, coughing and an elevated body temperature are the cardinal signs of transit-related respiratory disease. Veterinary attention is essential under these circumstances as pleuropneumonia can progress rapidly and result in severe damage to the respiratory system or even prove fatal. The sooner such cases receive appropriate treatment, the greater the likelihood of a successful outcome.

Because 30–40% of horses with travel sickness will not have an elevated rectal temperature until two or sometimes even three days after arrival, the rectal temperature should be recorded twice daily for at least two or three days after completion of a long journey.

Horse owners should check the bodyweight of their horses, either by using a commercially available measuring tape or preferably by weighing scales. Those persons who are professionally involved in horse sport should consider weighing their horses before and after longer journeys. The rate of return to pre-transport bodyweight may be a useful indicator of well-being, as well as a useful assessment of the post-travel recovery period.

DRIVING TECHNIQUE AND TRAVELLING TIPS

Good driving technique goes a long way to reducing the stress on the horses and hence the incidence of travel sickness. The horses' safety, your safety, and the safety of other road users depend on careful, competent driving. Towing is much more stressful than normal driving and therefore requires more knowledge and skill.

Plan the route

Be prepared, as it will take you longer to get to your destination than if you were not towing and you will use more fuel. Make sure that you know your route from start to finish. Plan your travelling times to avoid peak hour traffic congestion. Also plan your route to include straight roads where possible because even if this adds slightly to the length of the journey the horses will be less stressed than travelling around many bends in a shorter time.

Driver experience and technique

If you are an inexperienced driver do not attempt to tow until you have acquired more general driving experience. If you are simply inexperienced at towing, make every effort to travel with an experienced driver when he or she is towing a float or driving a truck with horses on board. Learn from their experiences and watch their driving techniques. Before you even consider transporting a horse yourself, practice with an empty float. You will need to be competent with all the float-towing driving manoeuvres When you are ready to tow with horses on board, preferably start with short trips until you have gained experience

Poor driving techniques are a major cause of travel stress. As soon as a float moves in any direction, a horse has to readjust its balance. The constant muscular effort required can be very tiring for the horse because it does not have arms with which to hold itself up. Acceleration and braking should be smooth and gradual to minimise excessive forward and backward movement of the horse. Avoid braking hard unless it is an emergency. Corners, bumps in the road, gusty winds, excessive speed, and road camber contribute to sideways movement, which a horse (being longer from nose to tail than from side to side) will find difficult to counter.

Cornering

Turn wide to allow for the fact that the float will 'cut in' on corners and curves. Remember you cannot drive around corners at the same speed as you would without a float in tow. Taking corners too fast will result in the horses falling over or scrambling as they struggle to stay upright. This soon becomes a habit and before long the horse will scramble every time you go around a corner. To prevent this happening, gently slow down before the corner, do not accelerate at all as you go around the corner and wait until you see that the float has fully straightened out (use your mirrors) from the corner before you start to pick up speed again.

Bumps and camber

Bear in mind that the shock absorbers of the float are probably not as good as the ones on the car, therefore the horse may be having a less comfortable ride. Avoid, if possible, running the left-hand tyres off the edge of road surface because that will lessen the likelihood of punctures, blow-outs and swerving, as well as giving the horse(s) a smoother ride.

Use your gears

Select a low gear when proceeding down a steep decline to slow acceleration without excessive braking and change down your gears when climbing before the engine starts to labour.

Traffic lights

When approaching traffic lights, keep your speed low and if the lights change to red as you are approaching, roll slowly towards them. If the lights change to green again, you will then not have to stop completely, but can slowly accelerate. This is particularly useful when transporting horses that start to fidget and kick as soon as they feel the float actually stop.

Driver comfort and safety

You will become fatigued more quickly when towing so plan for more rest stops. Frequent stops are vital for driver concentration. It is safer to have another person travel with you, preferably someone who is experienced enough to share the driving.

Drive defensively

Remember that the float will affect the handling of your vehicle in windy conditions and when trucks pass. Remember to allow plenty of breaking distance from cars in front and to look even further ahead than normal. Allow plenty of space between you and the vehicle in front of you and behind. When stopping or overtaking, be aware of the extra length and weight of your outfit and calculate accordingly. If there is any doubt about overtaking, don't do it! Only pull over to allow any traffic caught in a queue behind you to pass when it is safe to do so and you can easily re-enter the traffic flow.

Driving speed

Do not drive faster than you should regardless of pressure from other road users. Drive smoothly and at a steady pace. Avoid sudden acceleration, braking and swerving unless in an emergency because such actions can cause the horse to slide around or fall inside the float. Drive at a speed that takes into account the road surface, side winds, and the weather and traffic conditions. Experienced horse people usually agree that even though you can legally travel faster, 80 km per hour is the maximum speed for safe towing.

Bad travellers

Some horses 'scramble' when travelling. Scrambling is an attempt by the horse to balance itself by spreading its legs as the car travels around corners. If the horse finds it is unable to do so, it may panic and begin to thrash around. Even if the float does not have a full-length central partition that prevents the horse keeping its balance, you will often find with these horses that it is only either left turns or right turns that upset them, depending on what side of the float they are on. For example, if the horse is on the right-hand side of the float it will scramble around right-hand turns as it tries to spread both right legs further out than the body. The outside walls of the float stop the horse from doing this, so the horse panics and starts to scramble. Solving this problem can sometimes be as easy as taking out the central partition in a double float, which gives the horse room to move away from the wall and spread out all four legs. However, this is not an option if you have to transport two horses. Scramblers will usually travel well in a truck or angle-load float because they can then spread their legs out on both sides. There is also a specially designed, forward-facing two horse float available that has flared outer walls to allow the horse to spread its legs (JR Easy Traveller™).

Some horses like to look out the front window; it helps them to get their bearings. Others, however, find looking out the window frightening.

Some horses enjoy travelling on a certain side of the float. While it is generally accepted that the solo traveller or the bigger of two horses travels on the high (right-hand) side of the road behind the driver, some horses will not travel quietly on a certain side.

If things go wrong

If the float starts to sway, stay calm; do not slam on the brakes. If you have electric brakes lightly apply the float brakes only; if you do not and you are only travelling slowly you can apply the accelerator very lightly. Swaying can occur when travelling too fast or it can occur at any speed with an incorrectly loaded outfit. In either case try to slow down as gradually as possible. Stop at the soonest opportunity and check the horses.

If you hear very loud crashing and banging noises coming from inside a float use your rear view mirrors to check for ears in the front window of the float. If you can see them, the horse is still standing. If you cannot, the horse may have fallen. Even if you are sure your horse is in trouble ignore your instinct to stop immediately and check the horse. On a busy road, that course of action is very dangerous because your horse may get loose and cause an accident or be injured.

Drive until you can pull over safely and then carefully look in over the ramp or through the windows. You need to assess the situation before you open any doors or the ramp.

Every situation is different, you may have one or more horses in the float and you may or may not have a helper. Inside the float is a very dangerous place to be in this situation. Even if the horse is lying still for the moment it will probably start to panic again, either when it gets its breath back or when it is released. You will have to try to release the horse(s) without going in to the float i.e. through the door. Keep talking to the horse(s) in a calm manner. If there is another horse in the float, try to unload it to give the prone horse more room to manoeuvre. In most situations the prone horse will be able to get up. If it cannot you will need to call for help. You may need a vet as well.

The horse may have turned around rather than fallen over. If the float has a cover or doors over the ramp the horse cannot escape, but if not the horse may try to jump out. The horse will try to come out as you are lowing the ramp, so stand to one side of the ramp. You will need to reload the horse, making sure that it cannot get loose again.

A combination of common sense and experience will reduce the chances of things going wrong while transporting horses. If you are nervous take every opportunity to travel horses with someone more experienced than yourself until you gain confidence.

WHERE TO GO FOR MORE INFORMATION

National Road Trailers Australia (NRTA) can provide roadside assistance for vehicles with a float. Phone 1300 550 161 or go to www.nrta.com.au

The Land Transport Safety Authority in New Zealand has information sheets on trailer towing requirements (see www.ltsa.govt.nz).

A good website for information about floating is www.cyberhorse.net.au/safetowing. In particular, there is a comprehensive list of the website addresses of State transport departments.

The Rural Research and Development website (www.rirdc.gov.au/programs/hor.html) has a copy of the Code of Practice for the Land Transport of Horses (see also Appendix 1).

CHAPTER 10

WORKING IN THE HORSE INDUSTRY

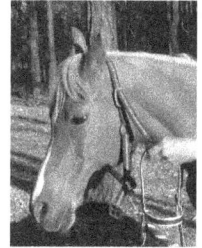

EMPLOYMENT OPPORTUNITIES

There are many jobs in the horse industry, both in the city and the country, locally and overseas, with breeding horses, pleasure horses, working horses, in the racing industry or the related services sector.

Racing

Both Thoroughbred and Standardbred racing require jockeys/drivers, strappers, trackriders and trainers. There are also jobs related to the administration and management of racetracks.

Breeding

Horse studs range from small private organisations to large commercial ones. On the large studs, usually Thoroughbred, it is possible to specialise; for example, as a stallion manager, brood mare manager or in yearling preparation. There are also employment opportunities in the administration of a horse stud.

Equestrian

Equestrian stables specialise in many areas, such as show jumping, endurance and Western riding, showing etc. Riding schools and horse agistment are often run in conjunction with one of these disciplines. Tourism (i.e. trail riding and/or farm stays) also falls into this category.

Other opportunities

Apart from the many opportunities in the variety of services supplying the horse industry, people can adapt their skills and qualifications gained outside the industry in order to work with horses. For example, an accountant can specialise in horse businesses or a human physiotherapist can do additional training to become an equine physiotherapist. Examples of careers with horses include:

- Administration
- Alternative medicine
- Dealing/brokerage
- Dentistry
- Equine consultancy
- Farriery
- Freeze branding
- Horse breaking/starting/training
- Insurance

- Journalism
- Manufacture of products
- Massage
- Mounted police
- Nutrition
- Pharmaceuticals (equine)
- Photography and art
- Physiotherapy
- Public relations
- Research and development
- Retail (saddlery and feed)
- Sales (bloodstock)
- Teaching
- Transport
- Veterinarian
- Veterinary nursing

Remuneration

Employment in the horse industry has been traditionally regarded as poorly paid, but that is changing as the industry becomes more professional. However, in the entry level jobs the pay may still be low, partly because working with horses requires skills that take time and practise to learn and employers are not willing to pay well for unskilled labour. There may be compensation for a low salary in the form of accommodation for you and/or your horse, transport, opportunities to compete and travel, etc.

EDUCATION AND TRAINING

Working with horses is an exceptionally rewarding career if you have qualifications and experience. There are many courses available, ranging from certificates from colleges of technical and further education (TAFE) or polytechnics, to university degrees, either full-time or part-time study, on campus or by distance education (home study). Training in subjects such as horse massage and alternative medicine is usually offered by private colleges that advertise in trade magazines.

Courses change regularly, so obtain the most up-to-date copy of a guidebook to educational courses in your local area, State or Territory. For courses in Australia, have a look at the website www.thegoodguides.com.au. For New Zealand courses, have a look at the New Zealand Register of Quality Assured Qualifications website, www.kiwiquals.govt.nz

Distance education

Many colleges or polytechnics offer distance education from certificate level to advanced diploma or higher. Some courses require that you attend for 'residential' components and others allow you to be assessed in the practical aspects of the course at your own workplace or on a work placement. Contact the different institutions and compare what they offer.

Entry level courses

Many TAFE colleges and polytechnics offer Certificate-level general horse courses that will equip you to work with horses at an entry level (e.g. as a strapper). If you wish to work with horses in any capacity, but have limited experience, these courses are a good place to gain practical skills. A basic horse course will cover subjects such as:

- horse handling

- horse identification
- horse care
- horse health
- stable skills
- communication skills

Basic riding or driving is usually included as is a short practical placement in the industry.

You may also receive accreditation toward some subjects in other courses, such as apprentice jockey training, but you should discuss this with the course coordinator of the institute that you are planning to attend.

Horse management courses

Directed at the educational needs of people seeking a professional career in horse stud or training stable management or associated services, these courses are generally offered at a TAFE college, polytechnic, or university as 2–3 years of full-time study on campus or as an external course (distance education).

As well as business skills, students are taught topics such as exercise physiology, nutrition, behaviour, reproductive physiology and pasture management. Practical skills are taught in breeding, breaking, handling, racing and equestrian sports.

Higher education

Some universities offer postgraduate study in equine subjects. Contact the individual universities for more information.

Specialist courses

There are some specialist courses available, such as equine dentistry at Glenormiston, Victoria, equine physiotherapy at Gatton College, Queensland and equine chiropractic at RMIT University, Victoria. These courses require an existing level of expertise and are aimed at people who intend to work as professionals.

Jockey

Adult jockeys should have a small body size and weigh less than 50 kg, so a prospective apprentice should have a medical assessment of mature weight and body size before starting the training.

Most States of Australia offer jockey training in conjunction with an apprenticeship/traineeship to a trainer. The instruction is provided by the Registered Training Organisation and covers a wide range of topics, such as personal diet and weight control, basic horse anatomy, communication skills, riding instruction, racing rules and administration, and financial management. There is usually at least one riding clinic conducted by former jockeys.

The national Thoroughbred racing website with links to the various States and Territories is www.australian-racing.net.au. The New Zealand Thoroughbred Racing website www.nzracing. co.nz has links to racing clubs, as well as information on the New Zealand Equine Academy

Farrier

Becoming a farrier usually involves an apprenticeship with a qualified farrier and attending a college for certain periods of time each year. In Australia an apprentice farrier is indentured through the government training authority of each State to a master farrier for a period of four years in order to obtain a trade certificate. Those presently practicing as farriers may be eligible to undertake a refresher course and/or trade test in order to attain a trade certificate. In New Zealand national certificates and a diploma in farriery are available (see www.nzqa.govt.nz/framework for details).

Farriery is a growing industry, limited only by the population of horses. The skills of the trade also allow farriers to work in wrought iron, hot forge work, and restoration work. It is possible to establish a small business with a relatively low capital investment compared with many other trades.

National Master Farrier Association
PO Box 72
Greensborough
Vic 3088

NZ Farriers' Association
PO Box 523
Rotorua
New Zealand

The Australian Farriers and Blacksmiths Association has a website, www.hypermax.net.au/~tcourt//home.html, from where you can contact the relevant State secretary to find out about farriers in your area.

Harness racing training

Specific training in the skills of harness racing and driving is available. The Australian national harness racing website with links to the various States and Territories is www.harness.org.au. The Harness Racing New Zealand website is www.hrnz.co.nz.

Riding instructor training

Training in riding instruction is part of the Australian Sports Commission's National Coaching Accreditation Scheme, and the Equestrian Federation of Australia (www.efanational.com) offers three levels of accreditation. In New Zealand, the New Zealand Qualifications Authority offers two levels of coaching accreditation through the National Qualifications Framework: Equine Community Coach or Grade 1 Coaching. See the website www.nzqa.govt.nz/framework for course details and providers.

Australian Horse Riding Centres (AHRC) also offer two qualifications for riding school instructors through the Australian Sports Commission's National Coaching Accreditation Scheme: AHRC NCAS Level 1 (Coaching) and AHRC NCAS Level 1 (Trail Ride). The web site, www.horseriding.org.au, has more information.

The Association for Horsemanship Safety and Education (AHSE) also runs clinics to acredit people who are already working in the industry as instructors and trail ride guides. (www.ahse.info/).

CONTACT DETAILS FOR AUSTRALIAN UNIVERSITIES AND COLLEGES THAT OFFER HORSE COURSES

Northern Territory

Northern Territory Rural College-Katherine
Phone: (08) 8973 8311
Email: ntrural@cdu.edu.au
Website: www.cdu.edu.au

Western Australia

Western Australian Turf Club
Phone: (08) 9277 0777
Email: apprentices@waturf.org.au
Website: http://racingcareers.waturf.org.au

Murdoch University
Phone: (08) 9360 6000
Email: admit@central.murdoch
Website: www.murdoch.edu.au

Swan TAFE Equine Training Centre
Phone: (08) 9390 0322
Email: Jo.Rogers@swantafe.wa.edu.au
Website: www.swantafe.wa.edu.au

South Australia

Torrens Valley TAFE Cheltenham Horse Skills Centre
Phone: (08) 8347 2182
Email: cheltgen@tv.tafe.sa.edu.au
Website: www.tvtafe.sa.edu.au

Tasmania

Launceston Institute of TAFE
Phone: (03) 6336 2761
Email: liz.curtis@tafe.tas.edu.au
Website: www.tafe.tas.edu.au

Queensland

Gatton College
Phone: (07) 5460 1260
Email: mah@sas.uq.edu.au
Website: www.uq.edu.au

Dalby Agricultural College
Phone: (07) 4672 3000
Email: training@dac.qld.edu.au
Website: www.dac.qld.edu

Emerald Agricultural College
Phone: (07) 4982 8777
Email: eacinfo@eac.qld.edu.au
Website: www.eac.qld.edu.au

Longreach Pastoral College
Phone: (07) 4658 4699
Email: jsarnold@lpc.qld.edu.au
Website: www.lpc.qld.edu.au

Burdekin Agricultural College
Phone: 1800 888 710
Email: aginfo@acta.qld.edu.au
Website: www.acta.qld.edu.au

Queensland Race Training
Phone: (07) 3869 0100
Email: qritc@racetraining.qld.edu.au
Website: www.racetraining.qld.edu.au

Victoria

Glenormiston College, The University of Melbourne
Phone: (03) 5557 8200
Email: glenormiston@landfood.unimelb.edu.au
Website: www.glenormiston.landfood.unimelb.edu.au

Marcus Oldham College
Phone: (03) 5243 3533
Email: courses@marcusoldham.vic.edu.au
Website: www.marcusoldham.vic.edu.au

RMIT University Chiropractic Unit
Phone: (03) 9925 7519
Email: victoria.janman@rmit.edu.au
Website: www.rmit.edu.au

Goulburn Ovens Institute of TAFE
Phone: (03) 5723 6137
Email: foliver@gotafe.vic.edu.au
Website: www.gotafe.vic.edu.au

Box Hill Institute
Phone: (03) 9286 9290
Email: h.wilson.bhtafe.edu.au
Website: www.bhtafe.edu.au

Harness Racing Industry Training Centre
Phone: (03) 54493590
Email: john.randles@bigpond.com
Website: http://hrtc.britafe.vic.edu.au

The Racing Victoria Education and Training Centre
Phone: (03) 9258 4620
Email: rvetc@racingvictoria.net.au
Website: www.racingvictoria.net.au

Northern Melbourne Institute of TAFE, Epping Campus
Phone: (03) 9269 1042
Email: info@nmit.vic.edu.au
Website: www.nmit.vic.edu.au

McMillan College, The University of Melbourne
Phone: (03) 5622 6000
Email: mcmillan-info@unimelb.edu.au
Website: www.horses.landfood.unimelb.edu/mcmillan.htm

New South Wales

Hawkesbury Agricultural College, University of Western Sydney
Phone: 1800 897 669
Email: hawkinfo@uws.edu.au
Website: www.uws.edu.au

Orange Agricultural College, The University of Sydney
Phone: (02) 6360 5511
Email: study@orange.usyd.edu.au
Website: www.orange.usyd.edu.au

Western Institute of TAFE, Orange Campus
Phone: (02) 6391 5780
Email: lindy.wellsmore@tafensw.edu.au

Charles Sturt University
Phone: (02) 6933 2121
Email: admissions@csu.edu.au
Website: csu.edu.au

CB Alexander Agricultural College
Phone: 1800 025 520
Email: info@tocal.com
Website: www.tocal.com

TAFE Hunter Institute – Scone Equine Studies Centre
Phone: (02) 6545 1808
Email: michael.thew@tafensw.edu.au
Website: www.tafensw.edu.au

TAFE websites

New South Wales: www.tafensw.edu.au

Queensland: www.tafe.net

South Australia: www.tafe.sa.gov.au

Tasmania: www.tafe.tas.edu.au

Victoria: www.tafe.vic.gov.au

Western Australia: www.tafe.wa.gov.au

Employment agencies

www.stablemate.net.au

www.workhorserecruitment.com.au

www.grooms4u.com

CONTACT DETAILS FOR NEW ZEALAND UNIVERSITIES, POLYTECHNICS AND ORGANISATIONS THAT OFFER HORSE COURSES

NZ Equine Industry Training Organisation
Phone: 0800 84 11 11
Website: www.equineito.co.nz

New Zealand Equine Academy
Phone:(07) 823 7999
Website: www.nzracing.co.nz/nztr/Academy.aspx

Massey University
Phone: (06) 356 9099
Email: enrol@massey.ac.nz
Website: www.massey.ac.nz

National Trade Academy
Phone: (03) 343 4901
Email: admin@nta.co.nz

Kyrewood Equestrian Centre
Phone: (06) 355 9148
Email: info@kyrewood.co.nz
Website: www.kyrewood.co.nz

Telford Rural Polytechnic
Phone: (03) 418 1550
Email: enquiry@telford.ac.nz
Website: www.telford.ac.nz

Waikato Institute of Technology
Phone: (07) 838 6399
Email: info@wintec.ac.nz
Website: www.wintec.ac.nz

Equine eLearning (website-based courses)
Phone: (06) 370 2494
Email: helensmith@xtra.co.nz

Bibliography and further reading

Avery A. (1997). *Pastures for Horses: A Winning Resource.* RIRDC: Sydney.

Bennet D. (1988) *Principles of Conformation Analysis*, Vol. 1. Fleet Street Publishing: Gaithersburg, MD, USA.

Bennet D. (1989) *Principles of Conformation Analysis*, Vol. 2. Fleet Street Publishing: Gaithersburg, MD, USA.

Bennet D. (1991) *Principles of Conformation Analysis*, Vol. 3. Fleet Street Publishing: Gaithersburg, MD, USA.

Brega J. (1995) *The Horse: Physiology.* JA Allen: London.

Brega J. (1995) *The Horse: The Foot, Shoeing and Lameness.* JA Allen: London.

Budd J. (1996) *Reading the Horse's Mind.* Howell Book House: New York.

Budiansky S. (1998) *The Nature of Horses.* Orion Books Ltd: London.

Bureau of Animal Welfare, Victoria. (2001) *Code of Practice for the Land Transport of Horses.*

Cleugh H. (2003) *Trees for Shelter: A Guide to Using Windbreaks on Australian Farms.* RIRDC: Sydney. (An edited version can be downloaded free from the RIRDC website: www.rirdc.gov.au/reports/AFT/02-162sum.html).

Country Style Magazines (date unknown). *First Aid for Horses.* Federal Publishing Company: Alexandria, NSW.

Crandell K. & Huntington P. Dietary Fat: Friend or Foe? Kentucky Equine Research (unpublished; available on website).

Crandell K. & Huntington P. Why is This Horse So Skinny and How Can I Put Some Weight on Him? Kentucky Equine Research (unpublished; available on website).

Crandell K. & Llewellyn M. Don't Pass on Processed Feeds. Kentucky Equine Research (unpublished; available on website).

Dawson J. (1997) *Teaching Safe Horsemanship.* Storey Communications: Vermont, USA.

Department of Food and Agriculture, Victoria. (1989) *Nutrient Requirements of Horses.* CSIRO, KER database.

Duckworth J. (2001) *They Shoot Horses, Don't They? The Treatment of Horses in Australia.* Robins Publications, Australia.

Ehringer G. (1995) *Roofs and Rails: How to Plan and Build Your Ideal Horse Facility.* Western Horseman Inc.: Colorado, USA.

Frandson RD. & Spurgeon TL. (1992) *Anatomy and Physiology of Farm Animals*, 5th edn. Lea & Febiger: Philadelphia.

Frape DL. (1986). *Equine Nutrition and Feeding.* Longman: UK.

Fraser AF. (1992) *The Behaviour of the Horse.* CAB International: Oxfordshire, UK.

Gardner S. & Huntington P. Will It Ever Rain Again: Feeding Horses in Times of Drought. Kentucky Equine Research (unpublished; available on website).

Gordon J. (2001) *The Horse Industry: Contributing to the Australian Economy.* RIRDC: Canberra.

Gower J. (1999) *Horse Colour Explained.* Kangaroo Press: Sydney.

Gray P. (1994) *Lameness.* JA Allen: London.

Gray P. (1994) *Respiratory Disease.* JA Allen: London.

Gray P. (1995) *Parasites and Skin Diseases.* JA Allen: London.

Harris SE. (1993) *Horse Gaits, Balance and Movement.* Macmillan: New York.

Hayes MH. (1987). *Veterinary Notes for Horse Owners*, 17th edn. Stanley Paul: London.

Hill C. (1990) *Horsekeeping on a Small Acreage: Facilities Design and Management*. Storey Communications: Vermont, USA.

Howey WP. (2000) *Supporting Learning in the Horse Industry: Formal Training Programs*. RIRDC: Canberra.

Huntington P. & Bishop C. Horses With Anhydrosis Require Careful Management. Kentucky Equine Research (unpublished; available on website).

Huntington P. & Cleland F. (1992) *Horse Sense: The Australian Guide to Horse Husbandry*. Agmedia: Melbourne.

Huntington P. & Crandell K. Taking Care of the Horse in the Golden Years. Kentucky Equine Research (unpublished; available on website).

Huntington P. & Jackson S. Feeding and Fitting the Halter and Sales Horse. Kentucky Equine Research (unpublished; available on website).

Huntington P. & Jackson S. Feeding and Fitting the Halter and Sales Horse. Kentucky Equine Research (unpublished; available on website).

Huntington P. & Pagan J. Developmental Orthopaedic Disease of Growing Horses. Kentucky Equine Research (unpublished; available on website).

Huntington P. & Valberg S. Tying up. Kentucky Equine Research (unpublished; available on website).

Huntington P. & Valberg S. Tying Up. Kentucky Equine Research (unpublished; available on the website).

Huntington P. *et al*. The latest on laminitis. *Equinews* Vol. 6, Issue 2. Kentucky Equine Research: Kentucky.

Huntington P. *et al*. The latest on laminitis. *Equinews* Vol. 6, Issue 2. Kentucky Equine Research: Kentucky.

Huntington P., Pagan J. & Duren S. Control Colic Through Management. Kentucky Equine Research (unpublished; available on website).

Huntington P., Pagan J. & Gardner S. Feeding the Broodmare. Kentucky Equine Research (unpublished; available on website).

Kiley-Worthington M. (1987) *The Behaviour of Horses*. JA Allen: London.

Kiley-Worthington M. (1997) *Equine Welfare*. JA Allen: London.

Leopold S. *Artificial Insemination with Frozen Semen*. Equine Centre, Horse Health Care–Reproduction (available from www.equinecentre.com.au).

Mackay B. (1999) *Commonsense with Horses*. NSW Agfacts (available through the website of the NSW Department of Agriculture).

McGreevy P. (1987) *Why Does My Horse...?* Souvenir Press: London.

McLean A. & McLean M. (2002) *Horse Training the McLean Way*. Australian Equine Behaviour Centre: Melbourne.

McLean A. (2003) *The Truth About Horses: A Guide to Understanding and Training your Horse*. Penguin Books, Australia.

Mills D. & Nankervis K. (1999) *Equine Behaviour: Principles and Practice*. Blackwell Science: London.

Myers J. & Cakebread P. (2001) *Horse Form and Function* Glenormiston College, The University of Melbourne.

Myers J. (2001) *Grooming*. Glenormiston College, The University of Melbourne.

Myers J. (2001) *Health Care*. Glenormiston College, The University of Melbourne.

Myers J. (2001) *Horse Care*. Glenormiston College, The University of Melbourne.

Myers J. (2001) *Horse Features*. Glenormiston College, The University of Melbourne.

Ottier D. *Embryo Transfer: The New Age in Breeding*. Equine Centre, Horse Health Care–Reproduction (available from www.equinecentre.com.au).

Pagan J. & Huntington P. Balancing Calcium and Phosphorus. Kentucky Equine Research (unpublished; available on website).

Pagan J. & Huntington P. Gastric Ulcers in Horses: A Widespread But Manageable Disease. Kentucky Equine Research (unpublished; available on website).

Pannam CL. (1986) *The Horse and the Law*, 2nd edn. The Law Book Company: Sydney.

Pollitt C. (2001) *Equine Laminitis*. RIRDC: Sydney.

Rees L. (1984) *The Horse's Mind*. Stanley Paul: London.

Roberts T. (1988) *Horse Control Series*. Greenhouse Publications: Elwood, Victoria (4 books).

Rossdale PD. & Ricketts SW. (1980) *Equine Stud Farm Medicine*, 2nd edn. Bailliere Tindall: UK.

Rossdale PD. & Wreford SM. (1993) *The Horse's Health from A to Z*. David & Charles: Devon.

Smythe RH. & Goody PC. (1993) *Horse Structure and Movement*, 3rd edn. JA Allen: London

Stashak TS. (1987) *Adam's Lameness in Horses*, 4th edn. Lea & Febiger: Philadelphia.

Stubbs A. (1993) *Healthy Land, Healthy Horses: A Guidebook for Small Properties*. RIRDC: Sydney.

Stubbs A. (1998) *Sustainable Land Use for Depastured Horses*. RIRDC: Sydney.

Taylor A. *Safety Around Horses*. (Horse Health Care: Codes and Guidelines; available on www.equinecentre.com.au).

Vehicle Standards Bulletin. (1999) *National Code of Practice: Building Small Trailers*. Department of Transport and Regional Services (DOTARS), Canberra.

Waring GH. (1983) *Horse Behaviour*. Noyes Publications: New Jersey.

West G. (ed). (1988) *Black's Veterinary Dictionary*. A&C Black: London.

Appendix 1

WELFARE ISSUES COVERED BY CODES OF PRACTICE IN VICTORIA, AUSTRALIA

The welfare of horses in Australia is protected by legislation and there are also Codes of Practice, which are not law, but are designed to provide detailed information about various methods of managing horses. The Codes of Practice set the minimum standards necessary to ensure that horses are given humane care. The legislation and Codes of Practice covering horse welfare vary between the different States of Australia, but there are also National Codes of Practice for Animal Welfare, published by the Primary Industries Standing Committee (PISC, formerly SCARM) that apply to each State unless a State Code of Practice has been created, in which case the State Code takes precedence over the National Code of Practice. The Codes of Practice are updated as necessary to address current issues and should be followed by all people involved with horses. The following is a list of horse welfare issues and the relevant Victorian Codes of Practice that cover these issues.

Feeding

The basic nutritional needs of horses, such as readily accessible food and water to maintain health and vigour, are explained in the Code of Practice for the Welfare of Horses. The appendixes in the Code provide more detailed information about water and food quantities appropriate for horses of different sizes and under different workloads. The provision of roughage, such as pasture or hay, as a large portion of the horse's diet is emphasised.

Overfeeding and founder

Overfeeding may result in founder (laminitis) in some horses and ponies. Animals at risk should be exercised and have their intake of feed reduced to minimal maintenance requirements. Veterinary attention is recommended when lameness from founder occurs, or when the owner is inexperienced in managing horses with the condition. Reduction in feed should only be undertaken by an experienced person or under instruction from a veterinary practitioner to ensure that all components of the diet essential for growth, health and vitality are readily available.

Short- and long-term confinement

Space requirements for standing, stretching and lying down, and the exercise requirements of confined horses are outlined in the Code of Practice for the Welfare of Horses. Additionally, the horse's need for social contact with other horses, people or other herd animals is emphasised, particularly for those horses kept in long-term confinement.

Health

Preventing disease and injury from occurring is explained in the Code of Practice for the Welfare of Horses. Good hygiene standards are recommended and some of the most common symptoms of ill horses are listed, as are those of conditions requiring immediate veterinary attention. Routine vaccination against tetanus is desirable, and under veterinary advice, vaccination against other diseases may also be recommended. To assist in prevention of paddock injuries, appropriate forms of fencing and enclosures are described.

Supervision

The Code of Practice for the Welfare of Horses recommends that horses kept under intensive management be inspected, fed and watered at least twice daily. Horses grazing under more extensive conditions should ideally be inspected daily and it is recommended that mares in late pregnancy be inspected at least daily for signs of impending foaling.

General care and management

Regular worming of horses is recommended in the Code of Practice for the Welfare of Horses to prevent intestinal damage and loss of body condition. Frequent paring of the hooves by a farrier is recommended to prevent overgrown or split hooves. Rugging should be appropriate for the weather conditions, particularly in extremes of heat or cold, to prevent the horse becoming overheated or chilled. Rugs should fit the horse well, be in good repair and checked daily, and removed at least once each week for inspection of the horse. Horses' teeth should be checked at least once each year by a veterinary practitioner or person experienced in equine dentistry, to ensure that teeth are healthy and functioning properly. Poor dentition may result in loss of body condition.

Tethering

The Code of Practice for the Welfare of Horses, the Code of Practice for the Tethering of Animals, and the Code of Practice for the Welfare of Horses at Horse Hire Establishments specify that long-term tethering is inherently dangerous and therefore tethering should only be on a short-term basis with extreme care. Suitable equipment and locations for tethering are described in the Codes.

Training

The Code of Practice for the Welfare of Horses recommends that persons engaged in educating and training horses should be experienced, or under direct supervision of an experienced person. Abnormal physiological and behavioural responses to training and confinement should be recognised and measures taken to correct them. It is an offence under the Prevention of Cruelty to Animals Act 1986 to beat or abuse a horse, and methods or equipment involving cruelty must not be used. Education of young horses should be minimal to reduce risks of injury and growth abnormalities, and immature horses should not be given strenuous training. In order to reduce the risk of injury horses in training should be conditioned before being subjected to strenuous exercise.

Transporting horses

Transporting a horse can cause stress, fatigue or injury, particularly on long journeys. Horses travelling to the abattoir are potentially at the highest risk because they often travel long journeys and are not required to perform at the destination. Welfare problems that may occur for horses in transport include fatigue, injury, insufficient water or food, extreme temperatures (e.g. too hot or too cold), inappropriate ventilation, and escape. The Code of Practice for the Land Transport of Horses (Victoria) outlines the required periods of rest for longer journeys, and explains the best practices that should be employed to prevent injury, dehydration or other welfare problems. These include pre-preparation of the horse for the journey, segregating horses of different sizes, stallions, groups of unbroken horses, mares in advanced pregnancy and mares with foals, maintenance of the vehicle or trailer to reduce the risk of injury to the horse, and safety equipment that may be suitable to protect the horse from injury. This Code applies when horses are in Victoria; however, the Model Code of Practice for the Welfare of Animals – Land Transport of Horses provides standards for horse transport in every State, and applies unless a State Code of Practice has been developed.

Overuse and injury of working horses

It is an offence under the Prevention of Cruelty to Animals Act 1986 to override, overdrive, overwork or abuse a horse. The Code of Practice for the Welfare of Horses in Horse Hire Establishments outlines the minimum standards that proprietors of such establishments should maintain. The Code of Practice for the Welfare of Rodeo and Rodeo School Livestock in Victoria outlines the minimum standards for operators of such events, to ensure that the horses and cattle are not overworked and that adequate rest and measures are taken when animals are injured. The Code of Practice for the Welfare of Film Animals outlines the measures that should be taken to reduce the risk of injury during the making of the film. Horses that are used to perform in films and are at risk of injury must be supervised by a veterinary practitioner and animal trainer. Bush or mountain racing may also result in stressful exertion and potential injury to the horses involved. The Code of Practice to Protect the Welfare of Horses Competing at Bush Race Meetings requires that a veterinary practitioner be in attendance at such races to inspect the horses, and the Code requires that horses are suitable for the event in regard to age, fitness, soundness or otherwise.

Contact: Ms Naomi Gleeson
The Equine Policy and Legislation Officer
Bureau of Animal Welfare
Department of Primary Industries Victoria

Appendix 2

SAFE RIDING ON THE ROAD: CODE OF CONDUCT FOR HORSES ON VICTORIAN ROADS

THIS CODE OF CONDUCT IS ONLY A GUIDE TO SAFE RIDING OF HORSES ON THE ROAD

1. Be sensible about your choice of riding area. Although horses are considered a vehicle, and hence you must ride on the road, do not expect motorists to slow down to a speed appropriate for horses on certain roads. Where possible, ride on designated trails or tracks away from traffic.
2. When riding in traffic, observe the road rules. Keep to the left. If you are with other riders, ride two abreast (never single file) and keep the inexperienced horse to the kerbside (furthest away from traffic) (Figure 1). In a larger group, surround the inexperienced horse to shield it from the distractions of the passing traffic (kerbside with horses in front and behind it) (Figure 2).
3. When riding two abreast in a marked lane or line of traffic, you are entitled to ride 1.5 metres apart. This means you will fully occupy the lane and will discourage motorists from trying to overtake you in the same lane, which should decrease the chance of a motorist coming so close that your horse is frightened or that you or your horse is actually hit by the vehicle.
4. Use the recommended hand signals to convey your intentions to other road users. ALWAYS thank motorists who pass you wide and slowly with a wave or a nod!
5. Increase your conspicuity and visibility. Ride where motorists can easily see you. Wear light-coloured, very visible clothing. Wear reflective material at night. Also consider what you can put on your horse to increase your visibility (e.g. light-coloured leg or tail bandages).
6. ALWAYS wear an approved safety helmet AND use the correct, well-fitting tack in good condition.
7. Ride slowly past pedestrians and cyclists as many people are nervous and unused to horses.
8. When riding across bridges always ride in a lane on the road, never on the footpath. Most rails on bridges are designed to keep pedestrians from falling over, not horses. If you horse does shy or move towards the edge of the bridge this will give you more time to react before you get near the railing. Because horses are top-heavy, if they do hit the railing when shying sideways, they may topple over the railing and off the bridge to the river, railway or road below. If in doubt of motorists ability to see you on the bridge, dismount and lead your horse along the footpath.
9. When riding on roads take care over metal grates, utility access covers, tram tracks or markings painted on the road. These are often slippery for horses to walk over, particularly if the road is wet
10. Always ride at a sensible pace suitable to the conditions.
11. Be environmentally aware and considerate. Avoid damage and do not litter.
12. Do not leave horses unattended near busy roads.

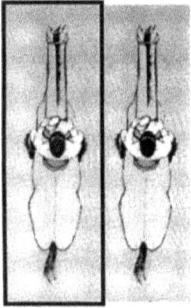

Figure 1. Riding with one other rider.

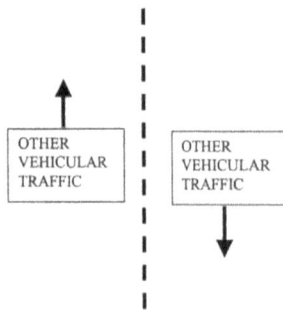

Figure 2. Riding with a group of riders.

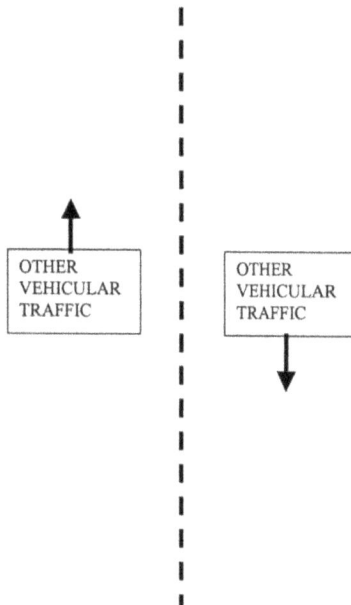

13. If riding on your own, leave details of your intended route and approximate departure and return times with a responsible person.

14. Do not ride an inexperienced horse on a busy road, especially by itself. Introduce the horse progressively to the noise and volume of traffic it will experience on roads. Always use an experienced horse to help train an inexperienced one to adjust to the road conditions and traffic noise.

15. In the event of an accident or unlawful driving by a motorist always reported the incident to the police. To help collate information on horse and motor vehicle road incidents (including near misses), please complete a copy of the Victorian Horse Council's 'Road Incident Report' and forward it to the Council. We cannot address any problems of horse and motorist interaction unless we can show that there is an issue requiring further attention by all stakeholders involved.

Happy riding!

Victorian Horse Council Incorporated
The Voice of the Victorian Horse Industry
11 Earlstown Road
Oakleigh Vic. 3166
Phone/fax: (03) 9569 0047
Mobile: (0411) 402 557
Email: vichorsecouncil@ozemail.com.au
Website: www.vicnet.net.au/~vichorse/

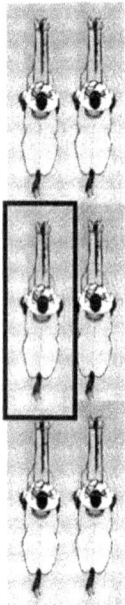

Appendix 3

TICK CLEARANCE POLICY: QUEENSLAND, AUSTRALIA

Horses are livestock and therefore local, State and Federal regulations cover their management, movement, housing and disposal. Horse owners in Queensland and northern New South Wales need to be familiar with the requirements set down by the Department of Primary Industries Queensland (DPI) regarding movement of horses from and between cattle tick-infected areas. These requirements also apply to horses from other States coming into Queensland for competition or breeding purposes.

Cattle ticks occur in some parts of Queensland and not others. Horses are potential carriers of cattle ticks and so they are included in the regulations designed to prevent the spread of the cattle tick. Livestock travelling from tick-infected to tick-free areas (i.e. over the 'tick line') are required to present for inspection and treatment at tick clearance facilities.

Updates of regulations details and requirements are available from the Department's website (www.dpi.qld.gov.au) and should be reviewed by anyone who keeps horses in tick-affected areas or who is anticipating moving horses into such areas.

The DPI clearance policies that are most relevant to horses cover movement of animals for competition and non-competition purposes from cattle tick-infected areas to cattle tick-free areas. Competition horses are divided into two categories.

Category 1 horses

Category 1 covers horses entering cattle tick-free areas for competition purposes at approved venues for less than 5 days. These animals do not require inspection or treatment at the tick line. Owners of such horses can apply for a yearly 'Multiple Movement Permit' that will permit the movement of their horses to and from approved show grounds or competition grounds. To qualify, the horses must meet the following criteria:

- manageable and trained to lead
- receive regular grooming and are free of ticks
- genuine competition horses travelling to scheduled events
- return to the cattle tick infected area within 5 days.

A waybill must be completed each time the horses move into the tick-free area and accompany the travelling horses.

For a single visit or for travel to non-scheduled events or training, a separate 'Travel Permit' is required.

Category 2 horses

Category 2 horses will remain in the cattle tick-free area for longer than 5 days, but less than 14 days. These horses must undergo a clean inspection at a tick line clearance centre en route. Horses staying in the area for more than 14 days will also require treatment at the line clearance centre.

The DPI stock inspector will issue a permit. A waybill must be completed and should accompany the travelling horses.

Non-competition horses

These horses will need to be inspected and treated at facilities located on the major highways from cattle tick-infected areas to the tick-free areas. Horse can be inspected from 8 am to 5 pm on weekdays, but an appointment should be made with the relevant centre prior to arrival. The following horses are exempt from inspection and treatment:

- competition horses (as detailed above)
- racehorses and pacers travelling to and from race meetings
- work horses participating in an approved regular 21 day treatment program
- horses for export from Queensland
- horses consigned for interstate movement and authorised in a travel permit issued by a stock inspector.

Before presentation for clearance at an inspection facility, the following criteria must be met.

- Ensure the horse is manageable and trained to lead.
- Groom the horse daily for at least 7 days immediately prior to travel. Coats must be free of loose hair and scurf.
- Carefully inspect the horse for ticks.
- Complete a waybill for the full journey. A travel permit will be issued at the inspection centre.
- Give at least 48 hours notice to the line clearance centre.
- Check the location of the centre (refer to the DPI website).
- Allow time for loading and unloading the horse at the centre.

If cattle ticks are found on the horse it will have to undergo spray treatments.

Glossary

A

abortion   expulsion of underdeveloped foetus, may be spontaneous or induced

abscess   cavity containing bacteria and pus

afterbirth   foetal membranes, placenta

aged horse   usually a horse that is nine or more years old

agistment   providing feed, shelter and water for payment

all-purpose saddle   suitable for jumping and flatwork riding

Anglo-Arab   recognised cross of Thoroughbred and Arab

Animalintex   commercial poultice that applies heat

anoestrus   period of sexual inactivity, usually during winter

artificial insemination   service of a mare by semen inserted into the uterus by a human operator

at stud   stallion being used for breeding purposes

azoturia   severe form of 'tying up syndrome'

B

bagged up   expansion of the mare's udder before foaling or because the foal is not drinking

banding (mane)   grooming the mane into loose 'pony tails' to keep it laying flat

bars   part of the upper and lower gums that does not bear teeth and where the bit of the bridle is inserted; also the part of the hoof running from the heel alongside the frog

Barb   an ancient, oriental breed that has influenced the development of many modern breeds

bareback   riding without a saddle

barrel   rib cage and belly of the horse

barren   a mare that is not pregnant

bay   red, brown or yellowish horse with a black mane and tail and black limbs

bell boot   protective cover for the heels and coronary band used during training, travelling or competition

bit   piece of metal attached to the bridle and which runs through the mouth and over the tongue; used to aid control of the horse when handling or riding

biopsy   sample of body tissue used for pathological examination

black   horse with black pigment throughout the coat, limbs, mane and tail, with no pattern factor present other than white markings

blaze   broad, white marking covering most of the front of the face

blinkers   eye shields fixed to a bridle or head-covering to prevent the horse from looking to the side, also called hoods

bog spavin   swollen hock joint, often associated with lameness

bolus   a lump of food that has been chewed and is ready to be swallowed

bot   botfly larva that lives in the horse's stomach, also refers to the free-living fly

bowed tendon   inflamed flexor tendon of the forelimb behind the cannon bone

box   stable for a single horse

brand   permanent visible change in hair growth or colour in the form of a symbol and/or numbers to identify a horse

breastplate   a device attached to the saddle to prevent it from slipping backward

bridoon   one of the two bits (the snaffle) used with a double bridle; a bridoon rein is the rein attached to the bridoon bit

bridle   straps, usually leather, that comprise the part of a horse's saddlery or harness that is placed about the horse's head

brood mare   mare used for breeding

browband   bridle strap that runs across the forehead

brown   mixture of black and brown pigment in the coat, limbs, mane and tail

buck   arching the back and lifting all the legs off the ground

buckskin   golden or yellowish shades with black points, mane and tail

by   refers to the sire of a horse, e.g. a horse is said to be 'by' Sir Tristram' if Sir Tristram is its sire

C

calkin   raised, square thickening of the metal of the hind shoe at the outer edge of the heel to give increased grip

cannon   bone between the knee and fetlock

canter   a three-time pace that is slower than a gallop

capped elbow   fluid-filled lump on the point of the elbow; often caused by the hind foot hitting the elbow when getting up or from lying on hard surfaces

cast   when a horse falls or lies down close to a wall or fence and cannot get up without help, it is said to have 'cast' itself; also refers to the horse losing a shoe

castration   removal of the testicles, the male horse is then referred to as a gelding

cavesson noseband   standard type of plain leather noseband

chaff   hay or oat straw cut into short lengths for use as a feedstuff

cheekstrap/cheekpiece   adjustable part of the bridle running from the headpiece to the bit

chestnut (colour)   varying shades of a yellowish tan colour; darkest chestnut is called liver chestnut

chestnut   horny bump on the inside of both the foreleg and the hock

clean legs   not swollen or deformed

clench   part of a nail that, during shoeing, is left projecting from the wall of the hoof after the end of the nail has been twisted off

clip   removal of hair in horses that are worked through the colder months when natural hair coat is longer

cob   a heavy-boned, short-coupled, muscular galloway sized horse of quality, generally having the type of head and neck of a pony

coffin joint   joint between the second and third phalanges and the navicular bone

cold back   stiff back and gait when the horse is first saddled or mounted

colic   pain in the abdomen

collected frame   the horse is ridden in a shortened frame (as opposed to being ridden on a long rein)

colostrum   thick, high-protein milk secreted by a mare that has recently foaled; contains antibodies to protect the foal against infections

colt   young male up to four years old

combined training competition   competition involving jumping and dressage

condition scoring   a method of estimating the condition of a horse

conformation   anatomical arrangement and proportions of the parts of the horse's body

concentrates   grains etc. that make up the non-roughage part of the feed

corn   bruising of the heel from an ill-fitting shoe or grain fed to horses

coronet   sensitive band around the top of the wall of the hoof that produces the new hoof wall

cover   the act of mating, i.e. a stallion 'covers' a mare

cow hocks   conformation fault in which the hocks are too close together, causing the back feet to be widely splayed

cradle   a light frame that can be fitted around a horse's neck to prevent it from biting or licking wounds, bandages etc.

creep feeding   practice of giving a foal access to special supplementary feed and excluding the mare from that area

crest   the upper line of a horse's neck

cribbing   vice in which the horse grasps wooden objects with its front teeth; may be accompanied by windsucking

croup   the upper line of a horse's hindquarters from the highest point of the rump to the top of the tail

cryptorchid   a stallion, colt or gelding with one or both testicles retained in the abdomen; also called a 'rig'

curb   a thickening of the ligament of the hind leg just below the hock

curb bit   either used on its own or as part of a double bridle (with a 'bridoon')

curb chain   a chain fitted to the eyes of a curb or pelham bit and running under the jaw

curry comb   a piece of grooming equipment used to remove dirt and scurf from a bodybrush; it has a flat back and the front consists of several rows of teeth; made of metal, rubber or plastic

D

dam   female parent of a horse

dapple   small circles of colour contrasting with the rest of the horse's colour

dermatitis   inflammation of the skin

dioestrus   stage of the oestrous cycle after ovulation when a mare is not receptive to the stallion

dishing   a faulty action resulting from turned-in toes: when moving forward the horse throws its front foot or feet outward instead of straight ahead

dock   tailbone

double bridle   a bridle consisting of two bits, a curb and a snaffle (bridoon), that are attached by means of two cheekpieces; the two bits may be operated independently

drench   oral administration of medication or supplements

dressage   the art of training horses to perform all movements in a balanced, supple, obedient and keen manner

dry mare   not lactating; does not have a foal

droppings   faeces, manure

dun   light, faded brown colour with dorsal full-colour stripe

dystocia   abnormal labour, difficulty in giving birth

E

electrolytes   minerals and salts (e.g. sodium, potassium chloride and bicarbonate) that are important for maintaining hydration and blood pH status

embryo transfer   transfer of a 7-day-old embryo from the uterus of one mare to that of another mare that will carry the embryo to parturition

empty mare   non-pregnant mare

endotoxaemia   toxins released by bacteria when they die

enteritis   inflammation of the lining of the intestines (bowel), usually accompanied by diarrhoea

enterotoxaemia   toxins in the blood, which have been absorbed from the gut

entire   uncastrated male horse (stallion)

equine   horses are part of the family Equidae, which also includes asses and zebras

ergot   small horny area in a tuft of hairs behind the fetlock joint

ewe neck   the line along the mane is concave; usually seen in thin horses

event horse   a horse that competes or is capable of competing in a combined training competition i.e. jumping and dressage

extension   straightening of a part of the body, such as a leg; also refers to a gait in which longer, but not faster, steps are encouraged

F

farrier   person who fits horseshoes

favour   indicates lameness, i.e. the horse is putting less weight on the affected (favoured) leg

feathers   long hairs around the pasterns of most heavy breeds of horse and some ponies

femur   large bone between the hip joint (above) and stifle joint (below)

fetlock   foreleg or hind leg joint formed by the cannon, pastern and sesamoid bones

fetlock boot   protective boot that stops a horse damaging the back of the fetlock when stopping suddenly or going down on bumpers during galloping

fever   an elevated body temperature, >38.5°C

filly   young female up to four years old

flaxen   blond coloured; some horses have flaxen manes and/or tails

flexion   movement that bends a part of the body, such as a joint

float (teeth)   filing of the sharp molars

float (transport)   a special trailer for transporting one or more horses

float boots   protective gear for horses' legs while travelling

foal   young horse up to the age of 12 months

foetus   developing foal from day 40 of pregnancy to parturition

follicle   fluid-filled sac developing around the egg (ovum) during oestrus; eventually ruptures (ovulation) to allow the escape of the egg

forging   action when the toe of the hind foot strikes the bottom of the front foot of the same side as it leaves the ground

forehand   the part of the horse that is in front of the rider

forward moving   ranges from the horse moves freely (as opposed to sluggish) in a calm manner to the horse is 'on its toes' and ready to move faster with little or no signals from the rider

founder   see laminitis

frog   the part of the hoof that, as it comes into contact with the ground, acts as a buffer to absorb the impact and prevent slipping

full mouth   the mouth of a horse at 5 years old when it has grown all its permanent teeth

G

gait   sequence of leg movements, usually forward

gall   skin sore resulting from rubbing of saddle or girth

gallop   the fastest gait of a horse

galloway   horse from 14 to 15 hands high

gaskin   part of the hind leg between the stifle and hock, the 'second thigh'

gelding   castrated male horse

gestation   the period between conception and birth, approximately 340 days

girth   strap attached to both sides of the saddle and running under the horse's chest, behind the elbow

good doer   horse that gains weight or stays fat on minimal feed

grackle noseband   figure-of-eight, double noseband

greasy heel   skin infection, usually on the coronet, heels and pasterns; also referred to as mud fever

green horse   has had basic training and requires further training

grey   black and white hairs with black skin Coat grows lighter with age, may be quite dark when young, but muzzle will be grey

H

hack   any horse used for riding, usually casual rather than in a particular fashion or a term for a show horse

hackamore   a bitless bridle

halter   rope headpiece with lead rope attached, used for leading a horse

hand   a linear measurement equalling 4 inches or 10 centimetres used for giving the height of a horse from the ground to its withers

headcollar   a bitless headpiece and noseband, usually made of leather or nylon, used for leading a horse

headpiece   bridle strap that goes behind the horse's ears

heat   colloquial term for oestrus, i.e. a mare is 'on heat'

heavy horse   a draught-type horse

hobbles   leather, rope or chain used to restrain a horse by restricting its leg movement

hock   joint in the hind leg between the gaskin (second thigh) and cannon bone

hollow back   horse's back is unduly dipped

hood   a head cover with openings for ears and eyes

hoof pick   a hooked metal instrument used to remove stones and dirt from a horse's foot

hyperlipaemia   serious medical problem particularly affecting ponies

I

immunity   the body's natural response to challenge from foreign agents such as bacteria and viruses

in foal   pregnant

intravenous   injecting a substance into the blood stream via a hypodermic needle or catheter in a vein, usually the jugular vein

irons   metal stirrups

J

jenny   female ass

jog   slow, short-paced trot

K

knuckling   the forelegs of the horse are straight and the fetlock joints are intermittently or permanently flexed

L

lactating   a term describing a mare that is producing milk to feed a foal

lameness   unevenness of the horse's stride when moving; limping (see also favour)

laminitis   inflammation of the sensitive layers (laminae) of the hoof, characterised by heat and pain

left rein   to be 'on the left rein' means moving to the left

legume   a type of plant (e.g. clover and lucerne) that provides high-protein feed

lice   external parasite that causes severe itching of the skin, mane and tail

light-horse breed   Arab or Thoroughbred type breeds

loins   lower back area behind the saddle

long reining   a stage of breaking involving driving a horse in two reins

lunge   to exercise a horse in a circle on the end of a rope

lunging cavesson   headcollar with a strengthened noseband, suitable for lunging

M

maiden   a mare that has never been bred

manege   an enclosed area, usually rectangular, used for exercising a horse

mange   contagious parasitic skin condition

manure   faeces, also called droppings

mare   female horse that is 4 years or older

martingale   a device used to stop a horse from raising its head too much when ridden

mouthing   educating a horse to respond to pressure on the reins; also refers to determining the approximate age of a horse by looking at its teeth

muck out   to clean out a stable of manure and dirty bedding

mud fever   skin infection of the heels and pastern; also known as greasy heel

myositis   inflammation of muscle caused by injury, sprain or infection, usually associated with tying up syndrome or azoturia

N

near side   left-hand side of a horse

nipping   giving a small bite

numnah   a saddle cloth that is cut in the shape of the saddle

O

oestrogen   female hormone responsible for sexual receptiveness

oestrous cycle   sexual cycle of a mare

oestrus   state of being receptive to the stallion; service or insemination at this time can result in pregnancy; also known as 'in season' or 'on heat'

off side   right-hand side of a horse

over-reach   a wound that occurs when a horse strikes into the heel of a foreleg with the toe of a hind leg

ovulation   shedding of an egg (ovum) from the ovary

P

pace   a lateral gait in two time, in which the hind leg and foreleg on the same side move together

palomino   gold yellow or dark cream colour with white mane and tail

parasite   an organism that lives on or inside the horse

parrot mouth   short lower jaw

parturition   birth process; delivery of the foal from the uterus

pastern   the portion of the leg between the fetlock and the hoof

pelham   a type of bit designed to produce, with only one mouthpiece, the combined effects of a snaffle and curb bit (see double bridle); a pelham bridle is a bridle fitted with a pelham bit

performance horse   competes in events such as racing, dressage and jumping

physitis   inflammation of the growing ends (epiphyses) of the long bones of young horses

pigeon toed   conformation fault in which the hooves and lower leg point inwards

pirouette   dressage term for a turn executed within the horse's body length

placenta   the membrane structure that surrounds and supports the growing foal during pregnancy; also called the foetal membranes or afterbirth

pleuropneumonia   inflammation and infection of both the lungs and the pleural cavity, with a build up of fluid in the pleural cavity

pneumonia   inflammation and infection of the lungs

poll   highest point of the head, just behind the ears

pony   a horse of any breed up to 14 hands

poor doer   a horse that loses or fails to gain weight with normal feed supply

poultice   a means of prolonged application of heat to a damaged or inflamed area

prepuce   fold of skin covering the penis

progesterone   female hormone responsible for dioestrus behaviour and maintaining pregnancy (from 'progestation')

prostaglandin   hormone that destroys the corpus luteum and which can be used to bring a mare into oestrus

proud flesh   excessive granulation tissue that may develop between the edges of the skin during healing of an open wound; bulges above the surface of the skin and delays healing

puller   a horse that pulls on the reins and is hard to stop

pulling   removing hairs from the underside of the mane and the sides of the tail to improve their appearance

Q

quarters   the area of a horse's body extending from the rear of the flank to the root of the tail and downward to the top of the leg; also known as the hindquarters

quidding   collecting feed in the cheeks without swallowing it; often associated with teeth problems

R

rain scald   skin condition causing loss of hair in small tufts

rear   rising up on the hind legs

reins   pair of long narrow straps attached to the bit and used by the rider or driver to guide and control the horse

rein back   to make a horse step backwards while being ridden or driven

refusal   in racing, it is failure of a horse to attempt to jump a hurdle or steeplechase; in show-jumping and combined training, it is either the act of passing an obstacle to be jumped or stopping in front of it

rig   abnormally developed or improperly castrated male (see also cryptorchid)

ringbone   a disease of the pastern joints

ringworm   a fungal skin disease causing circular lesions of hair loss and scaly skin

RIRDC   Rural Industries Research and Development Corporation

roan   coat colour with a mixture of white hairs

roarer   a horse with partial laryngeal paralysis that causes it to make a roaring noise when breathing in

roller   a special girth with rings at the top through which long reins can pass or side reins be attached to

roman nose   description of a head conformation that is convex from the eyes to the nostrils

roughage   high-fibre feed such as pasture, hay or chaff

running martingale   y-shaped strap that runs from the girth to each of the reins, which run through a metal ring on each end of the straps

S

saddle cloth   padding placed under the saddle to protect the horse's back from galls

saddler   a person who makes or deals in saddlery and/or harness

saddle tree   the foundation on which a saddle is built

(in) season   term used to describe a mare that is receptive to the stallion; also known as 'on heat' or oestrus

safety stirrups   quick-release or specially shaped stirrups to prevent a rider's foot being caught in the stirrup in the event of a fall

sand roll   an area of sand for rolling after a horse is worked or washed

schoolmaster   a very experienced, well-educated horse that will enable a beginner to safely learn to ride

scouring   diarrhoea, passing of loose, watery faeces

seedy toe   separation of the sensitive and insensitive layers (laminae) of the foot; sequel of chronic laminitis or hoof cracking

service   the mating of a mare by a stallion (see also cover)

shy   jump or move sideways or backwards at an unexpected sight or noise

sickle hocks   a conformation fault in which there is too much angle on the hocks

sire   the male parent of a horse

sleepy foal disease   common sign of illness in a newborn foal caused by a range of problems

snaffle bit   the oldest and simplest form of bit, consisting of a single straight or jointed bar with a ring at each end to which one pair of reins is attached

snip   isolated white marking between or near the nostrils

soundness   state of health or fitness to carry out a particular function

sow mouth   opposite of parrot mouth: the lower teeth extend further forward than the upper teeth

speedy cutting   action by which the horse strikes the inside of one leg with the inside of the foot or shoe of the opposite leg

spelling   break from work, usually at pasture

splint   bony enlargement of the small bones ('splints') on the side of the cannon bone

spur   pointed device strapped on to the heel of a rider's boot and used to urge the horse onward

stale   to urinate

stallion   an uncastrated male horse aged four years or over

star   white mark or patch on forehead

staring coat   hair that looks dull, dry and brittle

stirrups   leather or metal holders for the rider's feet; metal stirrups are also known as irons

stirrup leathers   straps that connect the stirrups to the saddle

stirrup rubbers/treads   inserts into a stirrup iron to provide extra grip

stock saddle   heavy type of saddle with knee and thigh pads and a deep seat

strapper   person who works as a groom, usually with racehorses

strangles   an infectious and highly contagious disease caused by the bacteria Streptococcus equi

stripe   narrow white mark down face

stud (breeding)   property where horses are kept for breeding

studs   screw-in projections in the heel of a shoe to give traction

subcutaneous   under the skin

supplement   extra feed other than pasture

surcingle   a belt of webbing or leather that passes over the saddle and girth to keep the saddle in position

sweat scraper   curved metal blade with a handle used to scrape sweat or excess water from a horse

T

tack   saddlery

tarsal bones   bones of the hock joint

teaser   stallion or rig used to find out if a mare is ready for mating

tendon boot   a boot that protects the tendons behind the cannon during jumping or work

tetanus   an often fatal bacterial disease related to deep, penetrating wounds

throatlatch   a strap that is part of the headpiece of a bridle

thrush   inflammation of the frog of a horse's foot, characterised by a foul-smelling discharge

tinker's grip   a restraining hold: a fold of skin on the neck is gripped to settle a fractious horse

trackrider   person who exercises racehorses

trot   a pace in two time, in which the legs move in diagonal pairs

tush   in the male horse, a tooth behind the corner tooth on each side of the upper and lower jaws, sometimes present in mares

twitch   a restraining hold placing pressure on the horse's upper lip using either a twisted rope or a hand

tying-up syndrome   stiffness and soreness of muscles; severe form also known as exertional myopathy or azoturia

U

ulna   smaller of the two bones of the foreleg

unsound   any defect that makes the horse unable to function properly

ultrasound   mechanical method of imaging internal tissues and organs using sound beams

V

vaccinate   to inject vaccine to stimulate immunity, e.g. for tetanus

vice   bad habit, such as windsucking

W

walk   a slow, four-beat pace

waxing   waxy plug at the end of a mare's teats, which indicates that foaling is imminent

weanling   young horse that no longer suckles milk from its mother

weaving   rocking from side to side, sometimes while lifting each foot in turn

wet mare   mare producing milk and feeding a foal

windgall   swelling of a joint or tendon sheath, usually not associated with lameness

windsucking   habit of gulping and swallowing air; may be accompanied by crib biting

withers   top of the shoulders between the neck and back

wolf tooth   a small vestigial pre-molar tooth that is often removed

Y

yearling   horse between one and two years old

Index

abnormal behaviour 16–17
abnormal movement 82–84
abortion 249–50
Aconite 226
adult horses,
    feeding 144–45
    water consumption 298
afterbirth 253, 254
age, brand 107
age, terms 102
aging by teeth 109–17
aggressive horse 60
agistment 2
alkaloid poisoning 226
allergic reactions 180
alsike clover 227, 228
American Miniature Horse 97
American Saddlebred 85, **87**
anatomy 11–13, 67–72
Andalusian 85, 87
angle of incidence 111
Anglo-Arabian 85
anhydrosis 163, 188–89
antibiotics 305
anti-inflammatory drugs 229
Appaloosa 85, **87**
appetite 177
appetite loss 139, 142, 143, 149, 303
apprentice jockey 311
approaching a horse 42
Arabian 85–86, **88**
arsenic 173
arterial bleeding 179
arthritis 78
artificial insemination 244–45
Australian Miniature Pony **88**, 98
Australian Pony 86
Australian Riding Pony 86
Australian Saddle Pony 86
Australian Spotted Pony 86, **88**
Australian Stock Horse **88**, 95
Australian stringhalt 162, 186–87
Australian Stud Saddle Pony 95
Australian White Horse 95
avocado 228

back at the knees 77
back conformation 74–75
back feet 21
back protection 47
bacterial abortion 250
bacterial acne 208, 212
bad travellers 307
bagging up 249, 251
bald face 105
bandaging 232–34. 296–97
barley 132
base narrow 78

base wide 78
beans 133, 170
bedding 283
bee stings 212
beet pulp 138
behaviour 177–78
behavioural characteristics 13–17
Belgian Draught 95
bench knees 77
bicarbonate 173
big head 165–66, 192
big leg 212
birdsfoot trefoils 224
Birdsville horse disease 225
Birdsville indigo 225
bit 49, 50
biting problems 62–63
black nightshade 224
blaze 105
blisters 209
blocked tear ducts 200
blood typing 109
body condition scoring 20, 138–39, 157, 177
bolting of food 27
boots (horse) 54
boots (riding) 47
boots, travelling 296–97
bots 208
bottle feeding 259–60
botulism 138
bow legs 76, 78
box stalls 282
box walk 17
bracken fern 225
braiding the tail 34
bran 132, 133, 168
branding 107–108, 266
breaking in 56–57
breast bars 292
breastplate 52
breeching door 292–93
breeding 237–67
breeding crush 281
breeds 6, 85–101
bridle 48–51
bridle teeth 109
bridling 61
bronchitis 196–97
brood mares 150–52
bruises 194, 201
brushing 83
bucked kneed 77
bucket feeding 260
bucking 63–64
Buckskin **89**, 95
buckwheat 227
buffalo fly bites 212
bulbo urethral glands 240

burr medic 227
buyer's agents 7
buying a horse 4–9

caffeine 173
calcium 126–27
calf knees 77
Californian poppy 226
camped out behind 78
camped out in front 77
canine teeth 109
cannon markings 106
canola meal 133, 170
canter 81
capillary fill time 177
Caslick's operation 250
Caspian 95
castor oil plant 225
castration 266
catching 42
certificate of identification 101–102
chaff 134–35, 137–38
chains, safety 292–93, 294
chest pain 302
chewing disease 225–26
chewing problems 17
chewing rate 27
children and horses 41
chin spot 105
choking 179
chopped forage 137–38
chromium 128
chronic bronchitis 196–97
clenches 23, 25
Cleveland Bay 95
clipping 35–36, 63
clothing 47–48
clover 142, 227, 228
clubfoot 80
Clydesdale **89**, 95–96
coat 28–36, 177
coat colour 6, 102–104
codes of practice 320–22
coital exanthema **210**, 212–13, 245
colic 160–62, 170, 182–84, 254, 257, 302
colic-like conditions 183
colitis X 197
colostrum 258–59
colours 102–104
colts 102, 266
commercial feed mixes 121, 168
communicating 15
composting manure 286
concentrates 132–34
concussion 81–82
condition scoring 120, 138–39, 157, 177
confinement 320
conformation 6, 72–80

conjunctivitis   200
Connemara Pony   **89**, 95
constipation   133
contagious acne   208, 212
contracted foot/heels   79
contracted tendons   257
Copper   128
corn gromwell   226
corns   204
coronary band   71
coronet   105
corpus luteum   238
corrective shoeing   24–25
costs of keeping a horse   1–4
cottonseed meal   133, 170
coughing   196–97
coupling   74
cow's milk   260
cow hocks   78
creep feeding   152
Cremello   95
crib biting   17, 27, 110
crofton weed   228
cross-firing   83
crushes   280–81
cubes (forage)   138
Cuboni test   249
cup (teeth)   111
cyanide poisoning   224

daily supervision   19, 321
dams   270–71
Darling pea   225
Dartmoor Pony   96
Death-causing plants   222, 224
dehydration   176–77, 302
dental star   111
dermatophilus   216–17
developmental orthopaedic disease   163–65, 189–90
diarrhoea   197–99, 256, 302
dicalcium phosphate   126-27, 154
dietary changes   121, 142–44
difficult horses   60–64
digestion   139, 143, 147, 148, 149, 157, 160, 161
digestive tract problems   139–40
digital cushion   72
dimethyl sulfoxide   172
dioestrus   238
dipped back   75
discipline   39
disease in older horses   149
diseases, and weight loss   140–42
dished face   74
dismounting, safe   45
DNA identification   109
dogs and horses   41–42
drenching   121, 207
dressage horses   3–4
drinking water   283, 298
dropped sole   80
drought feeding of horses   167–72
drugs, prohibited   172–74
drugs in feed   229

ear mites   213
ear movements   13
eating   13–14
education and training   310–16
elbow hitting   83
electric fences   276–77

electrolyte supplements   129–30, 305
electronic identification   109
embryo transfer   245
embryonic loss   249
employment opportunities   309–10
energy supplements   122–25, 169
enzyme supplements   143
epididymis   240
equine coital exanthema   245
equine granuloma   220
equine herpes virus   249
equine viral arteritis   245
ergot of paspalum   226–27
evisceration   179
evolution of the horse   11–16
ewe-neck   73
exercise   20, 36–37, 120
exercise yards   284
extrusion of feeds   136–37
eye injuries   200
eye problems   199–200
eyelid injury   200

face markings   105
facing up   42
Falabella   97
farrier   311–12
farrier, preparation for   22–24
fat   146
fat supplements   124–25, 169
fatness   138
fattening   156
feed bins   121, 283–84
feed concentrates   132–34
feed costs   1
feed pellets   134
feed processing   135–38
feed quality   121
feed quantity   120–21
feed ration   119–20
feed requirements   122
feed supplements   131–35
feed when travelling   298
feeders   19, 283
feeding   20, 119–174
   adult horses   144–45
   basic principles   119–21, 320
   condition scoring   120, 138–39, 157, 177
   in drought   167–72
   foals   20, 152, 255–56, 258–62
   to gain weight   142–43
   lightly worked horses   145
   to lose weight   143–44
   mares   150–52, 241–42, 254
   older horses   147–50
   stallions   157–59, 246
   supplements   20, 122–31
   weanlings   20, 152–57, 265–66
   weight estimation   139
   when travelling   298
   working horses/performance horses   145–47
   yearlings   152–57
feeding principles   119–35
   digestion   119
   good feeding practices   120–21
   quantities   120–21
feet see hoof
fences   274–78
fetlock arthritis   78
fetlock deformities   258
fetlock markings   106
fibre   125, 142–43

fibroma   213
filly   102, 157
fire branding   108
fire prevention   286–88
fireweed   227
first aid   178–80
Fjord   96
flat-footedness   24, 79
floats   290, 291–95
fly worry   213
foaling   251–54
   afterbirth   254
   difficulties   179, 253–54
   induction   253
   presentation   253–54
foaling signs   251
foals,
   age   102
   bottle feeding   259–60
   bucket feeding   260
   colic   160–62, 170, 182–84, 254, 257, 302
   diarrhoea   198–99, 256, 302
   disease immunity   255–56
   diseases   256–57
   feeding   20, 152, 255–56, 258–62
   hoof care   258
   formula feed   260
   fostering   262
   hand rearing   259–61
   handling   263–65
   health   255–58
   lameness and joint swelling   256
   learning to lead and load   263
   limb deformities   258
   management   255–67
   and mares in heat   238
   newborn problems   256–57
   orphans   258–62
   parasite control   258
   prematurity   257
   retained meconium   257
   teeth   111
   tendon trouble   257
   training   263–65
   transporting   293, 301
   urination problems   257
   vaccination   267
   weaning   261, 265–67
   worming   207
foal heat diarrhoea   256
foals, orphan   258–62
follicle development   238
follicle testing   238
follicle-stimulating hormone   239
folliculitis   208, 212
foot abscess   203
foot care   20–26
foot conformation faults   78–80
foot flight patterns   79
foot, bandaging   232–33
forage, during drought   167–68
forages, processed   137–38
forelegs   75–77
forelimbs   75–77
forging   84
formula feed   260
foster mothering   262
founder/laminitis   24, 171, 184–86, 302, 320
fractured limbs   179
freeze branding   107–108
Friesian   **90**, 96
frog   71

gallop 81
Galvayne's groove 111–12
gastric ulcers 165, 170, 190–92
gates 278
gear 47–48, 296–98
geldings 16
geriatric diet 149–50
German Warmblood 100
Gestation 150–51, 251
girth galls 217–18
girths 52–53
going down on the bumpers 84
good doer 144
gossypol 133
grain mixtures 168
grass hay 135
grasses, hazardous 165–66, 192, 272
grazing management 273–74
greasy heel 216–17
grooming 32–35, 37–38
growths on the skin 209
gut sounds 177

habronemiasis **211**, 219
Hackney 96
Haflinger **90**, 97
hair excess trimming 34–35
hair loss 209
halter and sales horses 36–38, 155–56
halters 54
hand service 243–44
hand-rearing newborn foal 259–61
handling,
    basic principles 39–44
    breaking in 56–57
    common problems 60–61
    difficult horses 60–64
    foals 263–65
    lunging 36, 58–60
    riding 44–47
    safety 39–42
    weaning foals 266
handling of foals legs 263–64
handling yards 279–80
hard to catch horse 64
harness racing training 312
harrowing 273
hay 134–35, 137, 298
hay nets 121
hay racks 121
haylage 138
hazardous grasses 165–66, 192, 272
head confromation 74
head markings 105
head nodding 201
head tossing 27
headcollars 54, 264–65
headstall 64, 263, 299
health,
    assessment 175–78
    bandaging 232–34
    colic 160–62, 170, 182–84, 254, 257, 302
    coughs and colds 196–97
    diarrhoea 197–99, 256, 302
    drought risks 170–71
    eye problems 199–200
    foals 256–57
    hyperlipaemia 162–63, 187–88
    injections 230–31
    nutritional diseases 159–67, 180–92
    oral medication 229
    poultices 234–35

respiratory diseases 194–97
    skin diseases 208–21
    teeth and dental problems 148–49
    travel sickness 301–305
    treatment 229–35
    vaccination 267
    veterinary assistance 178–80
    when buying horse 9
    worms and worm control 204–208
    wounds 192–94
hearing 13
heart and lung disorders due to plants 228
heart rate 176
heat cycle 237
heaves 196–97
heel 23–24, 71, 72
heel bruising 71, 80, 204
heel markings 105
height measurement 102
helmets 47
hemlock 222
herbal supplements 130–31
herd instinct 14
Highland 97
hind limb 77–78
hind limb lameness 77
hindquarters 75
hitching vehicles 296
hock arthritis 24
hock markings 106
hock, bandaging 233–34
hocks 75, 77
holding yards 279
honey 169
hoof,
    anatomy 70–72
    bones 72
    corrective shoeing 24
    foot care 24
    hoof trimming 21–22
    horseshoe removal 25
    horse shoeing 21, 23–24
    insensitive structures 70–71
    lameness 20, 201–203
    laminitis 24, 171, 184–86, 302, 320
    problems 79–80, 166–67, 203–204
    sensitive structures 71–72
    splitting 71
hoof angle 78
hoof bars 71
hoof care 20–25, 258
hoof conformation 78–79
hoof dressings 21
hoof growth 166–67
hoof oil 21
hoof pick 20, 33
hoof trimming 20–22
hoof wall 70–71
hoof wall cracks 204
hoof wall rings 80, 228
hordenine 172
hormones 241
horse auctions 7
horse dealers 7
horse facilities 269–88
    buildings 269, 278–84
    crushes 280–81
    exercise yards 284
    feed bin 283–84
    fences 274–78
    fire prevention 286–88
    paddock design 271

property design 269–71
    paddock shelters 284
    riding arenas 282
    safety 40
    stables 282–84
    urban horse properties 288
    yards 278–80
horse industries training 310–16
horse shoe removal 25
horse shoeing 21, 22–24
housing 17–19
how horses learn 56
hydration status 176–77
hyperlipaemia 162–63, 187–88

identification 101–109
incisor teeth 109, 111
injections 230–31
insurance, vehicle 290
interference 84–85
intestinal worms 198, 204–208
iodine 128
Irish Draught **90**, 97
Irish Sport Horse 97
iron deficiency 127
isolation 196
itching 209

jockeys 311
Johnson grass 224
jumpers bumps 74
jumping 81

kicking 63
kidney damage due to plants 228
Kimberley horse disease 225
knee chips 77, 201
knee hitting 83
knee markings 106
knee narrow 76
knee sprung 77
knee, bandaging 233
knees 76–77
knock knees 76
knocked down hip 78
kunkers 220

laburnum 222
lameness 20, 24, 201–203, 256
laminar corium 72
laminitis 24, 171, 184–86, 302, 320
lateral cartilages 72
laxatives 305
leading a horse 42–34
learning ability 56
leasing a horse 4
leg anatomy 70–72
leg bandaging 233
leg conformation 75–80
leg mange 214
leg markings 105–106
leucaena dermatosis 228
lice **210**, 214
lifting feet 20
limb deformities in foals 258
linseed 224
linseed meal 133–34, 170
lip tattoos 108
Lippizaner 97
liver damage due to plants 227–28
loading horses 263, 298–301
locked stifle 78

locking patella   201
locomotory stereotypies   16,17
lucerne hay/chaff   134–35
lumps in the skin   209
lunging   36, 58–60
lunging cavesson   58
lunging rein   56, 58–59
lunging whip   59
lungworm   206
lupinosis   227
lupins   133, 170, 227
luteinising hormone (LH)   238
lying down   14, 251, 256

maiden mares   152, 251, 255
maize   132
mane plaiting   34
mane pulling   34
manganese   127
mange   213, 214
manure management   273–74, 284–86
mares   15
   breeding behaviour   15
   Caslick's operation   250
   feed requirements   241–42
   fertility assessment   243
   health care   242
   management   241–45
   milking   259, 262
   nutritional requirements when
      pregnant   122, 144, 151
   parasite control   242
   oestrus cycle   237–39, 242
   reproductive system   237–39
   uterine infection   245, 249, 250, 254
   vaccination   251
mares, barren/dry   152
mares, brood   150–52
mares, foaling   250–54
mares, lactating   151–52
mares, maiden   152, 251, 255
mares, pregnant   150–51
mares, wet   254
markings   105–107
marshmallow   226
martingales   51
meadow hay   135
meconium   257
medications   229–30, 305
melanoma   **210**, 214
metabolism   143
Mexican poppy   226
micronisation   137
milk teeth   110
mineral supplements   126–128
Miniature Horse   97
Miniature Pony   98
mixed feeds   121, 134
molar teeth   109
molasses   135, 169
monkshood   226
Morgan   98
morphine   173
mosquito bites   214–15
mouldy feed   121, 132, 134, 162, 167
mounting, safe   45
mouth, opening   110
movement of the horse   80–85
mowing pastures   273
mucous membrane colour   177
mud fever   216–17
multipurpose crush   280

muscle indentations   107
muzzle markings   105

nasal discharge   177, 195
native birdsfoot trefoil   224
navicular bond   72
navicular disease   78
nebulisation   230
neck imperfections   73
nervous diseases caused by plants   225–27
nervous horses   60
neurofibroma   215
New Forest Pony   98
newborn foals   255–56
newborn foals, feeding   152, 255–56
nipping   62–63
nodular necrobiosis   215
normal movement   80–81
nose twitch   65
nosebands   49
nutritional diseases   159–67, 180–92
nutritional value of feeds   123

oak   228
oaten hay/chaff   135
oats   132
oestrogen   238, 239, 249
oestrus cycle   237–39, 242
offset knees   77
older horses, feeding   147–50
oleander   224
onchocercal dermatitis   **210**, 215
open knees   76
oral medication   229
oral stereotypies   16, 17
orphan foal   258–62
over at the knee   77
over-reaching   84
overfeeding   143, 165, 320
overheating   31, 146
overshot jaw   73, 107, 110
ovulation   238

pace   80–81
padding, floats   293
paddling   83
paddock mating   244
paddock rotation   273
paddock shelters   284
paddocked horses   18
paddocks   271–74
Paint   98
Palomino   **91**, 98
papillomatosis   221
parasite control   139, 171, 242, 247
parasites   204–208
parrot mouth   73, 107, 110
partly coloured eye   107
partly coloured hoof   107
paspalum spp.   226–27
pastern   78
pastern markings   105
pastes   229
pasture   120
pasture management   271–74
pasture species   271–72
patella   78, 201
Patterson's curse   227
peas   133, 170
pecking order   14, 15, 18, 121, 148, 304
pedal bone   72
pedal osteitis   24

pedigree   6
pelleting   136
pellets   134
penis   240
Percheron   **91**, 98–99
performance horses, feeding   145–47
Perlino   **92**, 95
permanent teeth   110–11
phosphorus   126–27
phosphorus supplements   126–27
photodynamic agents   227
photosensitisation   216, 227
phycomycosis   220
picking the feet up   65
pigeon toes   76, 78
Pinto   **91**, 99
pinworms   206
plaiting (hair)   34
plaiting (movement)   83
plant poisoning   221–28
pleuritis   196
pleuropneumonia   302–303
PMSG testing   248–49
pneumonia   196
points of the horse   67
poisoning   221–30
poisonous plants   221–28
   causing big head   165–66, 192
   causing death   222, 224
   causing heart and lung disorders   228
   causing kidney damage   228
   causing liver damage   227–28
   causing nervous diseases   225–27
   causing photosensitisation   227
   causing skin disorders   228
poll protection   297
pollard   132, 133, 168
polysaccharide storage myopathy (PSSM)   160,
   181
poor doer   143
poor quality feed   19, 119, 121, 161
poppies   226
posts   277–78
potassium   129, 130, 285
poultices   234–35
pregnancy   248–51
prematurity   257
preparing for competition   34–35
problem loaders   300–301
processed forages   137–38
progesterone   238–39, 250
prohibited substances   172–74
property, access   270
property, buildings   269, 278–84
property, design   269–71
property, security   270
prophet's thumb marks   107
prostaglandin   238–39
prostate glands   240
protein   146–47, 154, 169–70
protein supplements   125
pulled mane   34
pulling back   61–62

quarantine   235–36
Quarterhorse   **92**, 99
Queensland itch   **210**, 216
quidding   27

racehorse   4, 27, 75, 146, 166, 190, 191
ragwort   225, 227
rain scald   **210**, 216–17

ramp (floats) 292
rapid breathing 257
rattlepods 225
rearing 63–64
rectal temperature 175
recurrent exertional rhabdomyolysis (RER) 160, 181
red-flowered birdsfoot trefoil 224
redworms 198
registration 101, 102
regular horse care 20–36, 321
reins 49
releasing a horse 44
reproduction 237–67
    abortion 249–50
    breeding crush 281
    foaling 251–54
    mare management 241–45
    pregnancy diagnosis 248–49
    serving the mare 243–45
    sexually transmitted diseases 212–13, 245
    stallion management 245–47
reproductive systems 237–41
respiration rate 176, 257
respiratory diseases 194–97
respiratory problems 257
restraint crush 281
restraint methods 64–65
retained meconium 257
rice 133
riding 44–47, 323–34
riding arenas 282
riding boots 47
riding gear 47–54
riding instructor training 312
riding safety 44–45, 323–34
ringbone 78
ringworm **210**, 217
roach back 74
Roman nose 74
rotational grazing 273
roughage 120, 134–35, 155
round yards 279–80
roundworms 205–206
rubber vine 228
rugging 28–32, 298
rump bar 292–93
Russian knapweed 225–26
ryegrass 225

sacroiliac ligament damage 74
saddle boils 208, 212
saddle fitting 53–54
saddle sores 217–18
saddles 52–54
sale registries 7
salicylate 172
salivation 27
Salmonella spp. infection 197
salt 135
Salvation Jane 227
sanding 197
sarcoid **210, 211,** 218
scalping 84
scouring see diarrhoea
scrotum 240
secondary hyperparathyroidism 126
security 270
security, vehicle 290
seedy toe 203
selenium 127–28

selenosis 228
selling a horse 9
seminal vesicles 240
senses 13
sensitisation due to plants 216, 227
sensitive laminae 72
sensitive sole 72
sensitve areas of the horse 41
septicaemia 256
serving the mare 243–45
sesamoid bones 83
severed extensor tendons 24
sex 6–7, 102
sex identification 102
sexes, differences between 15–16
sexually transmitted diseases 212–13, 245
shape of horse 72–80
shear mouth 25
shelter, paddock 284
Shetland 92, 99
shin soreness 201
shock 179
shying 323
sickle hocks 77
side bone 24
side reins 37, 58–59
sight 200
size of horse 7
skeletal deformities 126
skeletal system 69
skin diseases 208–21
    bacterial acne 208, 212
    bee stings 212
    big leg 212
    buffalo fly bites 212
    classification 209
    coital exanthema **210**, 212–13, 245
    ear mites 213
    equine granuloma (kunkers) 220
    fibroma 213
    fly worry 213
    leg mange 214
    lice **210**, 214
    melanoma **210**, 214
    mosquito bites 214–15
    neurofibroma 215
    nodular necrobiosis 215
    onchocercal dermatitis **210**, 215
    photosensitisation 216, 227
    phycomyosis 220
    pinworm infestation 220
    Queensland itch **210**, 216
    rain scald **210**, 216–17
    ringworm **210**, 217
    saddle sores/girth galls 217–18
    sarcoid **210, 211,** 218
    spider bites 218
    squamous cell carcinoma **211,** 219
    stable fly bites **211,** 219
    summer sores **211,** 219
    swamp cancer **211,** 220
    tail itch 220
    ticks 220
    urticaria **211,** 221
    warts **211,** 221
skin disorders due to plants 228
slashing pastures 273
sleepy foal disease 256
sloping pastern 78, 79, 84
sloping shoulders 74
small feet 80

smell 13
snaffle bridle 58
snakebite 179
snip 105
socialising 14–15
solar corium 72
sole 71
sole bruising 204
solid yards 280
sorghum 133
sorghum spp. 224
soursob 228
sow mouth 26, 73
soybean meal 133, 170
speeding cutting 84
sperm production 240–41
spermatic cord 240
spider bites 218
splay footed 76
splints 36, 37, 77
sprains 194
squamous cell carcinoma **211,** 219
St Barnaby's thistle 225–26
St John's wort 227
stable fly bites **211,** 219
stable vices 16
stabled horses 18–19
stables 282–84
stallion paddock 246
stallions,
    exercise 246–47
    feeding 157–59, 246
    fertility 240–41
    fertility assessment 241, 247
    handling 246–47
    health care 247
    management 245–47
    parasite control 247
    reproductive system 240–41
    serving the mare 243–45
    stable management 246
Standardbred **93,** 99
standing under behind 78
standing under in front 77
star 105
starch 123–24, 143, 146
steam rolling/flaking 137
stereotypies 16–17
stifle 77, 78, 190
stirrups 52
stock saddle 52, 54
stomach tube 229
stone bruises 201
storage feed bins 283–84
storksbill 227
straight back 75
straight hind legs 77–78
strangles 195–96, 232
strapper 309, 310
stress 148–49, 198
stripe 105
strongyle worms 204–205
strongyles 198, 204–205
stubborn horse 55, 60
stumbling 83
summer sores **211,** 219
sunburn see photosensitisation
sunflower seeds/meal 133, 170
supplementary feeding 122–31
supplementary feeds 131–35
surcingle 54

suturing wounds 193
swamp cancer **211**, 220
swan neck 73
sway back 75
sweating 37, 144, 302, 304
swelling, generalised 209
swelling, joints in foals 256

tail braiding 34
tail itch 206, 220
tail protection/bandaging 297
tapeworms 206
teasing 242–43
teasing yards 279
teeth 12, 109
teeth and age estimation 109–17
teeth care 25–28
teeth problems 25–28
    broken 27
    crib biting 17, 27, 110
    decayed 27
    examination 28
    extra teeth 26
    floating the teeth 28
    imperfect meeting of teeth 26
    incorrect conformation of the mouth 26
    in older horses 148
    parrot mouth 26, 73, 107
    sharp cheek teeth 25
    shear mouth 25
    signs 27–28
    smooth mouth 26
    sow mouth 26, 73
    step mouth 26
    teething problems 26–27
    wave mouth 25–26
    wind sucking 17, 27
    wolf teeth 26, 109
temperament 5
temperature 175
temporary teeth 110, 111
tendon trouble in foals 257
testes 240
tetanus 231
tethering 321
theobromine 173
thin horse 142
thorax conformation 74
Thoroughbred **93**, 99–100
threadworms 206
thrush 203
tick-free areas 325–26
ticks 220
tie rings 293
tie up 264
tied-in below the knee 77
time budget 14–15
tinker's grip 65
Tobiano **91**
toe out 76
topical medications 230
trace mineral supplements 127–28
training 55–65
    how horses learn 56
    breaking in 56–57
    difficult horses 60–64
    foals 263–65
    to load 300
    methods 57–60
tranquillisers 229, 305
transporting horses 289–308

basics 321–22
breeding stock 301
coping with problems 307–308
costs 1
driving safely 305–307
equipment 289–95
equipment maintenance/
    preparation 295–96
feed and water 298
floats 291–95
foals 263
health status after arrival 305
leg protection 296–97
loading the horse 298–301
pregnant mares 301
poor travellers 300–301
quarantine 235–36
regulations 290
safety measures 306–308
tail bandage 297
tick-free areas 325–26
vehicles 289–91
travel sickness 301–305
travelling boots 296–97
travelling gear 296–98
treating horses 229–35
trees 270
tropical pastures 165–66, 192
trot 80
tushes 109
twinning abortion 250
twins 248
twitching 65
tying up 159–60, 180–81
tying up a horse 43

ulcerative lymphangitis 212
ulcers (eye) 199–200
ulcers (gastric) 165, 170, 190–92
ultrasound examination 248
uncooperative horses 60
under run heels 80
undershot jaw 73
uneven heels 80
unloading horses 301
unresponsiveness to the bit 27
urban horse properties 288
urination problems in foals 257
urticaria **211**, 221
uterine infection 245, 249, 250, 254
uveitis 200, **211**

vaccination 231–32
    foal 258
    pregnant mares 251
    Salmonellosis 251
    strangles 196
    tetanus 231
    weanling 267
valgus/varus deformities 258
vegetable oil supplement 124, 169
vehicles 289–91
ventilation, floats 294
veterinary assistance 178–80
veterinary examination 178–80, 243, 247
viral infections 195
vision testing 200
vitamins 128–29, 154–55, 173
vulval tears 254

Waler **93**, 100

walk 80
walkabout disease 225
wall eye 107
Waratah horse disease 227
Warmblood **93**, 100
warts **211**, 221
water requirements 121
water supply 270
water trough 121
weaning 265–66
weanling 265–67
    branding 266
    castration 266
    feed requirements 152–57
    handling 266
    vaccination 267
weed control 272
weeping sores 209
weight estimation 138–39
weight gain 142–44
weight loss 27, 139–43
welfare 17–20, 320–22
Welsh Cob **94**, 100–101
Welsh Mountain Pony **94**, 100
Welsh Pony **94**, 100
Welsh Pony of Cob type **94**, 100
Western Australian blue lupin 227
wheat 132
wheaten hay/chaff 135
whip 59–60
white horses 95
white iron weed 226
white line 71
whorls 108
wind sucking 17, 27
winging in/out 83
wolfsbane 226
wood chewing 17, 171
working horses 322
worm control 206–207
worms,
    bots 208
    large strongyle 204–205
    lung 206
    pin 206
    prevention 206–207
    round 205–206
    small strongyle 205
    tape 206
    thread 206
    treatment 206–207
Wormtest kit 207
wounds 192–94

yarded horses 19
yards 278–80
yearlings 152–57
yellow burr weed 226
yew 222
yielding to pressure 57–58
young horses,
    energy feeds 154
    forging 84
    protein feeds 154
    roughage 155
    vitamin and mineral supplement 154–55

zinc 128

www.ingramcontent.com/pod-product-compliance
Lightning Source LLC
Chambersburg PA
CBHW041801280326
41926CB00103B/4766